土豆帝国

历史·文化经典译丛

土豆帝国

FEEDING the PEOPLE:
The Politics of the Potato

Rebecca Earle

[英] 丽贝卡·厄尔 著

刘媺 译

上海人民出版社

献给马特·韦斯顿（Matt Western）和丽萨·厄尔（Lisa Earle）

目 录

序　言　自斟一大杯杜松子酒　/ 1

第一章　外来的土豆　/ 25

第二章　启蒙的土豆　/ 63

第三章　自由市场的土豆　/ 101

第四章　全球的土豆　/ 135

第五章　资本主义的土豆　/ 176

第六章　安全的土豆　/ 209

总结　帕蒙蒂埃、农民和个人责任　/ 247

致谢　/ 261

参考文献　/ 265

索引　/ 320

图片目录

序言

1. 彼得·哈伯·赫茨伯格1774年出版的土豆种植指南的卷首插图 / 12
Peter Harboe Hertzberg, *Underretning for Bønder i Norge om den meget nyttige Jord-Frukt Potatos: at plante og bruge* (Bergen, 1774).

2. 描绘安第斯山区土豆收获的17世纪绘画 / 15
Felipe Guaman Poma de Ayala, 'June: Time of Digging up the Potatoes', 'El primer nueva corónica y buen gobierno', 1615–1616, drawing 391 (fol.1147), The Royal Danish Library, GKS 2232 kvart.

3.《长官与土豆》/ 18
John Frost, *The Book of Travels in Africa, from the Earliest Ages to the Present Time* (New York, 1848), 169.

第一章

4. 腓特烈大帝视察土豆收成 / 29
Robert Warthmüller, *Der König Überall* (1886), Deutsches Historisches Museum, Berlin.

5. 贝尔纳多·德·西恩弗戈斯的本草书（约1630年）中的土豆插图 / 35
Bernardo de Cienfuegos, 'Of the potatoes from Peru which in the Indies are called chuño', 'Historia de las plantas', c.1627–1631, Biblioteca Nacional de España, Mss/3357–3363, vol. 1, fol. 498.

6. 流动土豆小贩，选自18世纪英国民谣 / 43
'The Potato Man' (London, c.1780), Broadside Ballads Online, Bodleian Libraries, Ballad-Roud Number V20070.

第二章

7. 乔瓦尼·巴蒂斯塔·奥奇奥利尼1784年论著中的土豆版画插图 / 64
Giovanni Battista Occhiolini, *Memorie sopra il meraviglioso frutto americano chiamato volgarmente patata* (Rome, 1784), 39.

8. 波茨坦无忧宫，腓特烈大帝的牌匾上留下的土豆供品 / 66
Photograph by Matt Western, 2014.

9. 由英国农业委员会推广的土豆烘烤机 / 80
Board of Agriculture, *Account of the Experiments tried by the Board of Agriculture in the Composition of Various Sorts of Bread* (London, 1795), 28.

第三章

10. 拉姆福德伯爵在他特别设计的壁炉前的漫画 / 117
James Gillray, *The Comforts of a Rumford Stove* (London, 1800), British Museum, 1868,0808.6897.

11. 巴伦西亚经济学会第三版汤食谱 / 120
Junta Pública de la Real Sociedad Económica de Amigos del País de Valencia (Valencia, 1801), 60.

第四章

12. 毛利人在吃土豆 / 155

    James Coutts Crawford, *Watikini Eating Potato*, pencil and ink drawing, 1861, Alexander Turnbull Library, National Library of New Zealand, E-041–046.

13. "快乐的土豆家族" / 165

    Ilaria María Sala, 'Tudou for the Tuhao: Can "Sister Potato," a Singing Peasant, Convince the Chinese to Eat More of the Lowly Spud?', *Quartz*, 24 Feb. 2016, https://qz.com/622594/can-potato-sister-a-singing-peasant-convince-the-chinese-to-eat-more-of-the-lowly-spud/.

第五章

14. 讽刺漫画表现的受推荐的"面包替代品" / 181

    James Gillray, *Substitutes for Bread* (London, 1795), British Museum, 1868,0808.6492.

15. 土豆与资本主义理性 / 194

    'The Effect of Limiting the Supply of Food', *The Struggle* 78 (1843), 1.

16. 弗朗西斯科·尼蒂的图表：高土豆摄入量与疾病的关系 / 202

    Francesco S. Nitti, 'The Food and Labour-Power of Nations', *Economic Journal* 6:21 (1896), 56.

第六章

17. 土豆皮特 / 210

    Back cover of *Potato Pete's Recipe Book*, Ministry of Food ([London], [1940?]).

18. "吃土豆锄德皇"的橱窗展示　/ 216

'Potatoes in Iowa become "the Newest Fighting Corps" on the Domestic Front', photograph, c.1917–1918, Records of the US Food Administration, US National Archives, 283501.

19. 2008 年国际土豆年标志　/ 234

Food and Agriculture Organization of the United Nations.

20. 国际马铃薯中心出版物中的安第斯土豆农户　/ 237

Christine Graves, *The Potato Treasure of the Andes: From Agriculture to Culture*, International Potato Center (Lima, 2001), cover photo.

21. 安第斯土豆田的示意图　/ 240

Stephen Brush, Heath Carney and Zósimo Huamán, 'Dynamics of Andean Potato Agriculture', *Economic Botany* 35:1 (1981), 81.

总结

22. 拉雪兹神父公墓内装饰着土豆的帕蒙蒂埃墓　/ 248

Daniel Thierry/Getty Images

23. 梵高《吃土豆的人》（1885）　/ 253

Van Gogh Museum, Amsterdam (Vincent van Gogh Foundation).

24. 不满的臣民向腓特烈大帝扔土豆的土豆印花　/ 258

Christoph Niemann, *The Potato King*, Owlkids Books (Toronto, 2015).

菜谱目录

序言

四川炒土豆丝　/ 4

刘建成、胡廉泉、杨镜吾、舒孝钧：《大众川菜》，成都：四川科学技术出版社，1995年，第202—203页；雅各布·克莱因（Jakob Klein）译。

第一章

"松露"的四种食谱　/ 39

Lancelot de Casteau, *Ouverture de cuisine* (Liège, 1604) (trans. Daniel Myers), 94–95, www.medievalcookery.com/notes/ouverture.html.

第二章

西班牙土豆面包食谱　/ 84

'Carta del cura del Linares sobre el cultivo y aprovechamiento de las patatas', *Semanario de agricultura y artes dirigido a los párrocos*, Madrid, 5 Feb. 1797, 204–205.

第三章

供 25 名士兵吃的杂烩汤　/ 112

Manuscript Recipe book, English, c.1798–1826, 'Recipes for cures and cookery, c. 1802–1826', Schlesinger Library, Radcliffe College, AR297, fols. 85–86.

第四章

波斯土豆脆皮饭　/ 137

Najmieh Batmanglij, *Food of Life: Ancient Persian and Modern Iranian Cooking and Ceremonies*, Mage Publishers (Washington, 2016), 238–243.

第五章

俄罗斯烤土豆　/ 198

Elena Molokhovets, *Classic Russian Cooking: Elena Molokhovets' A Gift to Young Housewives*, ed. and trans. Joyce Toomre, Indiana University Press(Bloomington, 1992), 193.

第六章

肉冻土豆沙拉　/ 227

Jane Holt [Margot Murphy], 'News of Food: Jellied Meat and Potato Salad Furnish Cool, Attractive Main Dish for Dog Days', *New York Times*, 5 Aug. 1943, 12.

总结

美味填馅土豆　/ 256

US Department of Health and Human Services, 'A Healthier You', 2005, https://health.gov/dietaryguidelines/dga2005/healthieryou/html/sides.html#9.

序言　自斟一大杯杜松子酒

吃什么是我们的事，或者我们通常是这么认为的。我们讨厌被告知要多吃蔬菜、少吃盐、多吃扁豆，尤其是当这些建议来自政府的时候。美食作家戴安娜·亨利（Diana Henry）总结说："当政府告诉我们要少喝酒的时候，我想自斟一大杯杜松子酒。"[1]这倒不仅仅是因为我们喜欢作对。我们也不知道这种对我们私人生活的干预，是否违反了根本的民主原则。难道不允许我们犯一些饮食方面的错误吗？ 2012 年，纽约市长迈克尔·布隆伯格（Michael Bloomberg）试图禁止超大包装饮料的销售，然后为此付出了代价。该计划遭到失败，因为批评者认为它是对个人自由的冒犯。"纽约人需要的是市长，而不是保姆"，《纽约时报》整版的广告大声疾呼。英格兰北部罗瑟勒姆附近的一所学校，因在食堂里取消了油炸鸡肉卷和汽水，导致愤怒的母亲们群起而抗议——她们坚持认为孩子有权吃汉堡、薯片和其他不健康的食物。[2]我们觉得，自己的饮食，是该自己关心的问题。

与此同时，我们却依赖政府来保证我们的食品安全。遍及欧洲的马肉丑闻，将人们的注意力集中在监管体系出错之后果的问题上。2013 年 1 月，欧洲大陆的消费者得知，他们超市的意式"牛肉"千层面和墨西哥香辣酱，可能含有大量添加了苯丁二酮和其他危险化学物质的马肉，而部分原因，是政府缺乏检查程序。这令消费者十分震惊。我们希望政府能帮助我们吃得安全，如果它做不到，我们就会感到失望。有报道称，对糖的喜爱和对锻炼的不屑，正在造成一场代价高昂的公共卫生危机并阻碍经济发展，这也让我们感到烦心。英国报纸定期发出警告说，糖尿病和肥胖症正在让国民医疗保健制度走向破产，而研究人员计算了我们集体不合理饮食的经济成本。一项调查发现，仅美国，这个数字就大大超过500亿美元。[3] 其他人至少应该吃得理智些，因为他们不明智的消费习惯会影响我们所有人。

　　任何人，特别是政府，在塑造我们的饮食方面应该起多大作用？对此的不一致态度，引起了心理学家所称的认知失调——由持有不相容或相悖的信念及观念而引起的模糊不适感。我们很想斟一大杯杜松子酒，但又担心，如果每个人都这样做，将会导致社会和经济的灾难性后果。我们也不是第一代担心如何在饮食自由和公众福祉之间取得平衡的人。二战期间，美国军队在招募新兵的时候，大量的人因不良饮食习惯造成的健康状况不佳而遭到拒绝，此后，美国的饮食问题上升到了威胁国家安全的程度。为了解决这个问题，政府官员制定了大范围的饮食改革计划，目标不仅是对美国有限的食物资源进行管理，而且还是通过改变国人的饮食习惯来提高公共健康水平。然而，恰恰是负责实施这一政策的人，怀疑它与自由民主制根本不相容。令联邦饮食习惯委员会感到不安的是，它自己的计划所鼓励的那种对规定的服从，被它认为是所谓极权主义政权的特征。真正的美国人，将会，而且应该抵制这种对私人生活的侵犯。[4]

　　个人选择、公共福利、国家富强之间的这些矛盾，是在启蒙运动中诞生的。正是在18世纪时，日常饮食习惯成了政府担忧的问题。关于如何建立经济成功的国家的新理论，引出了个人饮食和国家适应力之间关系的新观点,换句话说,就是我们所称的粮食安全概念的出现。《土豆帝国》讲述了粮食安全概念的悠远历史,并对饮食如何成为现代政治的一部分进行了崭新的阐述。通过展示健康的饮食如何嵌入新自由主义的框架——它强调个人的责任和选择,而不是政府主导的干预——它也有助于解释我们自己与饮食指南之间令人焦虑的关系。

《土豆帝国》通过象征这种转变的食物——土豆的历史来讲述这个故事。如今,土豆已经成为全球的一种主食。据联合国统计,土豆在全世界的每一个国家均有种植。在全世界最重要的粮食作物中,它排名第四。中国是最大的土豆出产国,2016年收获近1亿吨。世界各地每天都在食用土豆,土豆泥、土豆炖花菜、土豆炖孜然籽、炸土豆、煎土豆饼等,土豆的吃法超过千种。目前,欧洲人是吃土豆的主力军,而在单个国家的人均食用量方面,土库曼斯坦则占据第一,每人每年的消耗量接近140公斤。[5] 土豆是典型的现代粮食。由于它的全球重要性和营养价值,联合国曾把2008年定为"国际土豆年"。(参见四川炒土豆丝食谱。)

对于一种16世纪之前完全不为大多数人所知的食物而言,这是一桩了不起的成就。在那以前,只有在从玻利维亚、智利的安第斯山脉向北穿过落基山脉的山脊沿线生活的人,才食用土豆。这些山脉是土豆的故乡,也是庞大的印加帝国的故乡。印加帝国在16世纪被西班牙征服者所覆灭,从此一股旋风将土豆刮到了爱尔兰、印度及更远的地方。土豆在欧洲和全世界传播的故事,勾勒出日常饮食习惯与现代国家之间关系的新历史。

大多数学者认为,民众的饮食习惯影响国家政治、经济安全的概念,形成于19世纪末至20世纪中期。历史学家们解释说,到那时,个人健康不再是私事,而是变成了一个重要的公共问题。[6] 许多国家的政治家和官员前所未有地担心不良饮食习惯对国家效率和实力的影响,他们以一系列创造性的新方案来应对,包括政府补贴学校伙食、开展健康饮食运动等。詹姆斯·弗农(James Vernon)①提到,饥饿和营养不良从个人的不幸变为国家的紧急状况,因为它们开始被视为以影响整个社会的方式,威胁到了"政治稳定、经济生产和种族效率"。它们需要的"不仅仅是慈善的干预,而且还需要国家管理"[7]。关于粮食安全的大量著作,都将两次世界大战期间的发展,与人们日益加深的充足饮食对国家和全球的稳定至关重要的观念联系在一起。20世纪40年代后期联合

四川炒土豆丝

中国种植和食用土豆的数量都超过全世界其他任何地方。土豆与辣椒、玉米、花生一起于17世纪从南美传入东亚,并缓慢渗透到当地的饮食中。这道现代的辣土豆丝食谱,将几种新食物转化为一种地道的川菜。

四川炒土豆丝

1. 配料

土豆(250克)

鲜葱(25克)

干辣椒(4颗)

植物油(40克)

盐(2克)

---

① 詹姆斯·弗农(1965—),英国历史学家。2000年起任加州大学伯克利分校历史系教授。他参与编写了"伯克利英国研究系列"丛书,研究领域包括:现代英国史、大英帝国史、帝国史与全球化的比较研究,著有:《政治与人民:1815—1867年的英国政治文化研究》《再读宪法:19世纪英国政治史的新解读》《饥饿:一部现代史》《大英帝国古怪的自由现代性》《英国剑桥史:1750年至今》等。——译注

> 2. 方法
>
> （1）将土豆洗净削皮，切成粗丝，放入冷水中淘洗几次以去除淀粉，沥干水分。鲜葱切成2.5厘米段。干辣椒去蒂，切成2.5厘米段。
>
> （2）将油倒入炒锅中，加热至中高温，快速放入干辣椒、鲜葱、盐，待干辣椒颜色变至褐红，放入土豆丝迅速翻炒至熟，起锅后便可食用。
>
> 3. 特色
>
> 土豆丝鲜脆，味道微辣。下饭下酒均佳。

国粮农组织等国际机构的成立，通常被认为是这一新观念的顶峰。[8]

有效的治理必须对民众的饮食习惯进行有效管理，这个观点，是现代性的一个重要组成部分。始于19世纪晚期的发展，改变了国家与食物关系的许多方面，政治家和官员们已经能够以一个世纪前难以想象的方式设计和实施雄心勃勃的计划。国家承担人口福利责任的程度也发生了重大变化。不过，日常饮食习惯决定一个国家政治和经济成功的最基本的现代观点，并不是出现在19世纪晚期，而是出现在比它早100年的启蒙运动时期。

这个时代顺序很重要。把这些观点放在18世纪的背景下，我们可以看到食物、政治经济、公共福利和有效治国之道，在启蒙辩论当中的密切联系。所有这一切，都决定性地塑造了今天的世界。在启蒙运动时期，日常饮食习惯获得了新的政治重要性，因为政治家、科学家、哲学家和慈善家们都愈加相信，饮食与国家实力之间存在着联系。18世纪也出现了一种信念，即保证经济运行良好和国家安全的方法，是引导人们，而不是要求人们对食物进行正确的选择。按照政治科学家伯纳德·哈考特（Bernard Harcourt）①的解释，关键问题是支撑今天的营养治国之道的逻辑"根植于自由经济理论的最初构想中"。⁹

饮食究竟是自己的事还是大家的事，我们可以把这样的矛盾情绪归咎于启蒙运动。今天的健康饮食餐盘、食物金字塔②和政府的饮食指南，都起源于18世纪。对更有力干预形式（如布隆伯格市长的苏打水禁令）的敌意，也同样起源于18世纪。追踪土豆从安第斯山脉到世界各地的旅程，是追溯这段历史的一种方式。土豆的历史，同样也为思考粮食安全，或更好地思考人们常说的粮食主权，开辟了另一种视角。粮食主权所强调的，是赋予当地人自主权来决定他们自己的饮食和农业行为的重要性。对土豆成为南美主食起到主要作用的是安第斯的村民，而不是印加帝国。将土豆种植传遍欧洲的先驱者是近代早期的农民和劳工。如今，联合国的分析家和农业专家日益意识到，小农户才掌握着可持续农业未来的钥匙。

跟随土豆的旅途有助于讲述这个故事，因为它异常清晰地揭示出这些交集。土豆让我们看到，我们的饮食观念如何与资本主义的出现及其对自由市场的颂扬纠缠在一起。土豆的故事也提醒我们，普通人以持续塑造我们生活的方式来创造历史。简而言之，土豆是思考现代世界起源的一条好途径。《土豆帝国》一书最后提出，如果不了解我们在分析中使用的语言和思想的起源，就无法缓解我们目前对粮食正义和安全的担忧。

---

① 伯纳德·哈考特（1963— ），美国政治理论家，现任哥伦比亚大学法学和政治科学教授。主要论著有《镇压革命：美国政府针对其公民的战争》《危害性原则的崩溃》《燃料与粮食管理》等。——译注
② 健康饮食餐盘和食物金字塔是指导人们食用不同食物种类及比例的常用指南。1992年美国农业部推出了最广为人知的食物金字塔，2005年推出修正版，2011年更替为"我的餐盘"。哈佛大学公共卫生学院也发布了著名的健康饮食餐盘，超过25个其他国家和组织也发表过食物金字塔。——译注

国家的营养

　　对于普通人吃什么，国家并不总是有兴趣。人们有饭吃当然是非常重要的，各地统治者长久以来都十分担心饥荒带来的政治后果。都铎时期的政治家威廉·塞西尔（William Cecil）曾说过，没有什么"比食品短缺更能导致人们的暴动"。[10] 在古代和近代早期的世界中，许多地区的政体都设有公共粮仓，它们能够在粮食短缺的时候将粮食分配给城镇居民。以罗马帝国为例，它曾投入大量资源为首都居民提供可靠的粮食供应。中国政府也与众不同地利用国家粮仓系统供养城市和农村人口。大约从公元7世纪开始，中国的统治者就对帝国许多地方的粮仓网络进行监管，他们的目的是防止饥荒，为所有臣民（不仅仅是生活在城市的臣民）维持稳定的粮食供应。国王们也对忽视自己军队粮食需求的危险保持着警惕。在安第斯山脉，印加帝国看管着庞大的仓储设施网络，它们用来为帝国和帝国的军队储存粮食和其他物资。[11]

　　此外，统治者早就认识到调控粮食价格的重要性。奥斯曼帝国果断地干预谷物周期，以确保伊斯坦布尔能收到每日所需的数百吨面粉。奥斯曼帝国制定法规以控制粮食的销售价格，并通过使用国有船队，常直接管控粮食到城镇市场的运输。[12] 在中世纪的许多欧洲城市，市政府对食品的价格和质量都进行了管制。行会体制同样也是为了确保市场上出售的食品符合健康和质量的标准。没有达到这些要求的食品供应商将被罚款。例如，英国在1379年的一项法令，就对伦敦出售含有"不合适、发臭、骗人的废料"肉饼的面包师进行了制裁。[13] 对这样的做法进行禁止，是一种合法的，实际上也是必要的治理行为。

　　市政当局有时也会对婚宴或其他节日聚会上膳食的奢侈程度进行监

管。立法可能会禁止某些菜肴或限制总预算。公元前2世纪的罗马《法尼乌斯法》(*Lex Fannia*)就限制了私人宴席上菜品的数量。这类法规可能还会详细规定哪些人可以、哪些人不可以吃某些食物。17世纪的日本农民被禁止食用多种食物,包括豆腐和白米。[14] 这些禁止奢侈的法律旨在防止浪费性支出或过度放纵,也致力于维护社会的差别。到16世纪,欧洲的法规更开始反映出一种担忧,那就是个人或集体暴食将会预示着向更普遍恶行的堕落,这将威胁到整个国家。亨利八世臃肿过度的身体,对法国天主教徒来说,既是过度放纵对身体的危害,也是英国宗教改革后的道德沦丧。因此,个人的饮食缺陷,影响并反映着一个国家的精神状态。[15]

在大多数宗教中,为饥民提供食物更是一种慈善的必要。对锡克教徒来说,为需要的人分发食物,是最为重要的宗教义务。在近代早期的欧洲和地中海地区,各种宗教慈善机构也向囚犯、贫民和其他饥饿的民众分发食物。[16] 给穷人提供食物与更大的宗教架构之间的这种关系,有助于政府将粮食供应作为一个伦理问题来关注。儒家哲学家孟子坚持认为,如果君主在饥荒时期不提供粮食,在道义上无异于谋杀。在莫卧儿王朝之前的孟加拉,国王也被要求分发大米给需要的人。饥饿给统治者强加了一种道义上的义务,而未能履行这一义务者,则有丧失合法性的风险。[17]

总而言之,对于大多数古代和近代早期国家来说,确保城市人口能够获得稳定、安全的粮食供应,是公认的治国方略的组成部分。统治者更是通过对穷人福祉的关注,在某种程度上来证明他们具有治理国家的道德素质。历史学家查尔斯·蒂利(Charles Tilly)① 发现,粮食供应

---

① 查尔斯·蒂利(1929—2006),美国社会学家,政治科学家。蒂利的研究主要集中于社会运动及现代民族国家的产生,并因其在这方面的开创性贡献而获得社会科学研究委员会颁发的第二届艾伯特·赫希曼奖。主要著作有《强制、资本和欧洲国家:公元990—1992年》《欧洲的抗争与民主》《身份、边界与社会联系》等。——译注

涉及的各种因素，"无论如何都不能构成一个和谐的整体。相反，它们实际上使得粮食政策成为激烈的政治辩论的主题"。[18] 但是，只要不因饥荒而人口消亡，不因铺张浪费而造成社会秩序大乱，不因罪恶的过度消费而招致神的愤怒，政治哲学家们就不会过多考虑普通人饭桌上有什么。君主们关心的是防止粮食骚乱，通常并不关心臣民日常饮食的具体特点。不管他们的粥是用小米还是大麦熬的、汤里有没有卷心菜、面包是烤的还是煎的，都没有政治上的重要性。这些事是牧师和医生的职责，不是政治家的。只有统治者的日常饮食，才具有无可置疑的政治重要性，宫廷医生为他设计个性化的饮食计划，他的健康，是长期存在的政治焦虑话题。[19]

由于普通人的饮食并非统治之术的一部分，因此，人们的日常饮食习惯，在近代早期欧洲的政治论述中并没有突出的地位。尼可罗·马基雅维利（Niccolò Machiavelli）① 并没发现这个主题与他关于治国才能的讨论有关。乔万尼·博泰罗（Giovanni Botero）② 在1589年的《论国家理性》(*The Reason of State*)一书中，对有效统治进行了开创性的、具有影响力的分析，以完全传统的观点来看待食物对于有效统治的重要性。他提醒读者，"经验不止一次地、而是多次地告诉我们，面包的缺乏比任何其他东西更能激怒普通百姓"，因此能够导致叛乱。他指出粮食供应对军事行动的重要性，还告诫统治者在所有事情，包括饮食上保持克

---

① 尼可罗·马基雅维利（1469—1527），意大利政治思想家和历史学家。其思想常被概括为马基雅维利主义。在中世纪后期的政治思想家中，他第一个明显地摆脱了神学和伦理学的束缚，为政治学和法学开辟了走向独立学科的道路。——译注
② 乔万尼·博泰罗（1544—1617），文艺复兴时期意大利思想家、牧师、外交家与诗人。他反对尼可罗·马基雅维利的思想，他的经济思想亦为后世的经济学家提供了理论来源。主要论著有《论城市伟大至尊之因由》《论国家理性》等。——译注

制的道德必要。然而，民众特定的饮食习惯并没有成为他治国方略的一部分。[20] 政治哲学家托马斯·霍布斯（Thomas Hobbes）①完全没有关注过这个话题。在他1651年的《利维坦》(Leviathan)一书中，霍布斯将饮食简单地视为人类的基本需求，而不是国家的要务。[21]《论国民的营养》一章，并没有考虑人们究竟是如何给自己提供营养的世俗问题。而"营养"（nourishment）只是代表着商业和产权。对于像霍布斯这样的政治哲学家来说，维持国家政体所必需的"营养"，是促进贸易的黄金、白银和其他商品，而不是面包和肉汤。[22] 因此，治国之道的关键，不是人们在厨房里做的事，而是商业、市场监管和粮食供应。

到18世纪晚期，这样的国内事务在欧洲已成为统治之术的一部分。在英国，连首相也向议会提出，有必要呼吁民众在日常面包中添加更多全麦和土豆。[23] 对这类细节前所未有的关注，反映出对统治的理解发生了变化。几十年之前，法国理论家米歇尔·福柯（Michel Foucault）②就发现了这些变化，他描述了17世纪晚期欧洲出现的对"人口"的一种新的政治认识。新的治国理论家们把人口视为除森林和工厂等禀赋之外的另一种需要管理的资源。[24] 饮食改革与支持生育的政策、消除懒惰的计划、公共卫生运动和其他旨在提高人口素质的事业一起，构成了18世纪的统治理念。从18世纪政治理论家的角度来看，人们每天吃什么

---

① 托马斯·霍布斯（1588—1679），英国政治家、哲学家。他提出"自然状态"和国家起源说，指出国家是人们为了遵守"自然法"而订立契约所形成的，是一部人造的机器人，反对君权神授，主张君主立宪。著有《论公民》《论物体》《贝希莫特》《利维坦》等。——译注

② 米歇尔·福柯（1926—1984），法国哲学家、思想史学家、社会理论家、语言学家、文学评论家、性学家。法兰西学院思想体系史教授。他对文学评论及其理论、哲学（尤其在法语国家中）、批评理论、历史学、科学史（尤其医学史）、批评教育学和知识社会学有很大的影响。福柯的主要著作有《古典时代疯狂史》（1961）、《词与物》（1966）、《知识考古学》（1969）、《规训与惩罚》（1975）、《性史》（1976—1984）等。——译注

与评估国体的实力和健康程度密切相关。到18世纪末,关于劳动者和其他普通民众的饮食习惯不再像早些时候那样仅有寥若晨星的报道,渴望评估国家整体健康状况的热衷政治的观察家对此进行了大量的评论。政府的管理技巧和现代的政权,都已经开始将日常饮食包括在内。有效的统治需要对民众的饮食习惯进行某种程度的细查,这种信念是现代性的一个重要组成部分,而这种信念的出现,是在启蒙运动时期。

国家与个人

许多18世纪的作家对政体富强与民众活力之间的新关系都进行了表述。具有哲学思想的军官沙特吕侯爵(Marquis de Chastellux)曾撰写过一部有关公众幸福的极具影响力的专著。他认为,英国劳工比法国人更健康,因为他们享受着优越的饮食。因此,且不论两国的相对人口如何,英国是强于法国的。于是,改善法国劳动人民的饮食习惯,成了法国政府的当务之急;[25] 找出并推广改善饮食习惯的食物,被誉为一项爱国事业。许多作家认为,土豆就是这项挑战的最佳解决方案。几粒盐、一点黄油、培根或牛奶,就足以把它变成健康而令人满意的一顿饭,一本法国农艺手册如此写道。它总结说,"没有一种庄稼如此有用"。[26] 在法国大革命之前和之后,法国官员通过有奖竞争、出版物和劝勉来鼓励土豆的消费,且他们也并没有孤军奋战。在整个18世纪的欧洲,许多个人都在尝试推行改善劳动人民营养健康的方案,因为人们认为,这样的计划能够增强一个国家的实力。(参见图1)

推广的方案虽都很好,但将新的政治信念转化为真正的行为改变,却并非易事。从日常饮食第一次引起政治理论家注意的那一刻起,争论就已经开始:谁负责确保民众遵守这些建议,如果他们不遵守又该谁负责？18世纪的科学家、政治家和社会评论家常感叹工人阶级吃不好,并争论到底是穷人的无能还是经济结构限制了他们的购买力。1776年,苏格兰医师威廉·巴肯(William Buchan)抱怨说,农民们"对他们的吃、喝极其不重视,而且常常由于懒惰而吃不健康的食物,尽管他们本可以同样的花费吃到健康的食物"。[27]巴肯推荐土豆,说它是一种极其适合和经济的食物。但有一些人对这样的批评提出了质疑,他们或者称赞工人阶级现有的饮食习惯非常健康,或者把责任转嫁给商人和官员,

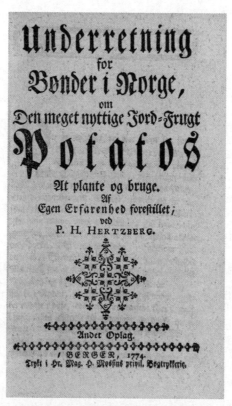

图1 彼得·哈伯·赫茨伯格(Peter Harboe Hertzberg)1774年出版的土豆种植指南的卷首插图。这本小册子为挪威农民提供了种植"非常有用"的土豆的指南。18世纪晚期,类似的文章在欧洲各地都有出版。把民众健康与国家富强联系起来的新思想,使土豆等营养丰富的主食受到政治理论家和爱国主义者的青睐。

认为是他们把粮食价格推到无法承受的高度。更激进的声音认为，权力和财富的剥削性分配才是劳动人民饮食可悲而不健康的真正原因。

饮食习惯不良主要是由"粗心大意"还是贫穷造成，这个分歧一直存在。在巴肯批评了农民饮食习惯的百年之后，营养学家马克斯·鲁布纳（Max Rubner）又批评了德国产业工人的"非理性"行为，说他们把有限的资源浪费在黄油三明治而不是更有营养的食物上。还有一些人，如写文提倡免费校园餐的一位布拉德福德的社会主义议员，则对不公正的经济结构进行了指责。约克的慈善家斯波姆·朗特里（Seebohm Rowntree）估计，该市许多穷人缺乏经济来源，连最低程度的合理膳食都吃不到。尽管如此，他相信，"如果人们对不同食物的相对价值有更多的了解，并能够通过这些知识获益，那么在不增加成本的情况下，他们的饮食可能会得到一些改善"。[28] 如今向社会保障金接受者提供烹饪课程的提议，就是这一场持久讨论的新篇章。[29]

把这些不同观点团结在一起的共同观念是，个人的饮食习惯不仅仅是有关个人的问题。当然，饮食对个人健康的影响早就得到承认，几千年来，医生们一直在指导人们吃什么才能保持健康。医生警告说，如果不遵循这些指南，无疑就是在自掘坟墓。[30] 然而，威廉·巴肯和布拉德福德市议员却看到了这种个人的失败所带来的更广泛的政治风险。巴肯认为，英国"真正的财富、人民的幸福和政府的稳定"与普通民众的日常饮食习惯息息相关。那位布拉德福德的社会主义者抱怨说，由于"水准较低和不合理的饮食"导致身体单薄，结果参加布尔战争的志愿者中，有40%的人太过于孱弱，连"当个靶子都不够好"。[31] 工人们因粗心或不负责任的饮食选择而影响自己的健康，他们就会削弱国家的安全和生产力。给孩子吃不适当食物的母亲，会招致批评说因此而削弱了孩子的身体和道德素质，而且还会增加劳工骚乱的可能性。为了应对这些弊害，18世纪的慈善家出版了许多经济食谱，旨在向穷困女性展示"以少量的钱带来更多益处的方法"。19世纪的社会分析家建议，应该向工薪阶层的女性灌输预算、普通烹饪和按营养进行菜谱计划的基本原则。到了20世纪，"家政学"已经进入了许多受国家资助的教育系统的课程。[32]

对于18、19世纪的政治作家而言，通过这样的教育运动来鼓励穷人接受推荐的饮食习惯，是合适的治国之道。然而，大多数人却并不同意国家负有保证最低饮食的全部责任的提法。[33] 政治哲学家、政治家埃德蒙·伯克（Edmund Burke）①于1795年曾说过："为我们提供必需品不是政府的权力。"[34] 19世纪晚期出现的福利主义哲学将这一责任归于国家，这标志着政治思想的新发展。然而，它并没有反映出人们对饮食对国家的重要性有了新的认识。个人健康与国家实力相关的观念诞生得更早，出现在18世纪。土豆作为启蒙时期超级食物的生涯，有助于讲述这个故事。

农民与土豆

早在欧洲的政治家们开始注意到土豆具有建设强健人口的潜力之前，南美的土豆就已经在数千年的时间里养活着许多的普通人。大约在1.2万年前，从智利到今美国犹他州，人们就在食用野土豆。它们的驯

---

① 埃德蒙·伯克（1729—1797），英国政治家、作家、演说家、政治理论家和哲学家。对法国大革命的反思使他成为辉格党里的保守主义主要人物（他还以"老辉格"自称），反制党内提倡革命的"新辉格"。他经常被视为英美保守主义的奠基者。主要论著有《为自然社会辩护：检视人类遭遇的痛苦和邪恶》《与美国和解》《对法国大革命的反思》《论崇高与美丽概念起源的哲学探究》等。——译注

化可以追溯到公元前7800年左右。[35] 安第斯山脉沿线的村民以种植土豆来维持日常生计，他们创造了大量的农艺知识和繁多的土豆品种，这些土豆以不同的用法给他们提供全年的营养。在欧洲人和西非人于16世纪30年代到达南美洲的太平洋海岸后，他们立即就注意到，土豆是一种必不可少的主食，在当地相当于欧洲的面包或西非的小米、高粱、山药粥。他们把土豆带到了欧洲，从而开启了它在全球的传播。殖民者还将土豆与种植和食用它的土著居民联系在一起，并推测，富含土豆的饮食在某种程度上解释了他们认为安第斯人所独有的鲜明特征。因此，

图2 描绘安第斯山区土豆收获的17世纪绘画。土豆与其他块茎植物、玉米和藜麦一起构成了安第斯人的主要饮食。这幅当地编年史家费利佩·瓜曼·波马（Felipe Guaman Poma）的绘画展示的是正在收获土豆的村民。这名男子使用的工具为chakitaqlla，即脚犁。类似的工具至今仍在使用，因它们有助于在陡峭不平的土地上进行耕作。

土豆，或更普遍而言，食用土豆的习惯，引起了近代早期殖民者的兴趣，他们依靠这类医学和人种学信息来了解和解释他们渴望统治的陌生民族和国家。[36]（参见图2）

在近代早期的欧洲，土豆很少引起国家政要的注意。由于对欧洲劳动者的日常饮食习惯缺乏关注，因此，很少有政治理论家评估这种新庄稼作为欧洲人食物的潜力，即使有，他们也只是抱怨土豆营养过多，并因此助长了懒惰。相比之下，普通民众却喜欢土豆，因为它具备比已有的食物更多的优势。即使在贫瘠的土壤中，土豆也能丰收，而且它还能成为营养丰富的大餐，特别是在与牛奶或其他乳制品配合时。而且，正因为国家对穷人的日常饮食习惯缺乏关注，过了好几十年，土豆才吸引了税吏和其他财政官员的注意。因此，农民和劳工因土豆得以逃避国家控制不受欢迎的一些方面，而正是他们，让土豆进入了欧洲人的饮食。本书第一章讲述的就是这段历史。近代早期农民和小规模栽培者在园艺和饮食创新方面的中心作用，对今天的农业改革纲领具有重要意义。本书的总结部分再次讨论了这一早期的农业和饮食变化在当代产生的共鸣。

土豆在政治上的隐身术结束于18世纪。曾经招致批评的营养性开始得到积极的对待。因此，土豆成为整个欧洲的科学和政治非常关注的对象。从挪威到那不勒斯，从圣彼得堡到日内瓦，官员、当地社团、农学家、牧师以及许多其他组织和个人，都在口头和行动上倡导土豆的消费。这场广泛的、遍及欧洲的土豆研究和宣传，导致了数百部书籍的出版，它们颂扬土豆作为劳动人民优秀主食的潜力，而劳动人民更多的消费，将有助于保证国家的实力和成功。本书第二章追溯了土豆在启蒙运动时期前所未有的政治地位。它受到的欢迎，反映了政治经济和治国新模式的出现，这种模式强调了健康、营养良好的人口对国家权力和财富的重要性。将土豆征服欧洲饮食的缓慢历史，与18世纪对土豆的狂热推广相结合，阐述了粮食在现代治国模式中所扮演的核心角色。

本书第三章将 18 世纪的土豆推广与同一时期出现的政治经济学新思想联系起来。正如亚当·斯密（Adam Smith）所认为的那样，允许个人追求自身利益，最终会导致经济繁荣，因此土豆热衷者（包括斯密本人）认为，构建健康人口的最佳方式，是通过知识和劝勉活动，赋予个人选择健康饮食的权利。启蒙主义土豆推广者相信，这种块茎食物的优点是如此巨大，单是个人利益就足以刺激消费的增长。因此，可让市场和公众健康以给个人和国家都能带来最佳结果的方式组织起来。要说明在这些思想最初形成的那一刻，这种私人利益和公共利益的融合如何发生，土豆就是一个具体的、日常的例子。

这段历史揭示，目前的新自由主义主张正是起源于 18 世纪，这种主张认为，健康饮食最好能被理解为是一种个人消费者的自主权，而这种自主权同时又能构建更强大的经济和政体。18 世纪哲学家伊曼努尔·康德（Immanuel Kant）认为，能决定自己的饮食，是智力成熟的基本特征，它如同其他形式的批判性思维一样，是典型的启蒙时期的特征。自主思考和行动的能力，无论是关于饮食还是其他事物，仍然是关于自由的一切开宗明义的基本组成部分。与康德相呼应，自由主义经济学家塞巴斯蒂亚诺·巴韦塔（Sebastiano Bavetta）、彼得罗·纳瓦拉（Pietro Navarra）和达里奥·迈蒙（Dario Maimone）在 2014 年的一项研究中坚持认为，"一个人能力的最高发展……源于能够进行自主选择"，包括个人消费的选择。[37] 有观点认为，作出选择会带来更大的个人幸福，从而带来更幸福、更公平的社会。正是通过这种理想化的趋同，个人的自由选择成了现代自由国家的理论基础。本书第三章探讨了饮食选择在这种自由主义治国模式中的作用。

第四章论述了土豆的全球历史。在近代早期，贸易扩张和殖民征服推动了土豆在世界各地的普及。但是，它成为全球主食的过程，不仅反映了这些力量，也反映出它在旅程中遭遇的各种不同的当地境况。欧洲的外交官、传教士和殖民官员，都庆幸自己把营养丰富的土豆带给了孟加拉和博特尼湾的落后居民，并把对土豆的接受视为当地所达到的文明整体水平的指标。对德黑兰的园丁和新西兰的毛利企业家而言，土豆则服务于其他的用途。随着这一美洲食物在1492年之后的全球传播出现的饮食习惯的转变，一直是当地环境、农业模式和土地所有、商业结构和已有饮食方式之间复杂交互作用的结果。第四章既解释了土豆在欧洲有关帝国主义论述中的地位，也关注了世界各地的普通人与这种新食物相互作用的方式。因此，对于第一章所关注的农民和小农户在将土豆融入当地饮食中的作用，本章提供了全球性的对比。（参见图3）

土豆在中国的地位变化，较好地体现了现代食物面貌的地方性及全球性。在好几百年里，土豆为不那么重要的地方的村民提供了重要的资

图3 《长官与土豆》（'The Aga and the Potato'）。从16世纪开始，土豆传遍了全世界。这幅19世纪的插图描绘了（毫无疑问是虚构的）一位土耳其长官正在品尝英国旅行者的行李中发现的生土豆的场景。

源，但政府几乎对其视而不见。然而，自21世纪初年起，土豆已被纳入了旨在增强中国粮食安全的广泛国家计划中。土豆政治角色的这一转变，与中国政府对市场经济的接受同步发生。因此，国家对土豆的大力推广，是对个人选择和个人利益的强调。一个鼓励人们食用土豆的广告宣传活动承诺："让我们都成为富有的土豆。"中国跟欧洲国家一样，资本主义、个人主义和个人饮食习惯与现代治国方式都紧密交错在一起。

产生于18世纪的土豆与政治经济之间的联系，赋予了土豆在政治和经济上的强大共鸣。因此，土豆深深融入了19世纪有关资本主义优点的争论中。第五章对这些争论进行了展示。政治激进分子谴责土豆是一种剥削工具。1830年，肯特郡的农业劳工们打出的横幅上写着："我们不能靠土豆生活。"[38]他们并不反对土豆本身，而是反对把土豆作为一种意识形态。"靠土豆生活"可简要理解为抗议者所反对的整个体制。土豆在19世纪备受争议的地位，在关于爱尔兰的讨论中显露无遗。在18世纪，爱尔兰曾是土豆在促进人口和经济增长方面具有有利作用的典范。到19世纪中期，它却成为令人担忧的经济闭关自守的例子。在历届英国政府眼中，爱尔兰的土豆不是促进了商业的繁荣，而是加重了对进入现代世界的拒绝。当国家需要的是一大批无产阶级化的劳动者时，"自给自足的吃土豆的民众"，便不再是一个具有吸引力的提法。

与此同时，随着城市化和工业化进程的推进，人们更加坚信民众的饮食习惯对国家政体的实质性影响。营养学的新语言为表达这种关系提供了词汇表。从19世纪中期开始，土豆在劳动者阶层的饮食中日益增加的重要性，并没有引来赞美，而是引起了营养学家和政治家的担心。他们谴责"土豆的懒惰血"与土豆食用者的破落户态度带来的共同影响：土豆在劳动者当中的普及，被谴责为经济低迷的原因。[39]谈论土豆，为劳动人民、科学家、经济学家和政治家提供了一种方式，以强调经济行为和民众日常饮食习惯之间的密切关系,来讨论那些重塑19世纪欧洲的巨大变化。

第六章的主题,是20、21世纪土豆与国家的关系史。自1900年始,土豆开始重新焕发出作为政治工具的光彩。营养科学领域的发展,使营养学家推翻了他们之前谴责土豆是劣质食物的说法。在营养学家对土豆的评价变得更为积极的同时,现代国家影响日常饮食习惯的能力大幅提高。在技术和制度的发展将其实现的过程中,一战和二战起到了尤为重要的作用。为满足民众的战时需要,欧洲各国政府在国家土豆育种农学方案的支持下,将积极鼓励土豆消费视为一种爱国工作。今天一些最著名的品种,就是这些国家支持计划的产物。

在20世纪,虽然土豆的商业育种和消费成为国家粮食安全计划的一部分,但在战后出现的经济发展模式中,土豆却很少受到关注,它成为更受忽视的发展中国家小规模、商业化程度较低的农业的一部分。直到最近,这种农业才被纳入国际粮食安全模式当中。将土豆种植向近代早期欧洲传播的农民的专业知识,在很大程度上一直不为人所知,同样地,让土豆保存其遗传多样性的小农专业知识,也只是刚刚才开始受到关注发展的国际组织的重视。联合国将2008年宣布为国际土豆年的决定,就反映了这种新的重视态度。土豆如今被誉为对抗饥饿的必要资源。它们还成为美食骄傲的理由,许多国家把特定的品种登记为一种国家遗产。土豆的当代历史,既概括了土豆能在国家安全中发挥作用的18世纪的信念,也概括了小农和农学家一样掌握着与建立可行粮食系统相关的专业知识的现实。

*

本书探讨了饮食与国家关系的特定思考方式的谱系,讲述了日常饮食变为现代政治一部分的过程。本书认为,产生于启蒙运动时期的思想,影响了我们对我们的饮食选择与国家实力和经济成功之间关系的看法。尤其是，我们认为吃什么既是个人问题又是公共利益问题的观念，它的根源来自18世纪。我并不是第一个注意到以下这一点的人——在欧洲，个人选择和公共利益之间的紧张关系在18世纪后期呈现出特别的力量，这些紧张关系成为政治理论的中心话题。而我所感兴趣的是，如果我们把食物与国家、政治和个人主义的兴起放在一起考虑，我们对这些紧张关系的分析会受到怎样的影响。《土豆帝国》在论述的时候，并不会佯称18世纪的思维模式与我们完全相同，或者18世纪的食物供应系统、农业或政治组织与今天一致。我也不认为跟随土豆的足迹是讲述这个故事的唯一方法。既然（我相信）这个更大的故事是正确的，那么用其他的食物或方法来追溯它也是同样可能的。不过，土豆恰好是一种有效的跟踪物，因为土豆渗透到普通民众的厨房的过程，并未与它在政治论述中占据突出地位的过程同步发生。它在18世纪政治论述中地位的突然升高，让现代治国方略的新特征成为焦点。而且，它是一种如此成功的全球食物，以至于它让我们能够描绘出日常饮食与现代国家之间关系的更具全球性的故事。

当然，关于土豆的政治讨论并非各处都相同。政治结构、食物体系和许多其他因素的差异，不仅会让单一主题产生极大的变奏，有时还会产生完全不同的基调。尽管如此，我一直在寻找潜藏在它们当中的和声，它让我们能够欣赏"某种信念是如何变得普遍，并且……这些信念如何影响我们的行为"。[40] 更重要的是，土豆的故事可提醒我们以其他方式思考民众、食物与现代国家之间的关系。使土豆变为全球食物的实际行动，在很大程度上是由普通人而不是政治哲学家们实施的。因此，以土豆为中心的历史，将现代国家对民众饮食时起时落的关注，与小农创新和农民专业技能的更强有力的持续低音进行了平衡。历史学家艾玛·斯帕里（Emma Spary）写道，一个单一的主题，"在不同的人手中有完全不同的含义和用法"。[41] 土豆兼具现代国家工具和普通民众历史代表象征的双重地位，就是一个很好的例子。《土豆帝国》讲述的，是饮食如何成为治国方略一部分的故事，同时它也以新的方式，记录了世界上最重要的粮食作物之一在全球的传播。

注释

1. 'Woman's Hour', BBC Radio 4, 17 Mar. 2014.

2. Martin Wainwright, 'The Battle of Rawmarsh', *Guardian*, 20 Sept. 2006; *New York Times*, 2 June 2012.

3. 例如，可参见 Rebecca Smith, 'Obesity Epidemic "Could Bankrupt the NHS"', *Telegraph*, 15 Oct. 2007; Amanda Platell, 'Sorry, Why Should the NHS Treat People for Being Fat?', *Daily Mail*, 27 Feb. 2009; Runge, 'Economic Consequences of the Obese'; Business in the Community, *Healthy People=Healthy Profits*; Harvard School of Public Health, 'Public Health and the US Economy'; Sarah Boseley, 'Obesity Could Bankrupt NHS if Left Unchecked', *Guardian*, 17 Sept. 2014。关于马肉丑闻，参见 Abbots and Coles, 'Horsemeat-Gate'。

4. Biltekoff, *Eating Right in America*, 45–79. Thaler 和 Sunstein 在 *Nudge* 讨论了这些紧张关系。

5. FAO, FAOSTAT.（确切地说，按照 FAO［联合国粮农组织］的数据，所有的国家都种植土豆）。

6. Rabinbach, *The Human Motor*; Harmke Kamminga and Andrew Cunningham, 'Introduction', *The Science and Culture of Nutrition*, ed. Kamminga and Cunningham, 2; Helstocky, 'The State, Health, and Nutrition'; Zweiniger-Bargielowska, *Managing the Body*, 22; Nally, 'The Biopolitics of Food Provisioning'.

7. Vernon, 'The Ethics of Hunger and the Assembly of Society', 699 (quote); Vernon, *Hunger*; Cullather, *The Hungry World*.

8. Maxwell, 'Food Security'; Shaw, *World Food Security*; Schanbacer, *The Politics of Food*.

9. Harcout, *The Illusion of Free Markets*, 44. "营养治国之道"一词，来自 Swislocki, 'Nutritional Governmentality'。

10. Walter, 'The Social Economy of Dearth in Early Modern England', 76. 另见 Davis, 'Poor Relief, Humanism and Heresy'; and Tilly, 'Food Supply and Public Order in Modern Europe'。

11. Murra, *The Economic Organization of the Inka State*, 121–134; Rickman, *Corn Supply of Ancient Rome*; Braudel, *The Mediterranean and the Mediterranean World*, I: 328–332, 570–605; Will and Wong, *Nourish the People*; Steel, *Hungry City*.

12. Murphey, 'Provisioning Istanbul'; Will and Wong, *Nourish the People*, 511. 关于法国谷物市场的监管的范例性研究，参见 Kaplan, *Bread, Politics and Political Economy*; Miller, *Mastering the Market*; Harcout, *The Illusion of Free Markets*。

13. 'Ordinances of the Pasteters, or Piebakers', 3 Richard II. A.D. 1379, *Memorials of London and London Life*, ed. Riley, 438 (quote); Pullan, 'The Roles of the State and the Town in the General Crisis of the 1590s', *Poverty and Charity*; Sharp, *Famine and Scarcity in Late Medieval and Early Modern England*.

14. Brundage, 'Sumptuary Laws and Prostitution in Late Medieval Italy', 344; Rath, *Food and Fantasy in Early Modern Japan*, 114–116.

15. Moyer, '"The Food Police"'.

16. 例如，可见 Pullan, *Poverty and Charity*; Safley, ed., *The Reformation of Charity*; Abreu, *The Political and Social Dynamics of Poverty*; Campbell, *At the First Table*。

17. Thompson, 'The Moral Economy of an English Crowd'; Kaplan, *Bread, Politics and Political Economy*; Khondker, 'Famine Policies in Pre-British India'; Davis, *Late Victorian Holocausts*, 286–288; Murphey, 'Provisioning Istanbul'; Will and Wong, *Nourish the People*, 2, 511.

18. Tilly, 'Food Supply and Public Order in Modern Europe', 431.

19. 关于食物对身体和精神健康的中心作用，参见 Gentilcore, *Food and Health in Early Modern Europe*。

20. Machiavelli, *The Prince*; Botero, 'The Reason of State' (1598 edition), *The Reason of State and The Greatness of Cities*, 73.

21. Hobbes, *Leviathan*. 另见 Pagden, *The Languages of Political Theory in Early Modern Europe*。

22. 因此，墨西哥和秘鲁源源不断地为欧洲提供"金、银的母乳"：Salinas y Cordova, *Memorial, Informe y Manifiesto*, 19r.。"如果胃不消化食物和分配它的益处，不仅身体的其他部位会挨饿和腐烂，胃本身也会。就像一位王子将他的臣民的财富拿走并消耗，如果不对它吸收并按照他们的需要分享，给他带来的破坏不会少于他的封臣"，博泰罗写道。'The Greatness of Cities' (1606 Robert Peterson translation), *The Reason of State and The Greatness of Cities*, 142.

23. *Times*, 4 Nov. 1795, 2.

24. Foucault, 'The Subject and Power'; Foucault, *Security, Territory, Population*; Coveney, *Food, Morals and Meaning*. 另见 Stern and Wennerlind, eds., *Mercantalism Reimagined*。

25. Marquis de Chastellux, *De la félicité publique*, II: 142–143.

26. Alletz, *L'Agronome*, II: 360. 另见 Spary, *Feeding France*。

27. Buchan, *Domestic Medicine*, 46.

28. Hartley, *How to Feed the Children*; Rowntree, *Poverty*, 240 (second quote); Vernon, 'The Ethics of Hunger and the Assembly of Society', 719–721; Treitel, 'Max Rubner and the Biopolitics of Rational Nutrition' (first quote); Biltekoff, *Eating Right in America*; Vernon, *Hunger*.

29. 'Jamie Oliver Bemoans Chips, Cheese and Giant TVs of Modern-day Poverty', Guardian, 27 Aug. 2013, www.theguardian.com/lifeandstyle/2013/aug/27/jamie-oliver-chips-cheese modern-day-poverty; BBC News, 'Tory Peer Apologises for Saying "Poor Can't Cook"', 8 Dec. 2014; Dowler, 'Food Banks and Food Justice in "Austerity Britain"', 164. 另见 Wheeler, 'To Feed or to Educate?'

30. 这句话来自 Smith, *Sure Guide in Sickness and Health*, 61。

31. Buchan, *Observations Concerning the Diet of the Common People*, 43; Hartley, *How to Feed the Children*, 2.

32. More, *The Cottage Cook*; Shapiro, *Perfection Salad*; Treitel, 'Max Rubner and the Biopolitics of Rational Nutrition'; BBC News, 'Cookery to be Compulsory', 22 Jan. 2008; Elliot, 'Cookery Examined'; Richardson, *The Political Worlds of Women*; Biltekoff, *Eating Right in America*; BBC News, 'Tory Peer Apologises for Saying "Poor Can't Cook"'.

33. Vernon, *Hunger*; Simmons, *Vital Minimum*.

34. Burke, *Thoughts and Details on Scarcity*, 2.

35. Ugent, Dillehay and Ramírez, 'Potato Remains from a Late Pleistocene Settlement in Southcentral Chile'; Louderback and Pavlik, 'Starch Granule Evidence for the Earliest Potato Use in North America'.

36. Earle, *The Body of the Conquistador*.

37. Kant, 'What is Enlightenment?'; Gray, *Liberalism*; Bavetta, Navarra and Maimone, *Freedom and the Pursuit of Happiness*, 43.

38. Cobbett, *Cobbett's Weekly Political Register*, 24 Mar. 1832, 786–787.

39. Moleschott, *Lehre der Nahrungsmittel*, 119.

40. Harcout, *The Illusion of Free Markets*, 50.

41. Serres, 'Theory of the Quasi-Object'; Spary, *Eating the Enlightenment*, 192 (quote).

第一章　外来的土豆

> 土豆,类似西班牙松露,只是大一些,也没那么美味。这种块根冻干后被称为"chuño"(冻土豆)。[1]
>
> ——韦森特·德·瓦尔韦德(Vicente de Valverde),1539 年

这是最早的对土豆的书面描述。韦森特·德·瓦尔韦德是一位多明我会修士,他跟随由弗朗西斯科·皮萨罗(Francisco Pizarro)和其他几个西班牙人率领的凶猛远征军,于 16 世纪征服了位于今日秘鲁的印加帝国。瓦尔韦德在绑架和谋杀印加帝国君主阿塔瓦尔帕(Atahualpa)的事件中扮演了重要(且不光彩)的角色。1532 年,在瓦尔韦德的命令下,西班牙军队将阿塔瓦尔帕抓获。西班牙军队占领印加首都库斯科(Cuzco)后,于 1538 年任命瓦尔韦德为首任主教。在秘鲁长时间的逗留,使得瓦尔韦德熟悉了当地的食物,包括被他比作松露或块根的东西:土豆。

数百年来,土豆一直是生活在安第斯山脉的许多人的主食。考古发掘表明,土豆可能是在公元前 7800 年,在秘鲁的卡斯玛山谷(Casma Valley),同时在更南的地方,或许以不同的驯化过程被培育出来的,而

野生品种被食用的时间则早得多。在一万三千年前，智利中南部蒙特佛得角（Monte Verde）的定居者就已经开始采集野生土豆。在同一条跨美洲大陆山脉的北部，位于犹他州（Utah）的一处遗址上，同样发现了早在公元前10900年就有野生土豆被食用的证据。山地居民用多种方式烹制土豆，犹他的先民似乎是把它们捣碎或磨碎食用。在瓦尔韦德到达秘鲁的时候，安第斯人已在食用土豆汤和炖土豆，配以玉米、藜麦、烤豚鼠和野味、鱼干，以及各种蔬菜、豆类和块茎，或还加上辣椒来增加味道。他们还发明了一种有效的冷冻干燥方法，可以把易坏的鲜土豆变成一种可长期存放的石头状的东西，叫做"chuño"。土豆以各种形式成为一种普通的日常食物。在15世纪就已经控制了南美大片领土的印加王朝，对待土豆颇有几分漫不经心，不像对玉米那样以庆典和仪式的年历来仔细监督其耕种。[2] 然而，土豆在世界其他地方却完全不为人们所知。

韦森特·德·瓦尔韦德不经意地描述了这种块根之后，其他征服者也有一些类似的报告，他们强调说，土豆是"印第安人的面包"——该地区的基本食物。来自西班牙巴斯克（Basque）的征服者帕斯夸尔·德·安达戈亚（Pascual de Andagoya）写了一部参加皮萨罗远征的回忆录，书中记录说，"这些地方的食物是玉米，还有一些块根，被当地人称为帕帕斯（*papas*），状如栗子或坚果"。[3] 在安达戈亚写下回忆录的十年后，另一个西班牙人佩德罗·谢萨·德莱昂（Pedro Cieza de Leon）写的书里也将它比作栗子或松露。没过几年，西班牙殖民者就开始收集土豆作为贡品，同时还有玉米和他们能从当地索取的任何有市场价值的东西。1587年，一名英国海盗在智利海岸停留时，发现了成箱的等待付款的土豆。他形容它们"非常好吃"。[4]

今天的土豆已经成为一种非常成功的全球粮食。它的年产量仅次于小麦、玉米和水稻，并且是世界上第五大最具价值的粮食作物。非洲马拉维的人均土豆消费量已经超过了秘鲁本土。总体而言，欧洲人如今是世界上最活跃的土豆食用者，平均每人每年消费 82 公斤。[5] 这种曾经陌生的块根植物，是如何成为欧洲人饮食的一部分的呢？

农民的保守主义

欧洲人如何学会吃土豆的标准故事，大致是这样：土豆最初于 16 世纪晚期传到欧洲，作为一种珍奇的植物吸引了人们的注意，并且，短暂地被精英阶层当作一种新奇的食物，但其他所有人都对它退避三舍。对于土豆的抵抗，或由于"农民的保守主义"，或由于对这种新植物可能带来的健康风险的恐惧，或由于不熟悉经块根而不是经种子培育植物所需的农业技术。植物学家很快就发现，土豆是茄属植物（Solanum）的一种，与致命的颠茄（nightshade）和有毒的天仙子（henbane）同属一科，人们经常提到这一点以证明他们的怀疑。这种块茎的球根形状，也引发了人们的担心，害怕它可能会诱发麻风病，因为得了麻风病之后，患者的四肢上都会留下类似的球根状赘生物。[6] 研究土豆的伟大历史学家瑞德克里夫·沙勒曼（Redcliffe Salaman）① 如此说道：

> 根本的麻烦在于，土豆从任何意义上来说都是一种新型食物，欧洲人在以前从未见过类似的食物。吃它，不仅是饮食上的一次冒险，也是对传统的大胆突破……除此之外，还有一个重要的事实

---

① 瑞德克里夫·沙勒曼（1874—1955），英国农学家、植物学家和马铃薯育种专家，他的里程碑式著作是他倾其一生的研究和实践在 1949 年出版的《土豆的历史和社会影响》，该书使营养史成为一种新的文学题材。——译注

是……土豆遭到明确的反对,是由于它在《圣经》中没有被提及,所以它不是上帝为人类设计的一种食物。[7]

由于这些原因,据说欧洲人在接下来的两百年里对土豆的各种魅力置之不理。而爱尔兰人却是例外,他们早期呈现出反常的热情。直到 18 世纪具有远见的精英人物,如普鲁士的腓特烈大帝(Frederick the Great of Prussia)等人开始支持土豆,它才开始被更广泛地食用,或传说中是如此。(参见图 4)

这些说法几乎没有任何历史依据。近代早期的欧洲人非但没有把土豆定义为一种需要陌生栽培技术的奇特植物,反而把它们简单地描述为类似于胡萝卜或欧洲萝卜的块根植物。它们共同的地下习性,比土豆培育自块茎、而其他植物培育自种子更让人在意。近代早期的植物学和园艺学文献在这个问题上也没有过多地留意。它们只是解释了如何培育土豆种,以及提出最佳种植方法的建议。至于土豆是茄科植物的一员而遭到排斥的说法,新世界的另一种舶来物,同样不为人所知的红辣椒也被确定为茄科植物,但如一位作家1590年所说,丝毫没有妨碍辣椒迅速成为"家喻户晓之物",并在伊比利亚半岛等地得到广泛食用。[8] 土豆与麻风病有关的唯一证据,来自 1620 年一部本草书(herbal)中道听途说的评论。该书的瑞士作者写道,他听说在 300 公里之外的勃艮第,土豆和耶路撒冷洋蓟都是遭到禁止的。[9] 没有迹象表明禁令真实存在,且这位草药师本人对土豆有益健康信心十足,因此提供了几种家乡巴塞尔的烹饪方法。而且,虽然卷心菜、菠菜、燕麦等许多其他作物在《新约》《旧约》中都无迹可寻,却并没有阻止它们成为欧洲大陆各处常见的食物。

认为农民保守主义是主要障碍的说法又如何呢?一种关于新世界

图4 腓特烈大帝视察土豆收成。腓特烈二世常被认为是说服持怀疑态度的德国人吃土豆的功臣。这幅19世纪的绘画描绘了仁慈的君主与卑微的农民在聊天,而他的侍臣们则狐疑地旁观。事实上,德国农民从16世纪就开始种植土豆。然而,直到18世纪,他们的活动才开始得到王室和官员的认可。

食物传播的权威历史认为:"让农民脱离行之有效的农业方法和行之有效的食物,也许只是一个时间问题,但在很多情况下,这是一个需要很长时间的问题。"[10] 本章的观点却与之相反:土豆进入普通人的日常饮食,并不总是需要很长时间。事实上,农民和劳工是最早接受土豆的人之一。尽管我们拥有的更多信息是有学识的人对这种新作物的反应,但劳动人民同样也种植和食用土豆。农民保守主义的说法,是对这些食用先驱的伤害。近代早期的欧洲人排斥土豆,并不是因为害怕它有毒,也不是因为不知道如何种植,更不是因为对它感到陌生。近代早期的土豆并没有取代谷物成为欧洲的主食,也没有被传播到欧洲大陆的每一个角落,这一点是毫不奇怪的,但是,在学者和王室说服持怀疑态度的民众食用它之前,它也并没有遭到不理解和怀疑。

很久以前

 1768 年，康沃尔的圣布里扬（St Buryan）教区卷入了一场关于土豆的纠纷。圣布里扬是一个古老的定居点，距离兰兹角（Land's End）大约 5 英里。土方工程表明，早在诺曼征服之前此地就有人居住。1768 年的争端与历史记忆有关。起因是当地礼拜堂的牧师认为，他该得的什一税应该比他从几个佃户种的土豆和其他"菜园产物"中收得的什一税更多。教会征税的惯例极其复杂，尚不清楚土豆是如何融入不断发展的判例法和正式法规体系中的。被告争辩说，他们应该支付较少的什一税，因为这些土豆是自己食用而不是出售的。此案凸显了不在农田而在菜园中小规模生长的作物的含糊状况，这一问题在什一税表中没有得到很好的解释。被告还坚称，无论菜园产物的确切法律地位如何，"很久以来"他们一直只为家庭消费而种植的土豆和其他作物缴纳少量税款，他们请求继续实行当地这种做法。[11] 这一案件因此取决于长期存在的地方习俗能否成为先例，以及，土豆事实上在圣布里扬种植了有多久。法院作出了有利于被告的判决，批准继续缴纳打折的什一税。以这样的做法，它支持了佃户们的说法，即他们许多代人都已经在为个人的用途而种植土豆。

 什一税纠纷为土豆在近代早期的欧洲传播提供了另一种历史。这些记录与关于农民保守主义和民众排斥的故事不同，它记载了土豆在欧洲许多地方的农舍菜园传播的过程，而这一过程的主角基本上是农业劳工、农夫和工匠家庭，而不是如腓特烈大帝这样的贵族。由于一种作物的什一税状态，部分取决于它是用于出售还是供个人使用，以及以这种方式种植的时间有多长，因此，关于土豆什一税的纠纷中，各方频频深入探究本教区土豆种植的历史和年表。有时被告坚持认为，土豆是新引入的东西，不具备征什一税的资格；在另一些场合，他们又认为，对它的什一税豁免，是世世代代的习惯和做法。土豆种植几乎总被认为是一项普通人为自己的生计而从事的活动。

在英国，这类纠纷最早可追溯到17世纪晚期，并在随后的几十年里变得更加常见。土豆第一次种植的时间在大多数情况下是非常明显的。以约克郡的柯克比马尔齐德（Kirkby Malzeard）教区为例，在1736年的一项调查中，明确向证人询问了土豆的引入时间，证人被问及："土豆在柯克比马尔齐德教区里生长、种植的时间，是在人们记忆之外吗？或者你知道或记得土豆第一次被引入的时间吗？"教区居民一致认为，土豆最早是在1680年到1700年间的某个时候开始在农田里进行种植的。在此之前，它们被种植在农舍菜园中供家庭食用。他们把土豆的大规模种植归功于一个名叫巴克的人的创新做法，而他们已不记得农舍菜园里没种土豆是什么时候了。[12] 在另一件案子中，兰开夏郡福姆比（Formby）、安斯代尔（Ainsdale）和瑞文摩尔（Raven Meols）的村民，如圣布里扬的居民一样，拒绝为土豆和其他菜园产物缴纳新规定的什一税，理由是他们种植土豆已经很长时间，从未被收取过什一税。他们坚称："为了保护我们自古以来所拥有和享有的财产和权利……（我们）不会付钱的。"[13]

克里斯蒂安·范登布鲁克（Christian Vandenbroeke）和埃洛伊·特伦（Eloy Terrón）等学者在法国、比利时、荷兰和西班牙发现了类似的案例。在这些国家，关于土豆承担什一税状况的争议，于18世纪早期开始出现。正如这些学者所指出，这种争议待到一种作物颇具规模时才会出现，因此它们出现本身就表明了作物的流行程度。这些资料表明，从17世纪70年代开始，在佛兰德斯（Flanders）、阿尔萨斯（Alsace）、加利西亚（Galicia）等地的一些地方，土豆被作为菜园作物进行种植。而在英格兰，教区居民常声称，对土豆什一税的豁免可以回溯到"远古以前"。[14]（牧师们常反驳说，即使他们选择不行使向土豆收税的权利，它也一直应该被征收什一税。）可以肯定的是，这些案例引出的历史故事是在法律纠纷的背景下出现的，在其中，土豆被引入的时间会带来直接的经济后果。[15] 不过，有一系列其他资料也证实了长期存在的土豆的本地化种植和食用。欧洲各地的劳动人民，早在18世纪的哲学家和国王们考虑鼓励他们之前，就开始种植食用土豆了。

了解土豆

欧洲关于土豆的最早讨论出现在本草书中。本草书本质上是食用和药用植物的百科全书，它们介绍有趣或重要植物的特征，并解释其用途。16世纪出现了数以百计的这类插图丰富的书籍。土豆的词条出现在16世纪50年代，同时出现的还有其他新大陆植物，如红薯、玉米和西红柿。这些著作描述了土豆的植物学特征，并试图在已有的科学知识和分类体系内对它进行定位。[16] 本草书揭示了有学问的作者对这种新植物的反应，同时也清楚地显示了它在欧洲各地更广泛的传播。例如，一本著名本草书的作者、佛兰德斯植物学家卡罗勒斯·克卢修斯（Carolus Clusius）在1587年就得到了一些土豆植株。据他说，他的土豆从意大利经由比利时而来。后来，他将它们分享给他在欧洲大陆其他地方保持通信的人。克卢修斯和他的伙伴们之间的信件往来清楚地表明，他们并未感觉从块根中长出土豆具有挑战性，而且也并没把土豆视为令人担忧的怪异植物。[17]

本草书还揭示出，不仅是植物学家种植土豆，而且它们在农舍菜园中也得到种植。例如，克卢修斯就说过，在德国和意大利部分地区，这些作物已经很是常见。不过，他并没有详细描述导致这种发展的农业实验者。（为了使这些南美块根适应欧洲较短的日照时间和较长的生长季节，试验是必要的。近年的研究强调，要取得可靠的收获，就必须进行谨慎的选种和评估。[18]）1588年，在比利时图尔奈（Tournai）的另一位有身份的植物学家在一篇文章中写道：几年前，

> 一位有名望的权威人士来到我的菜园里，看到了我种的（土豆）植株，他问我是否特别注重它们。我告诉他，我认为它们很稀罕。他回答说，他在意大利见过很多，有人用它来代替油菜根吃，有人把它和羊肉一起煮，还有人用它来养猪，各人想怎样都可以。[19]

显然，在16世纪的意大利，最早吃土豆的人就包括那些通常吃低贱的油菜根的人。油菜根是穷人常吃的一种蔓菁类块根。瑞士植物学家加斯帕德·鲍欣（Gaspard Bauhin）也指出，除了植物学家以外，其他人也在试验土豆种植。他在1620年提到，"很多人"更喜欢在秋天把它们挖出来，而不是把它们留在地里过冬。关于土豆的丰富地方农业知识已经出现了。[20]

这些记录清楚地表明，种植这些块根不仅仅是为了养猪，同样也是为了给人们提供粮食。克卢修斯和他的通信者们讨论意大利人如何把它做成炖菜，就像炖胡萝卜和芜菁一样。[21] 鲍欣记载，在巴塞尔（Basle），人们用黑胡椒烤土豆。[22] 17世纪的罗马植物学家托拜厄斯·阿尔迪尼（Tobias Aldini）把注意力太多地集中在如何烹制土豆上，以至于另一部本草书作者抱怨说，阿尔迪尼的记录更像是厨师而不是科学家的著

作。但这位批评者，西班牙医师贝尔纳多·德·西恩弗戈斯（Bernardo de Cienfuegos），却赞成阿尔迪尼的观点，即土豆有无数种烹饪方法：烤；像银鱼一样裹上面包丁油炸；裹上醋、盐和胡椒做成沙拉；切成薄片、加上柠檬汁和盐用油煎；像蘑菇一样用薄荷叶炒；像萝卜一样在肉汤里炖；做成砂锅菜……然而，在他看来，它需要一种被称为皮肯特（picante）的香辣调味汁，否则就会极其淡而无味。西恩弗戈斯还警告说，像所有的块根类蔬菜一样，它会引发胃肠胀气，因此会刺激性欲。[23]（参见图5）

16、17世纪健康手册的作者们同样也讨论了土豆。近代早期，在欧洲各地都出现了这些健康生活的实用指南，它们旨在诠释健康和福祉的原理。[24]它们的作者们一致认为，健康就是保持身体与周围环境的平衡。饮食是与外部世界的公开对话，因此需要特别小心地管理，以免身体乱套。身体乱套是一种长期的风险。所有的食物都以某种方式影响着进食者的身体，因为所有的食物都有可能改变体液的平衡，而体液代表着每个人的整体体质或"气色"。土豆因此和其他食物一样，具有一种内在的药用特性。西恩弗戈斯所谓的土豆和其他块根类蔬菜一样有引起胃肠胀气，因而也易刺激性欲的特点，是近代早期饮食学的典型观点。在他看来，土豆导致潮湿体液形成，因此用干辣的香料来烹制会有益处，这些香料可以平衡土豆的湿气。关于如何将土豆与恰当、正确的配料和烹饪方法进行搭配的介绍表明，土豆经得起做出健康而营养丰富的膳食所需的烹饪和药用技术的检验。这些手册还评论了土豆的味道特点，以及哪种烹饪方法能产生最美味的结果，有时还会写得很长。例如，1620年的一份意大利健康指南指出，土豆需要大量的盐和其他调味品，不是为了让它更健康，只是为了克服它的淡而无味。[25]当时，和现在一样，土豆为富有创造力的厨师提供了一张空白画布。

图 5　贝尔纳多·德·西恩弗戈斯的本草书（约 1630 年）中的土豆插图。贝尔纳多·德·西恩弗戈斯是阿拉贡（Aragón）的医师，著有多卷本草书，描述了数百种植物。他在"来自秘鲁的土豆"条目中列出了许多烹饪土豆的不同方法，并提到可以在马德里的蔬菜市场买到土豆。西恩弗戈斯明确表示，他讨论的是土豆而不是红薯，尽管他插图中的植物与后者很像。

称呼土豆……

但土豆究竟是什么呢？今天的生物学家解释说，"普通""爱尔兰"或"白"马铃薯（茄属马铃薯，*Solanum tuberosum*），完全不同于甘薯（*Ipomoea batatas*）和其他美洲的块茎植物，如菊芋（*Helianthus tuberosus*）（又称洋姜），也不同于非洲山药（*Dioscorea*）。"白"马铃薯是茄科（*Solanacea*）植物的一种，这一科植物还包括西红柿、辣椒和茄子。红薯与恼人的园林有害作物旋花（bindweed）是近亲，但与普通马铃薯只是远亲。北美菊芋和山药则完全是不同的种属。植物学如今对这些植物的明确区分，反映了与 16、17 世纪所使用的分类系统的较大区别。尽管近代早期的植物学家迅速将土豆归为茄属植物，但这并没有使它与其他在植物学上不同的植物完全区别开来。这些新的块根植物的重叠名称，反映出它们的相互关联，以及与欧洲现有食物之间的关联。

在近代早期的英格兰，"potato" 指的可能是红薯，也可能是土豆，甚至可能是菊芋。当作者想明确区分时，他们有时会把红薯称为"西班牙土豆"，而"加拿大土豆"通常指的是菊芋。"弗吉尼亚土豆"通常是指普通的土豆，因为许多英国作家认为这种植物起源于那里。不过，这三种都被认为是"potato"。西班牙语中的名称也是类似的重叠方式。西班牙人于 15 世纪 90 年代在加勒比地区第一次接触到红薯，它们在当地被称为"batatas"，这就是英文单词"potato"的来源。哥伦布称它是山药的一种，而山药这种非洲块茎植物早已被欧洲人所熟悉。西班牙人对美洲大陆的入侵，将指称红薯的新词汇引入了伊比利亚语词典中，也将在印加帝国通用的盖丘亚语（Quechua）中被称为"papas"的土豆引入了欧洲。当时和现在一样，说西班牙语的美洲人常以"batatas"与"papas"区分红薯与土豆；但在西班牙，"patata"一词被用来指其中任何一个，[26] 而"batatas"或"patatas de Málaga"（马拉加土豆）通常指的是红薯，因为早在 16 世纪，红薯就在马拉加附近以商业规模进行种植了，而"patatas de la Mancha"（拉曼查土豆）则指普通土豆。[27] 相反，菊芋也可能被定义为"pataca"。和英语一样，西班牙语中也有类似的词汇，强调了这些不同的新世界块茎植物之间的相互关联。16 世纪的德国作家也使用了一系列重叠的术语："Griblingsbaum"（松露植物）、"Erdäpffeln"（土苹果）、"Erd Artischocken"（土洋蓟）、"Tartuffeln"（松露），或直接称之为"Knollen"（根）。与松露的语义联系可以追溯到最早的西班牙语描述，并且在许多欧洲语言中很常见。例如，贵族黑森侯爵威廉四世（Wilhelm Ⅳ von Hessen）在16世纪90年代写道，意大利人称这些新块根为"Taratouphli"，还有些意大利人将其描述为一种"Tartufole"，即松露，[28]这就是德语"Kartoffel"（土豆）的起源。斯堪的纳维亚作家称之为"jordpäron"（土梨）、"jordäpple"（土苹果）、"artiskocker af Wirginien"（弗吉尼亚洋蓟）和"peruvianska nattskatta"（秘鲁茄）以及"potatoës"。[29]在16世纪晚期，植物学家克卢修斯和他的以多种语言通信的网络中使用了多个词汇，如"taratonfli""tartufy""papas americanorum""papos"和"papes"。

总的来说，近代早期作家有时会把土豆、红薯和菊芋区分开来，有时会花大量篇幅来解开那些造成混乱的复杂命名的乱麻。还有一些情况下，他们会顺利地从一种过渡到另一种，对它们既加以区别又加以结合。威廉·科尔斯（William Coles）的17世纪植物学手册首先列出了一系列适合"土豆"的食谱，然后简单地评论说，这些食谱对"弗吉尼亚土豆"同样有效。正如他提到的那样，"名字和种类就这样混在一起了"。[30]这样混在一起的结果，就是我们根本无法断定，南安普顿市长1593年送给赫特福德勋爵（Lord Hertford）的十磅"土豆"，到底是红薯还是普通土豆，因为两种都已在英国得到种植了。[31]在进入18世纪之后，无论是在学术的还是在地方的分类中，土豆、红薯和菊芋都依然属于一类。18世纪20年代，一位法国植物学家解释说，红薯是一种和菊芋差不多的土豆。一位在西班牙旅行的英国人，同样简单地将红薯描述为另一种"普通土豆"。18世纪的词典编纂者并不承担比旅行者或植物学家更多的义务以将这些植物区分为单独的词典条目。[32]从这个近代早期词汇中最强烈地显现出来的，是共性，而不是差异。

这种灵活性不足为奇。这三种植物都起源于美洲，并都作为"哥伦布大交换"（Columbian exchange）的一部分而传播到世界各地。"哥伦布大交换"是指 1492 年哥伦布到达西印度群岛之后，动植物随之发生的世界范围内的转移。因此，认为这些植物紧密相关是有许多理由的。它们相互重叠的名字，让我们想起了这段共同的历史。如果 16 世纪的欧洲人不能清楚地区分这些不同的蔬菜，那是因为他们认为它们是相同的。这三种都是来自新大陆的新的块根植物，而且欧洲人都把它们作为食物。显然，他们并没有被这些无《圣经》出处的新块茎所带来的分类学上的挑战吓住。[33]

给土豆削皮

想知道如何烹制这些新块根的读者，可以求助于新兴的印刷食谱。第一个疑是土豆的烹饪方法可以追溯到 1581 年。它出现在一本由马克斯·朗波茨（Marx Rumpolts）撰写并在法兰克福（Frankfurt）出版的烹饪书中。朗波茨曾是富有的美因茨大主教选帝侯的私人厨师，后来又为丹麦女王掌厨。他的烹饪书提供了大约 2000 个菜谱、30 多个菜单样本，以及采购和餐桌礼仪方面的建议。《新菜谱》(*Ein new Kochbuch*) 包括如何制作多层匈牙利挞、把糖做成一整块帕尔马干酪的形状，除此之外，它还包括一些新奇的美洲食材，比如芸豆等的菜谱。它还介绍了如何烹制典型的西班牙菜，比如"olla podrida"（什锦菜，英语中称为"olio"）。这是一种复杂的炖菜，要与新世界的火鸡一起烹制，后来它迅速成为欧洲各大宫廷中一道时髦的宴会菜。[34] 朗波茨显然熟知伊比利亚的烹饪传统，还熟知西班牙人在美洲殖民后引入欧洲的一些新食材。

他的烹饪书内还印了些"土苹果"的菜谱。也许他提到的"Erdtepffel"（德语"土苹果"）是土豆，也许是菊芋，也许是其他东西，我们无法肯定。不过，从另一些新世界和伊比利亚菜肴的菜谱来看，朗波茨熟知这些美洲块茎这一点似乎是可信的。他总是指示读者要"削皮、切小，把它们放在水里先煮到半熟，然后用细布好好压一下，切成小块，与同样切成小块的培根一起炒，再加少许牛奶一起煮，这样就美味而可口"。[35]

在朗波茨的土苹果食谱问世20年后，一本法国烹饪书介绍了如何烹制一种叫"tartoufle"的新松露。它推荐了几种不同的烹制方法，包括像烤栗子一样烤，或用黄油和甜酒调味。[36] 1596年英格兰的《好主妇之宝》(*The Good Huswife's Jewell*)介绍了一种"人们需要拿出勇气去做的果馅饼"食谱。实际上这种馅饼更像是布丁，是一种昂贵的食物，用上了黄油、鸡蛋、香料和"三四只公麻雀的脑花"配上"土豆"。[37] 这种甜土豆挞很快在欧洲各地大受欢迎。到了1611年，西班牙的烹饪书中已在批评这样的铺张和奢侈，认为是不惜一切代价都要避免的可憎烹饪行为。这个建议甚佳，因为第三代多塞特伯爵理查德·萨克维尔（Richard Sackville），就是在1624年狼吞虎咽"一块土豆馅饼"时不幸一命呜呼。[38]（参见"松露"食谱）

"松露"的四种食谱

16世纪晚期的欧洲的文献中，开始提到了一种新的"松露"。虽然很难确定，但它们很可能是土豆或其他一些美洲的块茎植物。兰斯洛特·德·卡斯特（Lancelot de Casteau）是一名专业厨师，曾为列日（Liège）的几个王室工作过，他在1604年的烹饪书中介绍了这种"松露"的四种不同的菜谱。他还介绍了如何制作奶酪香草乳蛋饼、爱尔兰风格的羊腿等新式美食。

> **"松露"的四种食谱**
>
> ***煮"松露"***
>
> 将洗净的"松露"放入水中煮,烹制时须削皮、切成小块,放入融化的黄油和胡椒。
>
> ***另一种"松露"***
>
> 将"松露"如上切成小块,放入西班牙葡萄酒、新鲜黄油和肉豆蔻一起炖。
>
> ***另一种***
>
> 将切成小块的"松露"与黄油、切碎的马郁兰(marjoram)、欧芹一起炖,然后取四五枚鸡蛋的蛋黄用少许葡萄酒搅拌,浇在煮沸的炖菜上,随后从火上移开,然后食用。
>
> ***另一种***
>
> 像烤栗子一样将"松露"放入热余烬中烤熟,然后将其削皮并切成小块,上面撒上切碎的薄荷、煮熟的小葡萄干、少许胡椒,淋醋,然后食用。

印刷和手抄的烹饪书,与取名的争议和本草书一样,都表明了在德国、英国等的一些地方,新世界的块根已得到小规模的种植,供家庭所用。正如撒克逊城市不伦瑞克(Braunschweig)1651年的一本烹饪书所指出的那样,"土洋蓟(earth-artichokes),或称为根……已经变得如此普遍,几乎每个农民在自己的菜园里都有种植"。[39] 到18世纪早期,德国的烹饪书中常已能明确区分不同的新大陆块茎,并清楚地指出,在某些地区，普通的土豆"已相当普遍"。[40] 随着 18 世纪继续向前发展，在欧洲许多不同的地方，更多的印刷和手抄的烹饪书中都出现了土豆食谱。[41]

也有其他资料表明，在人们认识土豆之前，它早就作为一种普通食材得到种植。值得注意的是，在法国、意大利和英国，17 世纪早期已有农业手册开始介绍如何种植土豆。法国园艺作家奥利维尔·德·塞尔（Olivier de Serres）在 1603 年的《农业讲堂与土地事务》(*Theatre d'agriculture et mesnage des champs*) 中，把土豆描述为园艺植物，而不是珍奇异草。英国植物学家约翰·帕金森（John Parkinson）在 1629 年的《阳光公园之土地天堂》(*Paradisi in Sole Paradisus Terrestris*) 中以同样的方式看待土豆。帕金森仔细地区分了不同种类的土豆，他报告说，"弗吉尼亚"土豆"我们很熟悉"，并介绍了如何进行种植和食用。[42] 他建议在热余烬中烤熟，或加酒和少许糖进行烘烤。约翰·伊夫林（John Evelyn）①在 1666 年的《园艺师日历》(*Kalendarium Hortense*) 中提醒园艺师们，2 月份应该"播种黄豆、豌豆、白萝卜、防风萝卜、胡萝卜、洋葱、大蒜等，并把土豆种植在最贫瘠的土地上"。因为即使在最边缘的土地上，土豆也能茁壮成长。几十年后，一本德国手册也以同样的方式介绍了如何培育土豆，并指出它们在奥地利的一些地区已经非常普遍。[43]

在海关记录和账簿上，也能追踪到这些新块根从家用的菜园作物变成商品之后，在欧洲各地的轨迹。直到进入 18 世纪之后，土豆在欧洲大部分地区仍是一种菜园作物，但从 16 世纪晚期起，新大陆的块根已开始在国际上进行贸易。海关记录显示，早在 16 世纪 70 年代，

---

① 约翰·伊夫林（1620—1706），英国作家、园丁和日记作者，英国皇家学会的创始人之一，曾撰写过有关美术、林学、宗教等的著作三十余部。伊夫林的文学成就主要体现在他的《日记》(*Diary*，1818 年第 1 版) 和《戈多尔芬夫人的一生》(*Life of Mrs Godolphin*，1647) 中。——译注

"patatas"就在加那利群岛进行商业规模的种植了，它们自该地被运往法国和荷兰。[44]我们无法确定这些资料指的是土豆、红薯还是二者皆有，但不管怎样，它们说明，在16世纪的欧洲，有一种新大陆的块根是长途商业贸易的对象。而我们可以肯定的是，17世纪20年代在马德里市场上销售的新大陆块根是土豆，因为报告此事的阿拉贡医师贝尔纳多·德·西恩弗戈斯仔细地将"papas del Perú"（秘鲁土豆）与其他类型的块茎进行了区分。他提到，返回的殖民者和本地人都在购买这些"papas"。秘鲁历史学家和法学家安东尼奥·莱昂·皮内洛（Antonio Leon Pinelo）在17世纪早期去过西班牙，应该对安第斯土豆非常熟悉，他也观察到了同样的情况。[45]税收记录显示，从17世纪30年代开始，一种叫做"jordpäron"，即土梨的东西，从葡萄牙和德国北部进口，经卡尔斯（Karlshamn）港进入瑞典南部的斯莫兰（Småland）。[46]到17世纪晚期，兰开夏和英格兰北部的其他郡，以及苏格兰和爱尔兰的部分地区，都有了专门的土豆市场，而且贸易立法详细规定了这种商品的关税。[47]土豆从维持生活的食物转变为商品，有助于解释土豆的什一税不断增多的争议，因为在土豆作物获得市场价值之前，它对什一税所有者的吸引力是有限的。18世纪佃户和牧师之间关于土豆什一税的许多冲突，很可能反映了土豆的商品化以及种植面积的增加。[48]其他种类的经济和法律数据，也表明了土豆作为一种商业作物而不断提高的地位。例如，1641年爱尔兰天主教徒反抗英国统治的起义之后，爱尔兰的新教徒要求对他们土豆产量的损失进行货币赔偿。这些资料进一步说明了土豆是如何被纳入现有的分类系统和消费体系的。1690年，法国维瓦莱（Vivarais）地区的官员在进行农业调查时，把土豆与栗子、芜菁和其他块根类蔬菜一起列在了清单上，因为所有这些食物都无疑为农民的膳食提供了淀粉来源。[49]（参见图6）

总之，在帕斯夸尔·德·安达戈亚不经意描述了这些栗子状块根后的70年里，土豆和其他新世界的块茎一起，出现在了不列颠群岛、西

图6 流动土豆小贩，选自18世纪英国民谣。18世纪的英国人对土豆和土豆商人都很熟悉。版画展示了一位与他的一篮子土豆和毛驴一起正在干活的小贩。在伦敦街头兜售土豆并不是这个土豆小贩唯一的收入来源。如这首民谣所述，他也是一个皮条客，其情妇也卖她自己的"好果子"。

班牙、德国、意大利和法国的部分地区。1577年，圣特蕾莎修女（St Teresa）在阿维拉（Ávila）的修道院吃了一个"patatas"，这是她收到的寄自塞维利亚（Seville）的礼物。[50] 在同一年，英国的讽刺文学嘲笑了它所谓的潜在刺激性欲的能力。[51] 去意大利和德国的旅行者也记录了当地人烹饪土豆的方法，植物学家也追踪到它在农舍菜园中的出现。在接下来的一个世纪里，土豆种植经弗朗契-孔德（Franche-Comté）和洛林（Lorraine）在法国进一步传播，也传播到佛兰德斯，并向东传播到勃兰登堡（Brandenburg）。在佛兰德斯，土豆得到大量种植，以至于在九年战争（1688—1697）期间，那里的士兵能够用从当地田地里掠夺来的土豆"极其丰足"地养活自己。[52] 这些片段，与什一税争议的证据和其他资料都是相符的，它们表明，至17世纪晚期，土豆已经出现在许多欧洲人的田间和桌上。

当然，土豆在整个欧洲还并非家喻户晓。在18世纪之前，几乎没有证据显示它在俄罗斯或芬兰出现过。虽然它作为农田作物较早被引入阿尔萨斯、弗朗契-孔德及周边，但却没有进入法国其他地区。它在德国土地上的不均匀分布，被历史学家冈特·维格尔曼（Günter Wiegelmann）制作成了详细的地图。此外，戴维·齐尔贝尔博格（David Zylberberg）还指出，当地价格实惠的燃料供应，是土豆进入英国人饮食习惯的先决条件。他解释道，土豆的消费"开始于以泥炭为燃料的社区，然后传播到以煤为燃料的社区，而1830年之前，在以木头为燃料的地区就不太常见"。[53] 对当地的这类仔细研究，对于解释土豆在特定地区的精确轨迹是有必要的。直到18世纪，土豆也并非重要主食。谷物，无论是烘烤成面包还是做成粥，一直才是欧洲大部分地区大多数人膳食的基础。然而，恐惧和不熟悉导致全盘拒绝的情形，与历史记录并不相符。欧洲人食用土豆已经有很长时间了。

土豆的传播

很难确定究竟是哪些人将土豆传播到近代早期的欧洲各地。历史学家威廉·麦克尼尔（William McNeill）猜测，西班牙、巴斯克的水手和渔民在其中扮演了重要角色。比如，很可能是这些人把土豆引进了爱尔兰西部，因为那里是从纽芬兰捕鱼回程的惯常停靠之地。如麦克尼尔所提出的，也有可能是西班牙船只将土豆带到了意大利南部的哈布斯堡领地和荷兰。水手也传播了其他新奇商品，如烟草，连同新的消费习惯。有理由认为，他们在美洲食物的传播过程中也扮演了类似的角色，只不过我们缺乏直接的证据。[54] 资料只是在极为罕见的情况下，才明确提到了传播的媒介。1606年之前的某个时间编纂的一本佛罗伦萨农业手册记载说，土豆是由西班牙、葡萄牙的加尔默罗修会（Carmelite）的修士引入托斯卡纳的。几年后的另一份意大利文文献将土豆引进皮埃蒙特（Piedmont）地区归功于"法国人"。有报道说，在18世纪早期，来自德国南部的玻璃吹制工等工匠将土豆带入了德国北部的梅克伦堡（Mecklenburg）地区。[55]

几十年前，历史学家道格拉斯·霍尔（Douglas Hall）注意到，近代早期的种植活动之所以如此难以重建，是因为它们的传播者往往是当时作家视线之外的普通人。[56] 那些在自家花园里种植土豆，并像煮胡萝卜一样煮土豆的劳工和小农户，他们的名字和生活，大部分都不为我们所知。我们所能做的，也只是注意到什一税法庭的判决结果。比如1766年的一次判决，判定土豆在海纳特（Hainaut）郡是由劳工最先在农舍菜园里进行种植的。[57] 不过，在18世纪末，一家保险公司的创始人、第二代马里兰准男爵弗雷德里克·莫顿·伊登（Frederick Morton

Eden），把第一位在苏格兰户外田地里种植土豆者的名字记录了下来。18世纪90年代，伊登根据合作者的调查、一些直接观察和自己对英国历史的广泛阅读，撰写了一部关于"英国劳动阶级"的多卷著作。伊登对土豆很感兴趣，认为它是极佳的食物。他认为应该鼓励大规模的种植。据伊登说，在拉纳克郡（Lanarkshire）基尔赛斯（Kilsyth）村的"托马斯·普伦蒂斯（Thomas Prentice），一个临时工"率先进行了这种创新。伊登指出，普伦蒂斯非常成功，以至于"每个农夫和佃农都以他为榜样，过了许多年，它已经成为主食品种"。[58] 我在基尔赛斯没发现普伦蒂斯的任何线索，也许伊登说的开拓者，是1701年出生在距该地8小时步行路程之外的卡尔路加（Carluke）的那位托马斯·普伦蒂斯。

很清楚的一点是，土豆的传播不均匀且分散，而它在18世纪早期已成为欧洲多地普通膳食的重要组成部分。爱尔兰情况如此，已属确凿无疑。按照17世纪50年代一位作家的说法，土豆是"穷人最主要的食物，每家都有土豆园"。[59] 同样，在17世纪的英格兰，按照牧师理查德·巴克斯特（Richard Baxter）的说法，穷佃户和农夫靠的是"黑面包、牛奶、豌豆馅饼、苹果馅饼、布丁、煎饼、燕麦粥、麦片粥、牛奶麦粥，甚至干面包、黄油、奶酪、卷心菜、萝卜、防风萝卜、胡萝卜、洋葱、土豆、乳浆、酪乳，和一些不起眼的喝的"为生。[60] 17世纪80年代，在奥德河畔法兰克福（Frankfurt an der Oder）东部的小镇勃兰登堡，一位博物学家发现，"在这周围，它们已经变得十分常见"。[61]

土豆可以在煤灰里烤，或与白菜和胡萝卜一起煮，因此在欧洲的许多地方，它很容易适应贫困家庭的烹饪方法。他们的膳食通常以淀粉为基础，来源于燕麦粥、黑麦面包、小米粥，还包括大量的以谷物为基础、由近代早期默默无名的穷人菜能手发明的菜肴。这样的食物构成了大多数人日常饮食的核心。劳动的人们，期许每天能吃到约一磅面包或等量的东西，这样的面包和汤汁通常会配上少量的奶酪、脱脂牛奶、蔬菜、豆类，以及其他馅料和开胃料。土豆可直接融入这样的菜肴中。在汤中，土豆可以代替萝卜，或与萝卜及其他块根一起食用，也可以加少量肉和洋葱与它混在一起，这样土豆就能吸收并增加它们的香味。在燃料便宜得足以在家中烘烤的地方，可以把土豆烤成薄饼或其他形式的面包。德国的烹饪书描述了如何把它们做成沙拉冷吃，以及如何配上洋葱、淋点醋一起炖。[62] 这样的食谱无需不熟悉的技术或新工具，所以它们能很容易地融入日常的烹饪做法中。

早期将土豆纳入欧洲饮食的主角们，似乎包括1651年在菜园里种植"土洋蓟或块根"的不伦瑞克（Braunschweig）农民、16世纪80年代把它们种植得"非常茂盛"的无名意大利人，以及憎恨将菜园作物上缴当地牧师的欧洲各地的村民们。[63] 为了让安第斯山区的栽培品种适应欧洲不同的日照长度和生长条件，这些人还进行了必要的试验。如历史学家马德琳·费里埃尔（Madeleine Ferrières）所说，是维瓦莱的农民在当地"发明"了土豆。即使于18世纪较晚才开始种植土豆的地区，往往也是村民和农民率先种植这种作物。以波兰南部的加利西亚为例，土豆在进入地主的厨房之前，都是生长在农民的菜园里的。[64] 欧洲工匠们为了生产现代早期消费者所渴望的新奇商品，必须发展新的技能。同理，农民们也学会了种植新的粮食。不过，与工匠不同的是，农民们往往能够享用自己的劳动果实。[65]

除了易适应现有的烹饪方法之外，土豆为什么能在农民和劳工中获得成功呢？它们在财政上的隐蔽性无疑是一种吸引力。正如什一税争端所显示的那样，教区居民可能数十年种植这种不起眼的块根，却不会引起当地教区牧师及其代理人的注意。此外，土豆将阳光、土壤和水转化为营养物质的方式特别高效。1公顷种小麦的土地一年可产出足以供养

7个人的蛋白质,而1公顷土地出产的土豆却能养活17个人。在热量上的对比则更加惊人:1公顷土地上种植土豆所产出的卡路里是种植小麦或燕麦的3倍。比起其他主要作物,土豆需要的水也更少。[66] 因此,土豆是以同样的农业投入养活更多人的极佳方法。它还有更多的优点:它能够在各种气候和条件下茂盛生长,而且富含维生素C和其他必要的营养物质。如果配上奶制品,加上家庭烹饪的燃料供应充足,土豆可以让每家人比几乎任何其他选择都更有效地养活自己。在18世纪的英格兰,每便士从市场上买的土豆提供的热量几乎是小麦面包的3倍,是肉的5倍。[67] 毫无疑问,正是由于这些原因,土豆在爱尔兰和加利西亚、阿斯图里亚斯(Asturias)等地特别受欢迎。英国对爱尔兰的殖民统治迫使农村人口的种植面积越来越小,而在加利西亚和阿斯图里亚斯等地,农民可以获得更多的土地,可以自由种植自己喜欢的作物。[68]

人类学家罗伯特·内廷(Robert Netting)研究了土豆对阿尔卑斯小村托贝尔(Törbel)的影响。在18世纪,该地居住的农民开始大规模种植土豆。[69] 村民自己如何解释对土豆的欣然接受,这个问题我们无法回答,但内廷的研究表明,在托贝尔,每公顷土豆提供的热量比该地区的主食黑麦要高3倍。而且,土豆能够生长在无法供谷物生长的过于陡峭或贫瘠的土地上。他的结论是,土豆有助于解释为什么托贝尔的人口在引入土豆后的世纪里翻了一番。[70]

在托贝尔,土豆最早种植于菜园的地里,这意味着它们最早是由妇女种植的。在近代早期欧洲的大部分地区,维护家庭菜园是女性的责任,因为她们被认为更适合要求手巧的劳动,如挖掘块根作物,以及其他不需要犁或长柄镰刀等大型工具的劳动。用锄头进行的小规模土豆种植,非常符合常由妇女从事的活动的特点,她们同时还可以养鸡,以及从事维持农家所需的种种其他农活。因此,在欧洲,女性农民很可能是土豆种植的先驱者。圣布里扬教区牧师要求的"菜园产物"很可能是村里的妇女们种植的。[71]

总之，这些特点表明，在近代早期的欧洲，土豆能被归入"逃避政府"的类别，"逃避政府"（state evading）一词由政治人类学家詹姆斯·C. 斯科特（James C. Scott）[①]发明，用来形容能从政府控制当中获得一定程度自主权的作物，因为它们可在土地边缘种植，也能为难以有效衡量并征税的复杂家庭供应系统作出贡献。与谷类不同的是，土豆不需要在一个明确的时间段进行收割，可以留在地里几个星期甚至几个月。收获之后，它们非常沉重，也难以运输到其他地方。从财政的角度来看，种植在家庭菜园里的土豆几乎不值得征税，因此，用斯科特的话说，这样就创造了一个"非政府空间"，特别是在没有建立商业框架来营销这些块根的地方。[72]这样的秉性，在一定程度上解释了土豆在加利西亚的小农、佛兰德斯的农户和英格兰的穷佃户中受到欢迎的原因。

相反，富裕的地主对土豆的优点却缺乏信心。18世纪早期，由于地主的敌意，瑞典鼓励种植土豆的努力没有取得任何成果。地主们怀疑，土豆的高产能会使农民更加自给自足，因此会不愿意为他们干活。[73]他们将这种庄稼视为对农田代价高昂的入侵，不然农田会出产更适销的商品。1727年，一个曾为几位英国公爵工作过的经验丰富的地产经理，建议他的客户禁止他们的佃户种植土豆，"除了少量他们自己用的"。非要在空地上种土豆的佃户，每耗费1英亩土地，应该被罚款10英镑。

[①] 詹姆斯·C. 斯科特（1936— ），耶鲁大学政治学斯特林讲座教授、人类学教授，农业研究项目主任之一，美国艺术与科学院院士。他于1967年获得耶鲁大学政治科学博士学位，研究兴趣主要集中在比较农业社会、东南亚、阶级关系和无政府主义理论等领域。主要著作有《农民的道义经济》《弱者的武器》《国家的视角》《逃避政治的艺术》等，在学术研究和政策制定方面都产生了巨大的影响。——译注

他们还应该有责任为这些土地施更多的肥，以补偿他认为的土豆从土壤中夺走的养分。[74] 如果佃户和劳工想吃土豆，那是他们的事，但是地主不应该允许他们把宝贵的土地用在这个目的上。在他看来，土豆是与落后、不经济的做法联系在一起的。他并不是唯一持这种观点的人。伟大的植物学家卡尔·林奈（Carl Linnaeus）一直对土豆的营养性持怀疑态度，他在思考18世纪40年代瑞典的本地饮食时，提出了一个疑问：究竟为什么仆人们"觉得那么有必要继续吃"土豆。[75] 在这些富人们的眼中，土豆是普通人种植和食用的普通食物，操心巨额财产管理的人不会对它产生兴趣，更不用说关心国家大事的人了。

根深蒂固的惰性

> 劳工们"盲目地重复着他们见过的他们父辈的所作所为"。[76]
>
> ——伯纳多·沃德（Bernardo Ward），1779年

> 农民的特点是根深蒂固的惰性，只有当有人无数次向他展示进步，迫使他摆脱无所作为状态时，他才会接受进步。只有到那时，他才会冒险亲自尝试。他有限的眼光，使他倾向于夸大他在过去成功种植的庄稼的优点。[77]
>
> ——阿梅代·德尚布尔（Amédée Dechambre），1877年

过了很长时间，人们才学会吃土豆。在一些地方，他们把土豆当做牲畜饲料，却不是供人食用。顽固的抵制需加以克服。人们担心土豆会引起各种疾病，如麻风病、腺病、肺结核或高烧。由于它与颠茄相似，一些人还担心它有毒。更有甚者，他们还不知道烹制

它的最好方法。⁷⁸

——伯纳德·亨德里克·斯利歇尔·范巴斯（Bernard Hendrik Slicher van Bath），1960 年

对于近代早期欧洲人是如何开始吃土豆的，本章提出了另一种解释。它并不在于数十次地耐心展示进步的专家们，而是在于农民和小农户。他们早在土豆引起什一税征收者和其他政府官员的注意之前很久，就开始种植这种块根作物。到 1700 年，在不列颠群岛和欧洲大陆的许多地方，村民、农民和劳工已经在家庭菜园里小规模种植土豆了。没有迹象显示，农民对土豆的块茎习性比植物学家更觉困惑，而且欧洲各阶层人民发明了许多烹制和食用这种淀粉块根的方法。就我们所能察觉到的传播方向而言，它与学者们的想象恰恰相反。

毫不奇怪的是，在近代早期的欧洲，农民和村民是农业和烹饪创新的先驱者。这类群体也是最早的玉米种植者。玉米是另一种美洲的输入物，在 16 世纪西班牙人将其引入欧洲和西非之后的几个世纪里，它同样给全球饮食带来了巨大变化。早在 16 世纪 40 年代，威尼托（Veneto）就有了玉米种植，该地的农民用它代替小米等便宜的谷物来烹制一种新的玉米粥。到 17 世纪早期，玉米粥已经成为该地区农民和穷人的主食。[79] 威尼斯的农民显然对玉米的新奇或《圣经》无记载并无不适。16 世纪 50 年代，当信奉新教的逃亡者将胡萝卜和其他块根类蔬菜引入的时候，英格兰东部的普通人也同样愿意接受种植它们所需的陌生农业技术。总的来说，如历史学家菲利普·霍夫曼（Philip Hoffman）在法国的例子中非常清楚地表明的那样，近代早期的农民绝不反对农业实验。[80] 此外，大量的研究也表明，世界其他地区的小农户在面对新作物和新经济状况时，也完全有能力采取农业创新的策略——这一点我们

将在总结中再进行讨论。

欧洲的农艺学也不是独立于小农农业而发展起来的。近代早期的各种科学知识，往往依赖于欧洲学者群体之外产生的信息、方法和理论。[81] 使17、18世纪英格兰和荷兰的商业性农业生产效率显著提高的新农艺措施，利用的是个体农民的农业实践。在这一时期，新作物、新技术的引进，加上有利于大规模农业的立法改革，共同创造了一种新型的农业资本主义。这种转变在极大程度上依赖于源于小农农业的技术。新农艺的一个关键要素，是苜蓿、三叶草和红豆草等绿肥作物的使用。将这些植物犁入土壤，是向土壤中增加氮和其他必要营养物质的有效手段。使用这类土壤增肥剂有助于减少为恢复失去的肥力而让土地休耕的需要。由此土地可以持续进行耕种，大大增加了耕地面积。这些作物还可以用来喂养家畜，家畜的粪便又进一步增肥土壤。不过，这整个过程须依赖于这些栽培品种种子的可靠商业供应。事实证明，以专业知识使其成为可能的人，是意大利的农民。他们对于这些庄稼的专门技术和实践知识是其他地方完全不具备的。结果，正如历史学家莫罗·安布罗索利（Mauro Ambrosoli）所言，"农民阶级种植的庄稼逐步进入了资本主义农业中"。[82]

当小农农业专业知识被农民之外的人所注意时，才能引起历史学家的注意。以红豆草和三叶草为例，这些作物对新农艺措施的重要性，将进步地主的注意力集中到栽培这些如今令人关注的庄稼的意大利农民身上。至于土豆，大多数16、17世纪的地主对它的商业潜力并不感兴趣。这在一定程度上解释了追踪土豆的早期欧洲史的困难。我们只能偶尔瞥见它在欧洲大陆各地的发展痕迹。这表明，小农户在其传播中发挥的作用，远比至今人们所认识到的更大。与我们所知的其他时期和地方的农民实践进行相互参照，也不难支持这一观点。例如，尽管学者们有时认为，土豆作为动物饲料的优越性向近代早期的农民证明了它"只适合作为动物饲料"，但这一结论与我们所知的欧洲或其他地方的农业实践并不相符。[83] 农作物能满足多种需求是一项积极的好处。墨西哥的农民长期以来一直很重视玉米兼为食物、饲料、燃料和建筑材料的能力。在欧洲本地，小麦极其重要的粮食和宗教象征的地位，并没有因小麦秸秆也可以充当低等家畜的草垫，或者家禽以小麦谷粒为饲料而被削弱。没有理由认为，近代早期的农民对土豆的多功能性有任何负面的看法。[84]

尽管小农户们在悄悄地种植土豆，希望不要引起收税牧师的注意，但土豆并不是地主和大庄园主特别感兴趣的对象。当然，土豆也出现在富人的餐桌上。食谱书介绍了如何烹制精致而昂贵的菜肴，如害死了多塞特伯爵的土豆馅饼；还有家庭档案记录了英国、法国等地大家族购买土豆和红薯的情况。在英格兰北部和爱尔兰等地也有商业规模的种植，有时还是在贵族的领地上。近代早期的精英并不排斥土豆，但也并不特别对它感兴趣。

土豆也没有被视为国家政策的重要组成部分。扩大土豆种植并不是官员或政治家的目标，除了极少数例外，也无人著文赞美土豆作为粮食被更多地食用会带来广泛公共利益的优点。[85] 相反，有政治头脑的评论家，对土豆可以帮助穷人逃避政府这一事实感到遗憾。作为英国殖民政府的代表曾在爱尔兰生活多年的哲学家和政治家威廉·配第（William Petty）① 便持有这种观点。配第于17世纪20年代出生在英国，因坎坷的生活而去过法国、荷兰，还去了牛津和伦敦，在那里教授音乐和解剖

---

① 威廉·配第（1623—1687），英国古典政治经济学之父，统计学创始人，最早的宏观经济学者。一生著作颇丰，主要有《赋税论》（写于1662年，全名《关于税收与捐献的论文》）、《献给英明人士》（1664）、《政治算术》（1672）、《爱尔兰政治剖析》（1674）、《货币略论》等。——译注

学,并帮助建立了英国皇家学会。17世纪50年代,他去了爱尔兰,爱尔兰当时刚被克伦威尔领导下的英国军队吞并。配第作为英国殖民政府的代表工作了很多年,而殖民政府的目标是从它的新领地中攫取尽可能多的利益。配第本人购入了几处大地产。配第认为,土豆妨碍了爱尔兰作为殖民者财政收入重要来源的发展。由于以土豆为生十分容易,因此爱尔兰人并不如配第所愿地辛苦劳作。结果,政府收取的税款,仅达到配第估计的更勤劳的人口所能产生的税收的一半。土豆让它们的种植者从英国政府那里获得了太多的自治权,在配第看来,如果爱尔兰人少吃土豆,英国政府就会更加富裕。[86]

随着18世纪向前推进,这种态度将会发生巨大的变化。从18世纪50年代开始,越来越多的政治家、政治经济学家、农学家、哲学家和地主开始拥护土豆。他们认为,每个人,尤其是穷人,都应该更多地种植和食用土豆。他们编写了土豆种植手册,尝试烹饪技术,并鼓吹土豆的许多优点。下一章探寻的是土豆在18世纪的巨大影响力,以展示有效治国之道和建立健康人口重要性观念的不断变化,是如何让普通民众不值一提的饮食习惯得到重视的。

18世纪这些土豆支持者的响亮声音,导致了本章开篇所总结的土豆历史的形成。这种历史认为,在18世纪之前,欧洲人并非普遍喜欢土豆,个体农民接受土豆的速度尤其缓慢。学者们倾听了18世纪的资料,并正确地听到了鼓吹人们应该吃更多土豆的18世纪舆论。如果我们仔细聆听,还可以听到其他故事。它们告诉我们,在18世纪哲学家开始推崇土豆之前,人们早就已经开始食用土豆了,并且,家庭菜园是寻找烹饪先驱和农业先驱最合适的地方。

注释

1. Vicente de Valverde to Charles V, Cuzco, 20 Mar. 1539, *Cartas del Perú*, ed. Porras Barrenechea, 314.

2. Murra, 'Rite and Crop in the Inca State'; Ugent, Dillehay and Ramírez, 'Potato Remains from a Late Pleistocene Settlement in Southcentral Chile'; Coe, *America's First Cuisines*; Naranjo Vargas, 'La comida andina antes del encuentro', *Conquista y comida*; Louderback and Pavlik, 'Starch Granule Evidence for the Earliest Potato Use in North America'. 卡斯玛山谷的土豆可能是野生品种，而不是驯化品种。

3. Pascual de Angagoya, 'Relación que da el Adelantado de Andagoya de las tierras y provincias que abajo se hará mención', 1545, *Pascual de Andagoya*, 138.

4. Cieza de León, *Parte primera de la Chronica del Perú*, chap. 40, 105; Zárate, *Historia del descubrimiento y conquista del Perú*, book 1, chap. 8; Santo Thomas, *Grammatica*, 159v; Molina, *Relación de las fábulas y ritos de los incas*, 62–63; 'Descripción y relación de la provincial de los Yauyos', 1586, 'Descripción de la tierra del repartimiento de San Francisco de Atunrucana y Laramanti', 1586, and 'Relación de la Provincia de los Collaguas', all in *Relaciones geográficas de las Indias*, ed. Jiménez de la Espada, I: 156, 234, 586; Purchas, *Hakluytus Posthumus*, II: 157 (quote); Acosta, *Natural and Moral History of the Indies*, 148, 201–202.

5. FAO, FAOSTAT: 'Food and Agricultural Commodities Production'; and Helgi Library, 'Potato Consumption Per Capita in the World'.

6. Salaman, *History and Social Influence of the Potato*; Slicher Van Bath, *The Agrarian History of Western Europe*, 267–268; Langer, 'American Foods and Europe's Population Growth', 53; Messer, 'Three Centuries of Changing European Tastes for the Potato', 104; Kiple, *A Moveable Feast*, 136–137 (quote); Toussaint-Samat, *A History of Food*, 646–653; Reader, *Potato*, 112–114; Galli, *La conquête alimentaire du Nouveau Monde*. 参见 Spary, *Feeding France*, 62, 86, 其评论具有启发性。

7. Salaman, *History and Social Influence of the Potato*, 116.

8. Las Casas, *Apologética historia sumaria*, III: 37; Acosta, *Natural and Moral History of the Indies*, 206 (quote); Monardes, *Joyfull News out of the New-found Worlde*, 20; Galli, *La conquête alimentaire du Nouveau Monde*.

9. Bauhin, *Prodromos Theatri Botanici*, 89–90.

10. Kiple, *A Moveable Feast*, 136–137.

11. Eagle and Younge, eds., *Collection of the Reports of Cases*, II: 228; Evans, *The*

Contentious Tithe.

12. Rev Charles Layfield v. Thomas Ayscough et al., Croston, 1686, LRO, PR 718; Decree in Chancery: Rev Charles Layfield, Rector of Croston, & Thomas Ayscough, Thomas Lathom, Peter Lathom, Thomas Crookham, Thomas Hodson, John Rutter, John Moore, Thomas Miller, Thomas Christophers, Henry Yate, Richard Tompson, Edward Disley, Richard Moore, & William Forshaw, all of Mawdesley & Bispham, 16 May 1686, LRO, PR 718; Elizabeth Save v. Henry Thwaites, Kirkby Malzeard 1736, TNA, E134/10Geo2/Hil3; Papers concerning the Rev. Richard Rothwell claim to Tithe Potatoes in Sefton parish, 1789, LRO, DDM 11/61 (quote); Resolutions of Inhabitants to resist the Rev. Glover Moore's claim to Tithe of potatoes, Seals, 3 Oct. 1791, LRO PR/ 284; Eagle and Younge, eds., *Collection of the Reports of Cases*, II: 91, 141, 149, 189, 258, 310, 313, 380–398, 552, 588–589, 648, 690–691; Salaman, *History and Social Influence of the Potato*, 452; Evans, *The Contentious Tithe*, 7, 47, 53; Thirsk, *The Agrarian History of England and Wales*, I: 64; Zylberberg, 'Fuel Prices, Regional Diets and Cooking Habits', 112.

13. Agreement by 162 tenants of Formby, Ainsdale, and Raven Meols to frustrate the attempt of the rector of Walton, to take tithe of Potatoes and 'Garden-Stuff', 24 Feb. 1789, LRO, DDFO 23/4.

14. Vandenbroeke, 'Cultivation and Consumption of the Potato'; Terrón, *España, encrucijada de culturas alimentarias*, 143–144; Ibáñez Rodríguez, 'El diezmo en la Rioja', 192; Olsson and Svensson, 'Agricultural Production in Southern Sweden', 117–139; Palanca Cañón, 'Introducción y Generalización del Cultivo y Consumo Alimentario y Médico de la Patata', 77, 240–250, 263, 266.

15. 比较 Eagle and Younge, eds., *Collection of the Reports of Cases*, II: 589 和 Aikin, *A Description of the Country from Thirty to Forty Miles Round Manchester*, 45–46, 204, 237–238, 285, 306, 362 中的叙述。

16. Cardano, *De Rerum Varietate*, 30–31; Michiel, *I cinque libri di piante*, 143, 447; Clusius, *Rariorum Plantarum Historia*, 80; Bauhin, *Prodromos Theatri Botanici*, 89–90; Gerarde, *The Herbal or General History of Plants*, 926–928; Besler, *Hortus Eystenttensis*, section on 'classis autumnalis', plate 27; Laufer, *American Plant Migration, part 1*, 27–69; Salaman, *History and Social Influence of the Potato*, 73–100.

17. 例如，参见 Carolus Clusius to Joachim Camerarius, Frankfurt, 18 Nov. 1589。这封信与其他许多信件被抄录于 Van Gelder, ed., *Clusius Correspondence*。

18. Clusius, *Rariorum Plantarum Historia*, 80; Gutaker et al., 'The Origins and Adaptation of European Potatoes'.

19. Jacques Plateau to Carolus Clusius, Tournai, 3 Sept. 1588, *Clusius Correspondence*.

20. Bauhin, *Prodromos Theatri Botanici*, 89–90.

21. Jacques Garet Jr. to Carolus Clusius, 19 Jan. 1589; Jacques Garet Jr. to Carolus Clusius, 28 July 1589; and Gian Vincenzo Pinelli to Carolus Clusius, Padua, 19 Sept. 1597, 8 Dec. 1597, all in *Clusius Correspondence*; Clusius, *Rariorum Plantarum Historia*, 80.

22. Bauhin, *Prodromos Theatri Botanici*, 89–90. 另参见 Gerarde, *Herball, or General History of Plantes*, 782; Clusius, *Rariorum Plantarum Historia*, 80; Sala, *De Alimentis et Eorum Recta Administratione Liber*, 12, 54, 65, 77; Zwinger, *Theatrum Botanicum*, 893。

23. Aldini, *Exactissima Descriptio Rariorum Quarundam Plantarum*; Cienfuegos, 'Historia de las plantas', BNE, vol. 1, chap. 88: 'De las papas del Perú, que en Indias llaman chuno, al pan que dellas se haze', fols. 498–505. 更多关于16世纪早期罗马和托斯卡纳地区新世界块茎植物的证据，参见 Michiel, *I Cinque Libri di Piante*, 143; Aldrete, *Del orígen y principio de la lengua castellana*, 110–111; Magazzini, *Coltivazione toscana*, 16; and Targioni-Tozzetti, *Cenni storici sulla introduzione di varie piante nell'agricoltura ed Orticoltura Toscana*, 38–39。也可参见 Gartner, *Horticultura* on 'papas indorum' or 'artofeler'。

24. Gentilcore, *Food and Health in Early Modern Europe*.

25. Benzo et al., *Regole della sanitá et natura de cibi*. 也可参见 Nuñez de Oria, Regimiento y aviso de sanidad, 41v; Elsholtz, Diaeteticon, 31–32; Albala, Eating Right in the Renaissance。

26. 玛丽亚·伊莎贝尔·阿马多·多布拉斯（María Isabel Amado Doblas）在其优秀的《黄金世纪文学中关于红薯/土豆的书目注释》('Apunte bibliográfico acerca de la batata/patata en la literaturedel siglo de oro'）中认为，在18世纪之前，"patata"一词总是指红薯，我认为她的这种把握没有理由。

27. Monardes, *Joyfull News out of the Newe Founde World*, 104, 注意到了贝莱斯-马拉加（Vélez-Málaga）地区广泛种植红薯。

28. 威廉四世描述道："一种近年来从意大利来到我们这里的植物叫做 Taratouphli，它在土壤中自己生长，有美丽的花朵、好闻的气味，在根部有许多鳞茎，拿来烹饪非常好吃。可先放在水里小火煮开后去皮，然后去掉水分用黄油炖。" Wilhelm IV von Hessen to Christian I von Sachsen, Kessel, 10 Mar. 1591, *Quellenbuch zur sächsischen Geschichte*, ed. Arras, 61. 另参见 Michiel, *I cinque libri di piante*, Libro giallo, 143, Libro rosso II, 447。

29. 例如，可参见 Rosenhane, *Oeconomia*, 130。

30. Coles, *Adam in Eden*, 33.

31. Jacques Garet Jr. to Carolus Clusius, 19 Jan. 1589, 9 Sept. 1589, 28 Aug. 1590, all in *Clusius Correspondence*; Gerarde, *The Herbal or General History of Plants. The Complete 1633 Edition*, 926; Salaman, *History and Social Influence of the Potato*, 426–439; and *The Book of Fines*, ed. Butler, 207. 汉普顿宫（Hampton Court）的园丁已经

成功培育红薯，从园艺上看，没有理由怀疑它们都已在都铎时代的英国得到种植，Historic Royal Palaces Blog, 'The History of the Sweet Potato'。

32. Labat, *Nouveau voyage aux îles de l'Amerique*, II: 400–401; Boissier de Sauvages, *Dictionnaire languedocien-françois*, 344; 'Pomme de terre, Topinambour, Batate, Truffe blanche, Truffe rouge'; Dillon, *Travels Through Spain*, 331.

33. 关于红薯和耶路撒冷洋蓟在欧洲的传播，可以参考 Monardes, *Joyfull News out of the Newe Founde World*, f. 104; Benzo et al., *Regole della sanitá*, 622–623; Parkinson, *Paradisi in Sole Paradisus Terrestris*, 518; Zwinger, *Theatrum Botanicum*, 400–401; Parmentier, *Traité sur la culture et les usages des pommes de terre, de la patate, et du topinambour*; Targioni-Tozzetti, *Cenni storici sulla introduzione di varie piante*, 46; Salaman, *History and Social Influence of the Potato*; Crosby, *The Columbian Exchange*; Hawkes and Francisco-Ortega, 'The Early History of the Potato in Europe'; Amado Doblas, 'Apunte bibliográfico acerca de la batata/patata'; Galli, *La conquête alimentaire du Nouveau Monde*。

34. Terrón, *España, encrucijada de culturas alimentarias*, 169; Muldrew, *Food, Energy and the Creation of Industriousness*, 46; Nadeau, *Food Matters*, 163, 165, 168.

35. Rumpolts, *Ein new Kochbuch*, 16b, 143b. Olla podrida 在 Nadeau, *Food Matters* 中得到了讨论。

36. Casteau, *Ouverture de cuisine*, 94–95.

37. 印刷出版的烹饪书，参见 Dawson, *Good Huswife's Jewell*, 20v (quote); Murrell, *A New Book of Cookerie*, 4, 80–81; Cooper, *The Art of Cookery Refin'd and Augmented*, 36–37; Woolley, *The Queen-Like Closet; Accomplished Ladies Rich Closet of Rarities*; Grey, *A Choice Manual*; Astry, *Diana Astry's Recipe Book*, 95。手稿收藏，参见 'Physical and Chyrurgicall Receipts', fol. 107; 'Collection of Cookery and Medical Receipts by Edward and Katherine Kidder', fol. 21; 'Recipe Book of the Godrey-Faussett Family of Heppington', fols. 19, 79; 'Anonymous Collection of Cookery and Medical Receipts', fols. 36–37; 'Cookery Receipts Collected by Johnson Family of Spalding, Lincs.', fols. 112r–v; and 'Manuscript Recipe Book', c. 1700, SL, Sophie D. Coe Manuscript Cookbook Collection, box 1, folder 1, fol. 4。另可参见 Thirsk, *Food in Early Modern England*, 111, 115; Pennell, 'Recipes and Reception'。

38. *The Letters of John Chamberlain*, ed. Mc Clure, II: 551; and Nadeau, *Food Matters*, 35.

39. Royer, *Eine gute Anleitung*, 104–105.

40. Elsholtz, *Diaeteticon*, 31–32 (quote); Helmhardt von Hohberg, *Herrn von Hohbergs Georgica Curiosa Aucta*, 387; and *Die Curieuse ... Köchin*, 563.

41. *Frauenzimmer-Lexikon*, cols. 1979–1981; Howard, *England's Newest Way in All*

Sorts of Cookery, 14; Hall, *The Queen's Royal Cookery*, 98, 100–101; Salmon, *The Family-Dictionary*, 390, 408; 'Anonymous Collection of Cookery and Medical Receipts', fols. 36–37; 'Recipe Book of the Godrey-Faussett Family of Heppington, Nackington, Kent', fols. 19, 79; Smith, *The Compleat Housewife*, 9, 132–133, 139; *Collection of Receipts in Cookery, Physick and Surgery*, 131–132; 'English Manuscript Cookbook', SL, American Institute of Wine & Food Recipe Books, box 1; Altimiras, *Nuevo arte de cocina*, 140–141; Buc'hoz, *Manuel alimentaire des plantes*, 485–486; Rigaud, *Cozinheiro moderno*, 396, 403; Pennell, 'Recipes and Reception'.

42. Serres, *Le theatre d'agriculture*, 513–514; Parkinson, *Paradisi in Sole Paradisus Terrestris*, 516 (quote); Magazzini, *Coltivazione toscana*, 16; *The Compleat Planter and Cyderist*, 245–247; Markham, *The Husbandman's Jewel*, 7; Turner, *An Almanack*, 17; Parker, *The Gardeners Almanack*, 33.

43. Evelyn, *Kalendarium Hortense*, 14, 19 (quote), 122; Helmhardt von Hohberg, *Georgica Curiosa Aucta*, 640; Switzer, *The Practical Kitchen Gardiner*, 217–219, 378.

44. Hawkes and Francisco-Ortega, 'The Early History of the Potato in Europe'.

45. Cienfuegos, 'Historia de las plantas', BNE, vol. 1, chap. 88: 'De las papas del Perú', fols. 498–505; Leon Pinelo, *Question moral si el chocolate quebranta el ayuno eclesiástico*, 63.

46. Lili-Annè Aldman, personal communication, 2015, 引用了瑞典瓦德斯泰纳国家档案馆（Landsarkiv in Vadstena）的资料, Vadstena: Drevs Församling: LIb: 1 (1635–1650), 19–149, 207–265; and Långasjö Församlin: LIa: 1 (1651–1667)。

47. Griffith v. Allerton, 1698–1699, TNA, C 6/414/31; Chamberlayne, *Angliae Notitia*, 40; *Act of Tonnage and Poundage, and Rates of Merchandize*; Salaman, *History and Social Influence of the Potato*, 224–225, 451; Hawkes and Francisco-Ortega, 'The Potato in Spain during the Late Sixteenth Century'; Hawkes and Francisco-Ortega, 'The Early History of the Potato in Europe'; Thirsk, *Agrarian History of England and Wales*, I: 64, 373; Clarkson and Crawford, *Feast and Famine*.

48. Evans, 'Some Reasons for the Growth of English Anti-Clericalism'; Barnard, 'Gardening, Diet and "Improvement"'.

49. Molinier, *Stagnations et croissance*, 266. 1722 年，斯塔福德郡卡夫斯维尔村（Staffordshire village of Caverswell）当局同样决定对"土豆、防风萝卜、胡萝卜和各种块茎"征收什一税。Evans, 'Tithing Customs and Disputes', 25. 关于栗子和块茎类蔬菜，也可参阅 Grieco, 'The Social Politics of Pre-Linnaean Botanical Classification'; Camporesi, *The Magic Harvest*。

50. Ávila, *Escritos de Santa Teresa*, II: 128, 158.

51. Rich, *True Report of a Late Practice Enterprised by a Papist*, sig. Bi. Salaman,

History and Social Influence of the Potato, 428ff., 对诸多类似著作进行了编目。

52. [Hamilton], *The Country-Man's Rudiments*, 31.

53. Wiegelmann, *Alltags- und Festspeisen in Mitteleuropa*; Morineau, 'The Potato in the Eighteenth Century'; Braudel, *Civilization and Capitalism*, I: 168–170; Ferrières, 'Le cas de la pomme de terre dans le Midi'; Zylberberg, 'Fuel Prices, Regional Diets and Cooking Habits', 119.

54. McNeil, 'How the Potato Changed the World's History'; Terrón, *España, encrucijada de culturas alimentarias*, 84–85; Lemire, '"Men of the World"'.

55. Magazzini, *Coltivazione toscana*, 16; Benzo et al., *Regole della sanitá et natura de cibi*, 622–623; Schuler, *Geschichte und Beschreibung des Landes Glarus*, 130; Boll, *Geschichte Meklenburgs mitbesonderer Berücksichtigung der Culturgeschichte*, II: 523. Berg, 'Die Kartoffel und die Rübe'; and Blum, *The End of the Old Order*, 271, 讨论了士兵和工匠在土豆传播过程中扮演的角色。

56. DeLoughrey, 'Globalizing the Routes of Breadfruit and Other Bounties'. 也可参见 Ploeg, 'Potatoes and Knowledge'。

57. Vandenbroeke, 'Cultivation and Consumption of the Potato', 19.

58. Eden, *The State of the Poor*, I: 508.

59. Symner, 'Notes on Natural History in Ireland' (quote); Petty, 'The Political Anatomy of Ireland', 1672, *Tracts; Chiefly Relating to Ireland*, 319, 355, 366, 374; Clarkson and Crawford, Feast and Famine; Barnard, 'Gardening, Diet and "Improvement"'.

60. Baxter, 'The Reverend Richard Baxter's Last Treatise'; Overton, *Agricultural Revolution in England*, 102.

61. Elsholtz, *Diaeteticon*, 31–32.

62. Royer, *Eine gute Anleitung*, 104; Helmhardt von Hohberg, *Georgica Curiosa Aucta*, book 5, chap. 42, 639. 近代早期农民的饮食，参见 Camporesi, *Bread of Dreams*; Capatti and Montanari, *Italian Cuisine*; Fox, 'Food, Drink and Social Distinction in Early Modern England'。

63. Royer, *Eine gute Anleitung*, 104–105; Jacques Plateau to Carolus Clusius, Tournai, 3 Sept. 1588, *Clusius Correspondence*.

64. Terrón, *España, encrucijada de culturas alimentarias*, 139–142; Ferrières, 'Le cas de la pomme de terre dans le Midi', 210 (quote); Miodunka 'L'essor de la culture de la pomme de terre au sud de la Pologne'.

65. 关于这一点，参见 Berg, 'Afterword: Things in Global History'。

66. Kaldy, 'Protein Yield of Various Crops as Related to Protein Value'; FAO, 'International Year of the Potato 2008: Potato and Water Resources'. 不过，一公斤小麦

的热量比一公斤土豆高得多。也可参见 Sauer, *Agricultural Origins and Dispersals*, 136–137。

67. Shammas, *The Pre-Industrial Consumer in England and America*, 137.

68. Terrón, *España, encrucijada de culturas alimentarias*, 139–143; Zylberberg, 'Fuel Prices, Regional Diets and Cooking Habits'.

69. Netting, *Balancing on an Alp*, 38. Bräker, *The Poor Man of Toggenburg*, 提供了瑞士山区土豆种植的第一手资料。

70. 关于土豆对人口增长之作用的持续辩论，参见 Komlos, 'The New World's Contribution to Food Consumption'; McNeill, 'How the Potato Changed the World's History'; Nunn and Qian, 'The Potato's Contribution to Population and Urbanization'。

71. Simonton, *A History of European Women's Work*, 19–33, 116–128. 关于女性和锄头文化的概述，参见 Boserup, *Woman's Role in Economic Development*。

72. Scott, *Seeing Like a State*; Scott, *The Art of Not Being Governed*, 195–207 (196 quote); Scott, *Against the Grain*, 130.

73. Hutchison, 'Swedish Population Thought in the Eighteenth Century', 91.

74. Laurence, *The Duty of a Steward to his Lord*, 30, 119; Thirsk, *The Agrarian History of England and Wales*, I: 56.

75. Linnaeus, *Skånska resa år 1749*, 7 June 1749; Koerner, *Linnaeus*, 148–149. 也可参见 Linnaeus, *Dissertatio Academicum de Pane Diaetetico*, 20; 感谢迪热尔·阿博（Desirée Arbó）提供翻译。

76. Ward, *Proyecto económico*, 72.

77. Dechambre, ed., *Dictionnaire encyclopédique des sciences médicales*, 229.

78. Slicher Van Bath, *The Agrarian History of Western Europe*, 267–268.

79. Zanon, *Della coltivazione, e dell'uso delle patate*, 71; Gasparini, *Polenta e formenton*; Galli, *La conquête alimentaire du nouveau monde*, 235–238. 玉米作为欧洲其他地方的农作物，另见 Forster, 'The Noble as Landlord in the Region of Toulouse', 241; Casado Soto, 'Notas sobre la implantación del maíz en Cantabria y la sustitución de otros cultivos'; Terrón, *España, encrucijada deculturas alimentarias*, 74–91; Gentilcore, *Food and Health in Early Modern Europe*, 147。

80. Hoffman, *Growth in a Traditional Society*; Thick, 'Root Crops and the Feeding of London's Poor', 294–295.

81. McClellan III and Regourd, 'The Colonial Machine'; Raj, 'Colonial Encounters and the Forging of New Knowledges and National Identities'; Chakrabarty, *Provincializing Europe*; Cañizares-Esguerra, *Nature, Empire and Nation*; Safier, *Measuring the World*; Smith, *The Body of the Artisan*; Raj, *Relocating Modern Science*; Roberts, 'Situating Science in Global History'; and McClellan III, *Colonialism and Science*.

82. Ambrosoli, *The Wild and the Sown*, 406.

83. Toussaint-Samat, *A History of Food*, 647.

84. 关于玉米的多功能性和农民的实践，参见 Brush, *Farmer's Bounty*。18 世纪挪威的农民吃土豆，而且熏制土豆叶以代替烟草。Wilse, *Physisk, oeconomisk og statistisk Beskrivelse*, 255–256. 阿蒙德·佩德森（Amund Pedersen）提供了挪威语翻译。

85. Forster, *Englands Happiness Increased*，就是一个例外。

86. Petty, 'Political Arithmetic' (1690), and 'Political Anatomy of Ireland', *Tracts*, 238, 319, 355, 366, 374. 或可参见 Brady, 'Remedies Proposed for the Church of Ireland', 166。

第二章　启蒙的土豆

"好几年来，报纸上几乎只讨论土豆。"1782年，法国历史学家皮埃尔-让-巴蒂斯特·勒格兰德·奥西（Pierre-Jean-Baptiste Legrand d'Aussy）如是说。医生们分析它们的特性，作家们推崇它们的优点，君主们鼓励食用它们——勒格兰德·奥西宣称，土豆已经成为启蒙运动的宠儿。[1]

有两个特点特别吸引18世纪土豆热衷者的想象力。首先，他们坚持认为土豆是一种健康且营养丰富的食物，每个人都可以愉快地食用。一本法国烹饪书宣称，土豆有三重优势，即"健康、美味、经济"，由此，它无疑是神圣的天意所赐。其次，土豆赞美者（一家报纸这样称呼他们）也提出，这种块根给普通人带来了特殊的好处。"对于拥有一大家子的穷人来说，一头奶牛和一片土豆菜园，是多么珍贵的东西啊！"著名的苏格兰医师威廉·巴肯如此赞颂道。[2] 在整个欧洲，医生、政治家、牧师和文学界的成员都一致认为，土豆对穷人来说是一种额外的资源，可以把他们从饥饿和贫困中解放出来。在土豆显著优点的鼓舞下，欧洲大陆许多地方的君主颁布法令，鼓励种植土豆，无数组织也提出了增加土豆消费的方案。勒格兰德·奥西说18世纪的公共领域被启蒙运

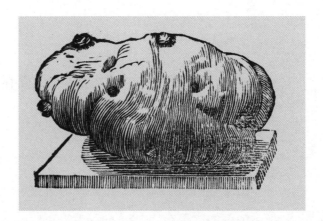

图 7　乔瓦尼·巴蒂斯塔·奥奇奥利尼（Giovanni Battista Occhiolini）1784 年论著中的土豆版画插图。罗马传教士乔瓦尼·巴蒂斯塔·奥奇奥利尼关于"通常被称为土豆的神奇美洲果实"的论著，针对的是农村劳工。他希望他对土豆农艺学的介绍能鼓励他们种植和食用更多的土豆。

动的土豆言论所占领也许有些夸张，但它确实在半个世纪里成了一个跨越欧洲大陆的强大主题。（参见图 7）

　　土豆从受人鄙视的怪植物转变为今天的主食，这个现象被一家受人尊敬的学术团体归功于 18 世纪的宣传。如我在第一章所指出的，这种转变的主力军据称是有远见的统治者和慈善家。一位食物历史学家声称："莱茵河两岸的人们都倾向于认为吃土豆会导致麻风病。"他解释说，在路易十六亲自资助了一场巧妙的促销计划之后，法国人才开始接受土豆。[3] 这项计划据说是由法国科学家、土豆推广者安托万·奥古斯丁·帕蒙蒂埃（Antoine Augustin Parmentier）设计的，他安排在王室土地上招摇显眼地种植这种块茎作物。帕蒙蒂埃的合作者朱利安-约瑟夫·维雷（Julien-Joseph Virey）在他的朋友及同事离世后出版的传记中描述道，当这些植物成熟时：

　　　　帕蒙蒂埃安排了宪兵进行守卫——但都是在白天。他的目的是让它们在夜里被偷走，老百姓确实也帮了忙。每天早晨他都被告以夜间的盗窃行为，他十分高兴，并慷慨地酬谢告密者，而告密者却对他那让人费解的喜悦感到吃惊。但公众舆论被征服了，从那一刻起，法国因一种持久的资源而富起来。⁴

据说在欧洲其他地方，要想征服公众的舆论也需要采取几乎相同的计策。在希腊，19世纪早期的民族主义者依昂尼斯·卡波迪思特里亚斯（Ioannis Kapodistrias），据说是以极其相似的策略，说服了持怀疑态度的希腊人接受土豆。卡波迪思特里亚斯

命令把土豆倒在纳普良的码头上并加以看管。谣言开始散布，说既然土豆得到那么好的保护，它们一定很贵重。不久就有人想偷这些"值钱的"土豆。由于守卫们被告知不要理会偷窃行为，因此没过多久整批货就消失了。土豆至今仍然是希腊菜肴的重要组成部分。⁵

这些故事在欧洲流行文化中广为流传。我与许多人交谈过，他们讲述了在学校学到的这些土豆英雄和他们克服顽固阻力的事迹。⁶（参见图8）

这些营养方面的慈善举措，通常是在饥荒或粮食短缺的特殊时刻典型呈现的应急响应。在普鲁士，据说是18世纪40年代反复发生的饥荒，促使腓特烈大帝颁布了一系列鼓励土豆种植的法令。支撑这些故事的假设是，政府推广一种受欢迎的粮食，是特定短缺时期的自然反应。可以肯定的是，18世纪的粮食短缺，将富人的注意力引向饥饿造成的人道主义后果和政治后果。在1750年到1815年的半个多世纪里，几乎

图 8　波茨坦无忧宫（Sans Souci Palace），腓特烈大帝的牌匾上留下的土豆供品。参观腓特烈二世无忧宫的游客在这块牌匾上放上土豆，他们进一步巩固了他作为德国第一位土豆推广者的声望。

持续不断的战争，以及不断增长的人口，给粮食供应带来了巨大的压力。不过，这样的事件，应该被放在更广泛的背景下。招募和装备庞大军队的需要、对粮食贸易自由化影响的调查、有用农业知识的传播导致的当地社会的形成、对农民和劳工酒精消费的束手无策、减少婴儿死亡率的计划，以及土豆推广，共同成为粮食与国家富强之间关系的更大的重组概念的一部分。本章将阐明这些关系。

以这种角度观察 18 世纪对土豆的着迷，就会发现个人饮食与启蒙运动时期出现的政治经济、公共卫生和有效治国等新观点之间的联系。认识到这些联系，可以让我们更好地理解粮食短缺与政府反应之间的历史关系。饥荒曾长期困扰着欧洲，"饥荒"解释了 18 世纪对土豆的巨大兴趣，但并不能解释为什么只有在 18 世纪，政府才开始通过推广特定的粮食来应对短缺。从饥荒有时夺走多达 40% 当地人口的 17 世纪，到 18 世纪，所发生改变的，并不是饥荒的发生率和严重性的增加——学者告诉我们，饥荒的频率实际上降低了。[7] 发生改变的是塑造健康人口的新治国模式的重要性。启蒙运动对土豆的着迷，反映的不是新食物的出现或新的饥饿程度的呈现，而是人口的健康和活力与国家富强之间关系的新观念。正是这一点，才将土豆的地位从农舍菜园和船舱货物的默默无闻，抬升到启蒙运动的论著之中。因此，土豆的小历史，揭示的是更大的历史变化，这些变化使得政府和它的理论家们关注到了普通人的日常习惯。

人口与粮食供应

启蒙时期的人们对土豆的讨论，与 18 世纪关于"人口"的辩论是分不开的。在整个 18 世纪，哲学家、经济学家、官员和文学界的成员，都致力于长期研究居住在一个地区的人口数量和其财富之间的关系。他们尤其思考的是，人口多是否是贸易和商业成功的根本动力，人口增长本身是否表明一个国家管理良好，以及人口是否有可能对于一个特定地区而言太过于庞大。这些问题都引起了激烈的争端和讨论。

自 16 世纪晚期开始，讨论好政府的论著开始提出，国家通常会从庞大的人口中获益。更多的人口提供了更多的工农业劳动力和更多的军队后备。它们反过来又会增加君主的权力。这种对人口规模与国家权力之间联系的兴趣，推动了 19 世纪数学领域，如测算人口增长所需的概率和统计等学科的发展。在 18 世纪，治国理论家们开始认为，人口不仅是君主的个人财产，而且是国家财富和权力的基石。这种政治信念，推动了越来越多的计划的制订，它们的目的是通过排干沼泽、管理医院，以及采取其他公共卫生措施来保护人口免于疾病和死亡。这种勇敢尝试的倡导者们，不仅强调推动他们努力的强大人道主义，而且还强调维持人口的政治重要性。英国 1750 年出台了《穷人预防注射诊疗所计划》(*Plan of a Dispensary for Inoculating the Poor*)，启动这项计划的依据是："由于国家的实力在一定程度上与它的居民数量是成比例的，因此每一次通过保护生命而增加人口的尝试，都会被认为是*爱国和人道的*。"[8]

到 18 世纪中叶，这些观点在欧洲许多地方已经司空见惯。西班牙政治家确信"人口是万物的基础"，因为没有人民，"就没有农业、工业、商业、艺术、权力和财富"。[9]瑞典皇家科学院秘书、数学家佩尔·沃占廷（Pehr Wargentin）赞同的是："一个公民社会的最强实力由大量的好公民构成，这是一种现在几乎无人怀疑的说法。"[10]庞大且（在理想情况下）密集定居的人口具有优势，这种信念贯穿了整个 18 世纪。当然，人口多也可能是不利因素的理论同样存在，而且，并不是每个人都赞同导致经济增长的原因是人口庞大。举例来说，重农主义，是强调土地和农业在创造财富中的中心地位的新经济理论，其倡导者们认为，人口增长是经济成功的结果，而不是原因。尽管如此，他们还是赞同，它是一个积极的迹象，是经济健康的象征。[11]

在 18 世纪的最后 25 年里，反对的声音开始抬头。一些作家认为，无限制的增长最终可能适得其反，会削弱而不是增强政体。托马斯·罗伯特·马尔萨斯（Thomas Robert Malthus）牧师成为这种观点最具影响力的倡导者。尽管如此，主流观点仍然强调人口众多的优点。1795 年，即马尔萨斯出版《人口论》(*Essay on the Principles of Population*)的三年前，由农学家亚瑟·杨（Arthur Young）主编的一家有影响力的英国杂志《农业年鉴》(*Annals of Agriculture*)刊登了一篇记者文章，它确信地论述说,"国家的富裕,无可争议地与它的居民数量成比例"。¹² 晚至 19 世纪 20 年代,詹姆斯·穆勒(James Mill)① 等作家,面对人口众多必然有益的执着观念,还在耐心地试图解释马尔萨斯的观点。¹³

正如米歇尔·福柯几十年前所言,这些人口问题的辩论标志着一种权力行使的新方式的出现,这种方式强调了让国家政策与更大的力量保持一致的重要性,而这些力量本身塑造了一个地区居民的活力、规模和生产力。¹⁴ 新的治国科学,不是简单的确保服从或施加权力的问题。它需要对资源进行管理,并为开发资源建立有效的制度。"人口",换句话说,远不止是居住在一个地区的个体的集合体,它是必不可少的资源。论述人口问题的作者们不断强调,对这个至关重要资源的成功管理,是政府效力的基本晴雨表。法国人口学家让-巴蒂斯特·莫霍(Jean-Baptiste Moheau)在一篇论述人口的文章中指出,居民数量的增加,既是整体公众福祉的表现,也是它几乎不可避免的结果。由于这个原因,人口的增长或不增长,"证明了政府的好或坏"。¹⁵

然而,仅人口多是不够的。有一篇关于如何重振西班牙疲软经济的论著,对需要更多地做些什么进行了详细说明。"任何经济体制的最基本要素,"它阐明,"都是确保人们得到*有效*就业。"¹⁶ 这篇论著的爱尔兰作者伯纳多·沃德(Bernardo Ward),曾受西班牙国王费迪南德六世(Ferdinand Ⅵ)委托,对西班牙的农业、商业和工业进行研究,为此他走遍了伊比利亚半岛。沃德后来担任西班牙皇家铸币厂厂长,有许多年一直在思考让西班牙人得到有效就业的最佳途径。他早先曾设计了一个

---

① 詹姆斯·穆勒(1773—1836),19 世纪著名的苏格兰历史学家、经济学家、政治理论家、哲学家、功利主义伦理学家和功利主义教育思想家。他与大卫·李嘉图一同是古典经济学后期的代表人物。他是英国机械联想主义的典型代表,联想主义心理学的主要传播者。他的儿子约翰·斯图亚特·穆勒也是著名哲学家。——译注

计划，打算将乞丐和无家可归的人安置在一个机构里，让他们"为了国家的利益"而工作。[17]

沃德对流浪者的关注表明，如果不积极从事生产劳动，庞大的人口也是无用的。在为费迪南德六世撰写的报告中，沃德建议对西班牙在欧洲和新世界领土的管理进行改革。他强调，西班牙人口的劳动潜力是一种必不可少的资源，而良好的治理恰恰在于释放这种潜力。人口可以通过个体数量的增加而"物理"地增加，但这并不是发展有效人口的唯一或最好的方法。将"一个不工作、对国家没有任何有用贡献的人"转变为一个勤奋的劳动者，才是更大的成就，因为这构成了人口的"政治"增长。[18]沃德认为，"当人们说，一个君主的财富由他的臣民的数量构成，他们的意思是'有用'的臣民，因为上百万的流浪者和职业乞丐不仅没有用处，反而是国家的障碍。没有他们，国家会更好更富裕。"[19]沃德对发展勤劳人口的重要性的评判，得到了广泛的赞同。1731年，瑞典作家、政治家爱德华·卡尔森（Edvard Carleson）指出，如果一个国家的人口没有在农业或制造业中得到有效的就业，它就"如同埋在地下的死宝藏"。[20]勤劳的居民，而不仅仅是普通大众，才是国家伟大和成功的核心。

民众必须身体健康才能勤劳工作，只有这样，国家才能繁荣昌盛。而民众健康所需的不仅是预防接种计划、沼泽排水方案，还必须包括充足的粮食供应。如土豆推广者帕蒙蒂埃所提出的那样，"食物的种类和选择极大地影响着人口，因此要确保人们吃得够，就不可能采取太多的防范措施"。[21]这些观点，并不仅仅是特定短缺时刻的反应，而是基于一种日益增强的信念，即强大的人口可以极大地增加一个国家的权力和财富。当然，战争和短缺的特别压力，会将官员的注意力集中在粮食供应上。1796年，数学家、官僚主义者约瑟夫-路易斯·拉格朗日（Jean-

Louis Lagrange）①设计了"营养微积分"，用于估算新法兰西共和国的食物需求。他的动机既包括对理论的好奇，也包括实际的担忧。[22] 从 1792 年到 1815 年，全球战争共动用超过 700 万兵力。从法国大革命爆发到滑铁卢战役的余波，这场战争将欧洲的实力消耗殆尽。再加上反复歉收，许多国家的粮食供应出现紧张状态。然而，即使在这样的战争年代，对劳动人民饮食的兴趣，也不仅仅是这些连续不断状况的反映。作家们一直强调确保"下层人民"得到充分营养的重要性，因为这样的观念在治国的讨论中已是司空见惯。[23]

从 17 世纪晚期开始，政治理论家们开始将粮食视为建立强大国家的重要组成部分。博学的德国外交官、政治理论家约翰·约阿希姆·贝歇耳（Johann Joachim Becher）将公民社会定义为"人口众多、营养丰富的社区"。英国律师威廉·培提特（William Petyt）写过许多论述政府治理的著作，他也提出了同样的粮食与人口规模之间的关系。[24] 由于发展规模庞大而活跃的人口对理解政治和经济福祉变得越来越重要，因此，确保人口吃好的政治和经济重要性也更加受到关注。18 世纪 30 年代，法国律师、政治哲学家让-弗朗索瓦·梅隆（Jean-François Melon）将粮食供应纳入他的国际贸易分析中。梅隆在一篇颇有影响的文章中，提出了一系列思维实验，以说明塑造商业交换的力量。他对三个假想岛国之间贸易关系的研究得出的结论是，一个国家的实力与其拥有"尽可能多的粮食"密切相关。[25] 在解释国家如何变得强大的更大的理论框架中，粮食是一个组成部分。

---

① 约瑟夫-路易斯·拉格朗日（1736—1813），法国著名数学家、物理学家，在数学、力学和天文学方面都有突出贡献。他的重要著作有：《分析力学》（牛顿之后的一部重要的经典力学著作）、《论任意阶数值方程的解法》、《解析函数论》和《函数计算讲义》等。——译注

从这个角度来看，确保有足够的、适合劳动者的粮食种类供应是至关重要的。当作家们谈到要保证人口的精力和勤劳时，他们内心所想的是劳动人民的精力和勤劳。英国慈善家乔纳斯·汉威（Jonas Hanway）曾明确表示："财富和权力的真正基础是穷人劳动者的数量。"[26]因此，国家的力量与财富所需要的，用诸多讨论该问题的小册子之一的话来说，是劳动人民要"吃得饱，吃得便宜"。[27]劳动人民的身体与国家实力和繁荣经济之间的相互联系，促使人们对劳动者的饮食习惯产生了新的兴趣。

将劳动者的饮食与国家政体相联系时，这些作者借鉴了在不同的写作体裁——健康手册中日益普遍的观点。正如在第一章中提到的，在17世纪，这类致力于介绍良好生活原理的各国的指南，出版的数量日益增多。饮食建议占据了突出的位置，因为每个人都认识到，好身体从根本上取决于吃合适的食物。这些书籍除了强调选择符合个人体质和气色的饮食的重要性外，还经常将特定地区的饮食与当地人口的特点联系起来。不同的饮食习惯，有助于说明诸如英国人和法国人之间的民族差异，以及欧洲人对自己和非欧洲人之间差异的理解。一位西班牙医生说，美洲印第安人在身体和性格上与西班牙人完全不同，"因为他们吃的食物不同"。[28]当地的历史学家也提出了类似的观点，他们也试图对一个地区的特色进行解释。威尔特郡（Wiltshire）居民忧郁的性格，格拉摩根郡（Glamorganshire）人精力充沛、和蔼可亲的性格，都被归因于当地的饮食。[29]个人的饮食习惯就这样塑造了地区和民族性格。这些概念，使得个人饮食习惯和整体人口特征之间的关系得以建立，18世纪的政治思想家正是对它们加以了利用。

一个国家福祉的许多方面，据说都会受到劳动人民饮食习惯的影响。观察人士认为，工业生产率是一个明显与之关联的领域。正如亚当·斯密所说："一般来说，人在吃得差的时候工作强过吃得好的时候……似乎是不太可能的。"[30] 而且，营养不良的劳动者不会养育精力充沛的孩子来从事制造业和农业劳动，人们也不能依靠营养不良的士兵来保卫国家。这两者——贫困儿童的健康和军队的强大——在任何情况下都是相关的。小说家亨利·菲尔丁（Henry Fielding）①想象着喝杜松子酒的母亲生下的病弱婴儿，担心"这些可怜的婴儿"成为"我们未来的水兵、士兵"的后果。他预言，如果英国依靠这些羸弱的人来保卫自己的领土、发展商业和农业，后果将十分可怕。[31] 欧洲各地的作家与菲尔丁一样，担心贫穷儿童的死亡或衰弱会造成"国家的政治损失"。在考虑该问题的众多文章中，有一篇认为，确保婴儿成长为健康而有活力的劳动者，对国家的"荣耀与繁荣"至关重要。[32] 论述该主题的书籍在它们的书名中明确地表达了这种关联。有一部西班牙的此类著作的标题是：《弃婴夭亡的具体原因：对这种严重不幸的补救方法，以及使他们成为有用的基督徒公民以使西班牙的人口、力量和财富显著增长的方法》（*Concrete Causes of Mortality in Foundlings during their Early Years: Remedies for this Serious Evil, and Method for Making them into Useful and Christian Citizens to the Notable Increase to Spain's Population, Strength and Wealth*）。它的作者是潘普洛纳（Pamplona）综合医院的一名牧师和托管人。他明确表示，为国家救活这些命运不幸的婴儿，将会增加士兵和劳动人口："将会有多少——我们现在缺乏的人——从事公共工作！有多少劳动者！多少诚实的士兵啊！"[33]

潘普洛纳的医院，是旨在通过照顾弃婴来增加诚实士兵和劳动者人

---

① 亨利·菲尔丁（1707—1754），18世纪最杰出的英国小说家、戏剧家。18世纪英国启蒙运动的最伟大代表人物之一，英国第一个用完整的小说理论来从事创作的作家，被沃尔特·司各特称为"英国小说之父"。——译注

口的众多机构之一。在这些工作中，食物是核心问题。喂养弃儿的难度很大。由于缺乏足够数量的奶妈，孤儿院的死亡率有时接近100%。正因为强壮而多产的劳动人口依靠的是它自身的再生能力，因此婴儿喂养与婴儿死亡率的关联引起了欧洲许多地方作家的关注。关于最佳人工配方奶粉，以及婴儿总体喂养的特别重要性之类的论著，被大量撰写出来，其中一些直接写给统治君王。这种情况说明，它们的作者们认识到了贫困儿童的营养与国家事务之间的关联。[34]

对国家安全、营养食品和劳动人口的关注，在海军补给需求中是最紧密交织的。英国政府供养皇家海军的工作清楚地说明了这种关系。由于人们认为这种"帝国赖以生存的人的身体"需要高营养的饮食，皇家海军的水兵所享受的口粮供应量远远超过商船上的水手。[35] 对它的保障并非易事。尽管供养任何一支军队都需要先进的组织，但船上生活造成的特殊障碍，却需要更复杂的基础建设。在不列颠群岛，供养海军的艰巨任务，是由1683年成立的食品储备局（Victualling Board）承担的，它负责监督所有与海军供应有关的事务。它的数百万英镑的预算，反映了英国政府对这项事业的重视。它的条例强调了对高质量食品的需求，1760年的条例规定："牛必须是最肥的，土豆、洋葱和饲料必须是能买到的最好的。"[36] 公务表中规定了水兵有权获得的供应品，同样也是为了确保劳动力的健康和工作能力。储备局一丝不苟的记录显示，条例最轻微的变更都可能招致审计，一方面是因为它可能意味着有欺骗财政部的企图，另一方面也因为不符合标准的口粮会削弱海军的战斗力。

其他国家虽然缺乏粮食储备局的强力制度，但也同样相信，水兵的饮食对于国家治理有直接的重要性。西班牙医生、西班牙加的斯皇家外科医学院教授佩德罗·马利亚·冈萨雷斯（Pedro María González）坚称，让水兵们因饮食不良和医疗不当而患病，对国家来说是巨大的损失，而对敌人却极为有利。[37] 其他地方的医生也同样强调为水兵提供营养食物的政治重要性。当医师安托万·普瓦索尼耶·德斯佩里埃（Antoine Poissonnier Desperrières）向法国海军提议一种激进的素食饮食方案时，他将它置于一种明确的政治背景下。普瓦索尼耶·德斯佩里埃是第戎科学院（Dijon Academy of Sciences）的成员、殖民医学早期文章的作者。他对海军补给中典型的对咸肉的依赖提出了批评，他（和其他许多人一样）指责它是坏血病的罪魁祸首，这种病是对水兵个人和海军整体的严重威胁。他提出将以大米和豆类为基础、用姜和腌洋葱调味的口粮作为替代。他承认，这种新饮食可能会让水兵不悦，但他坚持认为，它会让海军更健康、更有活力。由于它旨在保护"国家珍贵人群"的健康，因此他认为他的计划是一项"爱国和经济方案"。[38] 简而言之，考虑劳动人口的饮食问题，对关心政治的人而言是正当事业。

*

新的治国模式在于培养强健的劳动人口，由于其重要性，到18世纪晚期，普通人日常的饮食习惯已经变得与国家的整体成功休戚相关。在这些新的观念之外，政治家、地主等许多人依然在担心着饥荒和粮食匮乏给社会和政治带来的不稳定影响，而哲学家和牧师继续在哀叹着饥饿造成的苦难。这样的担忧并不是理论上的，因为在整个18世纪，食物短缺经常困扰着普通民众，有时还会激起长期困扰当权者的那种民众行动。1764年发生在那不勒斯的那场灾难性饥荒后来波及整个欧洲大陆，远至马德里都发生了严重骚乱。[39] 因此，对于18世纪的国家科学而言，食物供应既是公共秩序的问题，也是更大的政治经济模式的核心组成部分，这个模式将国家富强与劳动人口的活力联系在一起。

然而，如何确保劳动人民吃到合适的营养食物呢？除了最激进的政治哲学家，谁也不希望对经济秩序进行深刻的变革，将更多的财富输送给贫穷的劳动者。[40] 相反，人们的注意力集中在寻找现有饮食的廉价替代品上。在刚从上个世纪瘟疫爆发造成的人口损失中恢复的博洛尼亚，农学家、地主彼得罗·马利亚·贝格美（Pietro Maria Bignami）著文解释说，如果他的家乡拥有充足的粮食供应，其人口将会显著增长；如果人口增长，工业肯定也会随之增长。如果这成为现实，该地区无疑将成为"全意大利最富有、最幸福的地区之一"。[41] 为了实现这一目标，他认为有必要确定一种"新产物"，至少能部分弥补现有粮食的不足。他提出，也许土豆可以达到这个目的。

贝格美希望土豆能满足劳动人民对廉价、有助于人口建设的主食的需求，而他并不是唯一有这种想法的人。整个欧洲的政治思想家都把希望寄托在土豆上，把它视为培养强健劳动人口的一种工具。如爱尔兰人这样的长期食用土豆的群体，他们的吃苦耐劳和繁衍后代的能力，让人们看到了面颊红润的一家人靠自己种植的土豆维持生活的动人景象。苏格兰农学家、印刷商大卫·亨利（David Henry）在称赞土豆的营养特性时特别指出，除了美味之外，它还有更多优点："它对人口增长有利，因为人们已经注意到，在贫穷劳动者几乎只有土豆吃的爱尔兰西部，看到一家有六、七、八个或十个，有时甚至更多的孩子都不奇怪。"亨利赞许地提到了"挤满那些卑微之人的小木屋的健康后代"。[42] 还有人列举了苏格兰的高地，那里"吃苦耐劳的强壮居民"主要以土豆为生。[43] 在法国南部阿尔卑斯滨海省（Alpes-Martimes）的山村里，吃土豆的人同样因无疑代表着健壮的"显著肥胖"而受到称赞。[44] 伯尔尼经济协会（Economic Society of Bern）是18世纪欧洲建立的数百个传播有用知识的爱国组织之一，该协会主席塞缪尔·恩格尔（Samuel Engel）跟贝格美一样，写了一篇关于土豆各种优点的论著，其中阐述了土豆作为人口增长工具的巨大潜力。[45] 爱尔兰再次成为以土豆为主的饮食之优点的令人信服的例证。他相信,土豆要想在其他地方发挥魔力,只需劳动人民认识到它的潜力。

土豆与劳动阶层的饮食

许多评论家认为,很不幸的是,普通人现有的饮食习惯极大妨碍了大量的、精力充沛的劳动人口所带来的幸福结果。人们认为,与吃土豆的健康爱尔兰人不同,许多劳动者是被失策的、自以为是的饮食错误给害了。科学家安托万-亚历克西斯·卡代·德沃克斯(Antoine-Alexis Cadet de Vaux)抱怨说,乡下人具有的许多"饮食恶习",造成了他们糟糕的身体状况。[46]一位英国作家严厉地批评说,过度喝茶、吃糖和吃白面包是"影响人类劳动的所有罪恶"之根源。[47]"普通人大量饮用极燥热、极毒烈酒的习惯"遭到了特别多的批评。[48]慈善家们以不那么具有敌意的语气撰写了"友好的建议",旨在向穷人展示如何通过减少食物浪费、采用更经济的烹饪方式以及多吃蔬菜来增进健康。[49]马里兰殖民地总督之子弗雷德里克·莫顿·伊登(他考证了第一个在户外种植土豆的苏格兰人),在英格兰对劳动者和工匠的饮食习惯进行了大量的调查之后,认为"穿衣、饮食等个人花费中"的浪费,是加剧贫困的一个重要原因。伊登相信,只要稍加注意,一个劳动者就能"把在食物上的花费减少一半,而又不会减少食物的可口、营养和健康益处"。[50]

苏格兰医师威廉·巴肯撰写了几本书阐述这一主题。巴肯有一段时间在约克郡的一家孤儿院担任医务官,还在伦敦的一家私人诊所工作过。这些经历为他的畅销家庭医疗手册积累了资料。这本手册介绍了如何治疗耳痛,强调定期锻炼的必要,并提醒学生勤换衣服。《家庭医疗》(*Domestic Medicine*)还批评了"穷人"的饮食习惯,在巴肯看来,穷人对自己的健康不佳负有一定责任。他坚称,"农民对他们的吃、喝极其不重视,而且常常由于懒惰而吃不健康的食物,尽管他们本可以花同样多的钱吃到健康的食物。"[51]他还指责肉贩和杂货商诡诈的经营行为,这些行为导致变质和掺假食品常出现在穷人的饮食中。他强调,这些问题应该引起每个人的关注,因为变质的食物会引发流行病,同时也因为"穷人劳动者的命对国家十分重要"。[52]

巴肯在1797年的《平民饮食观察》(Observations Concerning the Diet of the Common People)中继续谈论了这些担忧。这本书撰写于18世纪90年代的饥荒之时。与法国的战争、粮食歉收以及政府的政策,共同导致英国反复出现粮食短缺的状况。《平民饮食观察》发展了巴肯早期对改善大众饮食的兴趣,同时它反映了粮食供应上的这些压力。巴肯写这两本书的目的,是向"普通人"展示如何通过对饮食作出更好的选择来"更便宜、更好"地生活。[53]正如他在《平民饮食观察》中所说的那样,大多数普通人吃太多的肉和白面包、喝太多的啤酒,却没吃足够的蔬菜。他指出,这不可避免地导致了健康不佳,并引起坏血病等疾病,严重损害劳动者和儿童的身体。他强调,如此一来就会破坏英国的贸易、削弱这个国家。

然而,如何确保人们营养充足呢?哪类食物能提供比啤酒和白面包更好的营养基础呢?威廉·巴肯鼓励以全谷物和块根类蔬菜为主的基础饮食,他坚信,这种饮食不仅比其他选择便宜,而且更加健康。他对土豆特别感兴趣。他相信,一个国家只有"从地下获取大量食物"才能变得"人口众多"。土豆和牛奶配在一起,能提供理想的营养:"我们所知道的一些最强壮的人,是牛奶和土豆养大的。"巴肯说。(就如法国农民,强壮的身体代表强健的体格。)即使没有牛奶,土豆也能成为完整的一顿饭。巴肯引用理查德·皮尔森(Richard Pearson)医生提交给农业委员会的一份报告来证明后者。⁵⁴ 巴肯坚持认为，地主很容易为他们的雇工提供土豆莱园，土豆也很容易被劳动者种植。这样的好处会惠及劳动者个人和他们的家庭，他们的身体会充满活力，由此也惠及国家。如果劳动者接受土豆，地主们支持这些新的饮食抱负，他确定将会带来双重性质的改观。"这是真正的财富和人口的源泉！"他强调，"人们将会生生不息，而贫穷，仅在挥霍无度的人当中才会出现。"⁵⁵ 土豆，将助英国走向富强。

事实上，这也是一开始英国农业和内部改进委员会（British Board of Agriculture and Internal Improvement）出版皮尔森博士关于土豆的报告的原因。该委员会成立于1793年，旨在调查和推广创新的农业方法。在3000英镑政府基金的支持下，该组织开展了一项多元化的研究和出版计划，主题包括改进磨盘、"消灭懒汉的秘密"、盐在粪肥中的作用，以及一种没有耳朵的"羊的特别繁殖"。⁵⁶ 它的第一位主席，是苏格兰农学家、乡绅约翰·辛克莱爵士（Sir John Sinclair）。他写了大量文章论述这些信息给英国公众带来的好处。委员会显示出对土豆的持久兴趣，定期报告与土豆有关的事件。它的成员听取了关于处理土豆叶片卷曲问题、土豆淀粉的制造和用干燥法保存土豆的报告，委员会认为它们可能对海军补给有用。⁵⁷ 委员会寻找新品种并安排试种，并委托学者对试验结果进行论述。它还提供额外的经费，并投放宣传，以鼓励改进种植技术。1794年，它成立了一家"土豆委员会"，询问面包师们以商业化规模生产土豆面包的可行性，并回答英国各地记者提出的有关土豆的问题。⁵⁸（参见图9）

委员会成员们深信土豆是一种优质作物，它收成稳定、种植容易、产量可观。辛克莱说："在粮食短缺和经济不景气的时候，什么物资都无法和土豆相比。"⁵⁹ 委员会的计算显示，一英亩土豆的出产一年可以

第二章　启蒙的土豆　79

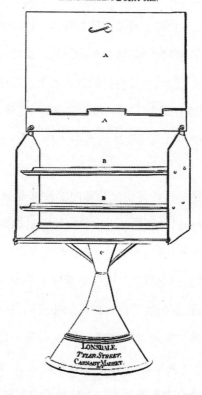

图9 由英国农业委员会推广的土豆烘烤机。从种植花椰菜到饲养安哥拉兔,18世纪农业委员会的成员们涉猎的话题颇为五花八门,但他们对土豆的兴趣是持续不变的,他们认为土豆对英国的福祉至关重要。这个土豆烘烤机出现在一本书中,该书描述了他们用燕麦、土豆等白面粉的替代品制作令人满意面包的努力尝试。

养活八到十人,远远超过同等面积的小麦。因此,"在委员会看来,增加土豆这种有价值的块根的种植,是最重要的目标之一,可以将它向英国农民大力推荐"。[60] 劝地主多种土豆的目的是让老百姓多吃土豆。按照委员会的解释,他们的目标是增加"下层阶级的常规消费"。[61] 为此,它出版了一系列小册子,其中包括煮土豆、烤土豆,尤其是土豆面包的食谱,土豆面包是使用高比例土豆的一种非常繁琐的做法。这些出版物用一种救世主般的口吻论述土豆的优点。委员会称，在苏格兰高地，土豆被认为是"现代赋予这个国家的最大福祉"。[62] 和威廉·巴肯一样，农业委员会也确信，土豆为穷人提供了改善他们饮食的绝佳机会。反过来它又是一个"在政治经济上具有重大意义"的问题。[63] 土豆的消费，尤其是穷人对它的消费，是直接的政治利益问题。

值得一提的是，到18世纪90年代，英国人吃土豆已经有悠久历史了。早在17世纪70年代，词典里就将这种块茎植物定义为"最初来自西印度群岛，但如今在英国菜园中很常见的一种产物"。[64] 根据园艺手册，一个世纪后，它的栽培"扩展到了英国的每一个郡"。虽然不是所有人都吃土豆，但在燃料便宜的地区，土豆已经成为许多劳动人民日常饮食中常见的品种。委员会在它的出版物中认可了这一点。[65] 他们对土豆的鼓励，并不是基于它在现有食物舞台上的缺失。相反，他们对土豆的兴趣反映了一种新的观念，即改善劳动人民的饮食"在政治经济中具有重要意义"。

土豆的赞美者

始于18世纪50年代中期、持续到18世纪末的土豆热席卷了整个欧洲，英国农业委员会的推广活动就是热潮的一部分。从西班牙的卡洛斯三世到瑞典的阿道夫·弗雷德里克（Adolf Frederick）等君主们，纷纷颁布鼓励种植土豆的法令。阿道夫·弗雷德里克受到了瑞典皇家科学院（Swedish Royal Academy of Science）开展的土豆实验的影响。[66] 那不勒斯医师菲利普·巴尔迪尼（Filippo Baldini）和瑞典商人乔纳斯·阿尔斯托姆（Jonas Alströmer）撰文宣传土豆的优点，芬兰官员、法国化学家等其他许多人也同样如此。[67] 哥本哈根的报纸刊登表扬信，鼓励读者种植土豆。在英国，进步地主、慈善家约翰·霍华德（John Howard）（后来他因监狱改革的成就而闻名）进行了新品种试验。霍华德希望能"为我们的同胞中最有价值的那部分人——这个国家的穷人劳动者"提供"极大的救济和帮助"。[68] 挪威和西班牙的牧师们就栽培技术发表布道，并劝告其教区的居民们种植土豆。芬兰南部阿西卡拉（Asikkala）的阿克塞尔·劳雷尔（Axel Laurell）牧师在教坛上就土豆的优点进行了长达4个小时的演讲，他还撰写了一篇宣传文章，其中包括种植建议和一些食谱。[69] 都灵的医师安东尼奥·坎皮尼（Antonio Campini）报告了土豆物美价廉的品质，他这样说道：

> 这是我们从美洲得到的最佳礼物之一。它为乡下的人们提供了良好、健康、令人愉快和美味的营养，它也同样有利于人口增长……我看到那些会烹饪土豆的士兵很少生病。[70]

二十年后，杜伊勒里宫（Tuileries）的花园被挖掉，取而代之的是按照一次国民大会狂热的决议而种植的土豆，决议希望以此来鼓励更广泛的土豆种植。[71] 假使有人想知道如何对待这些土豆，汉娜·莫尔（Hannah More）1795年的《小屋厨师，或琼斯夫人的便宜菜：价廉物美的烹饪方式》(*The Cottage Cook; or, Mrs. Jones's Cheap Dishes: Shewing the Way to do Much Good with Little Money*)中的每一份经济食谱都有它们的影子，而法国读者可参考同样以土豆为主的《共和国菜谱》(*Cuisinière républicaine*)。[72]（劳动人民对这种建议的反应将在下一章探讨。）

某些主题引起了持续的讨论，最佳栽培方法成为反复出现的话题，土豆到底是会消耗还是增加土地肥力，也引起了特别的争议。[73] 与存储和保存相关的问题也受到农学家和研究人员的注意,土豆易于腐烂的性质,造成了实用和技术上的挑战。研究人员评估了多种方法,包括将多余的土豆埋在沙子里,或更复杂的干燥等工序。[74] 在将土豆用作牲畜饲料这方面,许多研究人员一致认为,马、鸡等农场动物都能靠吃土豆茁壮成长,他们还提供了制作最佳土豆糊的方法。[75] 土豆的不同吃法也得到广泛宣传。爱尔兰小火煨土豆的方法被公认为最佳,农业委员会等制作的小册子也介绍说,用土豆代替小麦面包来烹制价廉味美的餐食是多么容易。[76] 科学家们做试验来提炼以土豆为原料的烈酒,把它作为谷物酒的替代品。[77] 最有代表性的是对令人满意的土豆面包配方的追求。把土豆"做成面包"的强烈愿望遍及整个欧洲。这样的面包,其健康性和营养性引起人们激烈的讨论,不过并没有得出结论。[78] 另外,土豆还被认为是极好的人工婴儿食品,因此它也有助于确保人口的增长。理想的是,当弃儿们长大成人之后,他们就可以开始在孤儿院的土地上种植自己的土豆。[79](参见土豆面包食谱。)

欧洲各国建立强国的渴望,加强了这种对流行饮食的新关注,因此,对土豆的兴趣超越了国界和语言。期刊和通信网络将启蒙时期的土豆讨论传遍了整个欧洲。法国科学家安托万-洛朗·拉瓦锡(Antoine-Laurent Lavoisier)① 曾在他位于布洛瓦(Blois)的庄园里试验种植土豆,他的探索结果,被西班牙一份由王室资助、成立于1797年的农业杂志《教区牧师农业艺术周刊》(*Parish Priests' Agriculture and Arts Weekly*)迅速报道。[80] 都柏林的植物园和农业协会都对土豆进行了实验,报纸也同样进行了报道。[81] 西班牙的卡洛斯三世还资助爱尔兰人亨利·道尔(Henry Doyle)出版了一本关于土豆种植的书籍。[82] 瑞士的土豆热

---

① 安托万-洛朗·拉瓦锡(1743—1794),法国著名化学家、生物学家,被后世尊称为"现代化学之父"。——译注

西班牙土豆面包食谱

1797 年，萨拉曼卡（Salamancan）的利纳雷斯·德尔·里奥弗里奥（Linares del Riofrio）村的教区牧师发明了一种面包配方，用 6 磅土豆配 3 磅小麦。他解释说，他的教区种了很多土豆，但小麦很少。他希望少吃谷物面包能给穷人带来好处。他说，实验结果非常出色。他说他第一次尝到面包的那天是他一生中最快乐的一天。"在我看来，"他写道，有了土豆，"饥饿和贫穷就会从地球上消失。"

土豆面包

"我做了一个实验，用了 6 磅洗干净的主要农作物白土豆。我把它们煮到皮裂开，也就是它们充分煮熟。然后，我把它们沥干、剥皮，在一个容器里捣碎，用一把牢固的抹刀尽可能地把它们捣碎，但不能把它们压碎，以免颜色变深。接着，我加了 3 磅小麦粉，还有一点溶解在大约 4 盎司水里的酵母和一大把盐，因为土豆本身是淡而无味的。我将它们充分揉匀，做成一个结实的面团，看起来和只用小麦粉做的面团没有明显的区别。它涨得更厉害，可以像小麦面团一样很快就放进烤箱。烤箱温度应该比烤普通面包的温度要高一些。"

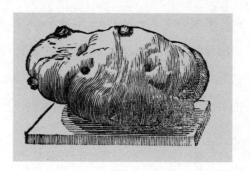

衷者参考了瑞典农学家以及英国、法国和德国关于土豆的论著。[83] 英国作家翻译了欧洲大陆关于实验种植方法的文章。[84] 瑞典人卡尔·斯基特（Carl Skytte）从土豆中提炼白兰地的工作，被狂热的业余化学家们在欧洲各地复制。然而，他的姓氏在这样的旅程中却难以完整地得到保留。[85]

土豆的倡导者们很是夸大其词。波兰作家、政治家朱利安·尤尔辛·聂姆策维奇（Julian Ursyn Niemcewicz）宣称土豆是仅次于洗礼的"上天的伟大赐福"。他认为，"我们更应该感谢美洲给予了我们土豆，而不是致命的贪婪工具：金、银这两种贵重金属"。[86] 马德里皇家植物园的总园艺师克劳迪奥·博特罗（Claudio Boutelou）表示，土豆"无疑是我们从新世界得到的最有价值的东西"，他还说，欧洲其他国家应该感谢西班牙引进了这种植物珍品。[87] 法国科学家安托万-亚历克西斯·卡代·德沃克斯把它比作以色列人在沙漠中维持生命的吗哪①，同样是仁慈的神送来的。[88] 一家瑞士报纸的编辑把法国制作土豆面包的进展描述为"本世纪最重要的发现之一"。[89] 另一位土豆的倡导者、美国出生的拉姆福德伯爵（Count Rumford）提醒读者，既然鼓励土豆消费的计划最终是为了改善穷人劳动者的福祉，那么它们当然是"开明政治家"所关心的。[90]

土豆倡导者采用了一套常见的技巧，通常包括赠品、免费提供土豆和派发操作手册等。奖励很常见。欧洲各地的爱国人士、爱国组织为最大的土豆产量、最佳土豆面包配方、最有效的土豆病害治疗方案提供奖励。1790年，来自利弗兰德（Lifland）的农民彼得·西卡尔（Peter

---

① 吗哪（希伯来语：מן；英语：Manna）是《圣经》中的一种天降食物，在古代以色列人出埃及时，在40年的旷野生活中，由上帝赐予。教会中现在便惯用"吗哪"来比喻"精神食粮"。——译注

Sirkal）在圣彼得堡自由经济学会（St Petersburg Free Economic Society）主办的土豆大丰收比赛上拔得头筹，他得到了10卢布，够买一匹马或两头公牛。[91] 在欧洲许多地方，类似的"经济学会""改善农牧、生产和其他有用技术学会"等纷纷建立起来。并且，从18世纪30年代起，它们也在欧洲的殖民地建立起来，其目的是传播有用的知识，尤其是传播在整个欧洲流行的、旨在使农业更有效率、利润更高的农业新模型。（由于这类协会很少包括像农民这样有实际耕作经验的人，因此它们往往反映的是作为协会成员的地主、牧师和官员的观点。其目的往往是向农民传播精英知识而不是相反。）[92]

　　土豆吸引了欧洲许多地方的这类组织的关注。埃尔福特（Erfurt）的学会成员被教授如何用土豆酿酒；还有一些德国的学会出版论著、赞助竞赛，并进行田间实验。所有的实验都表明，土豆甚至是一些高产谷物（如荞麦或黑麦）的极好替代品。[93] 芬兰经济学会（Finnish Economical Society）在图尔库（Turku）大力鼓励种植，免费分发土豆种，并向最成功的土豆种植者发放奖励。[94] 布鲁塞尔帝国皇家科学与艺术学院（the Imperial and Royal Academy of Scienceand the Arts of Brussels）、威尔士蒙茅斯郡农业学会（the Monmouthshire Agricultural Society）、苏格兰高地学会（the Highland Society），以及阿拉贡皇家经济学会（the Royal Economic Society of Aragón）也都采取了类似的做法。[95] 例如，在1796年，阿拉贡学会为那些能证明自己和家人食用土豆最多的劳动者提供了高达300里亚尔的奖励。[96] 皇家巴斯克学会（the Royal Basque Society）不仅提供奖品，而且还进行土豆种植和土豆面包制作实验，并翻译了一些关于土豆的农业论著。正如它在1786年的一份报告中介绍的那样，该学会长期以来一直将土豆当做"可以为国家带来好处的最重要的作物之一"。[97] 在遥远东部的喀尔巴阡

（Carpathian）山区，最初的特兰西瓦尼亚农业学会（Agricultural Society in Transylvania）也同样积极地鼓励土豆种植。[98]

这些工作的要点是对"经济"（*oeconomy*）的关心。"经济"是个人或国家对资源的理性管理，它包含了一系列以约束、节制为优先的行为和价值观。人们可以从健康、个人支出、烹饪，以及森林管理等各方面体现"经济"。"经济"将个人、家庭的行为——包括食物消费——与政府更广泛的关注点联系起来。它明确地表达了一种信念：即使很小的国内事务也会影响国家的福利。因此，英国农业委员会把它对土豆的推广描述为"在政治经济上具有重大意义"的事情。[99]

可以肯定的是，土豆并不是唯一一种引起这类关注的庄稼。由于国家的福祉需要充足的粮食供应，不少个人和组织评估了其他许多食品的潜力，包括莱比锡经济学会（the Leipzig Economic Society）在1771年调查的西伯利亚荞麦、巴伦西亚皇家经济学会（the Royal Economic Society of Valencia）评估的可能替代橄榄油的花生提取物等。来自世界其他地方的陌生食品引起了人们特别的兴趣，许多欧洲国家支持旨在将新型经济作物、新型主食引入欧洲的"经济植物"计划。植物学家约瑟夫·董贝（Joseph Dombey）建议把藜麦引入法国，作为大米的替代品。还有一些人支持野生水稻、面包果、西米。[100] 然而，土豆才是全欧洲受到关注最多、最长久的对象。

其原因，按照土豆倡导者们的说法，是它具有独特的优点。他们报告了它与谷物比较而言的高产量，这一点如今也得到联合国和其他关注世界饥饿问题之机构的重申。[101] 他们认为土豆耕种比较简单，因此但凡有一小块土地，任何人（如巴肯所说的一大家子的穷人）都能轻松地种植自己的土豆。它不需要特别肥沃的土壤，在各种天气和气候下都能生长良好。倡导者们坚称它营养丰富，不过，由于缺乏公认的评估食物营养潜力的科学方法，它只是停留在一种说法上，而不是确定的事实。[102] 支持者们也强调烹饪的简单性，并指出，多余的都可以拿去喂牲畜。据说孩子们都很喜欢它。最重要的是，推广者们强调说，土豆并不是新鲜事物，而是其他欧洲人常吃的熟悉食物。与野生大米或藜麦不同，土豆在欧洲有着悠久而成功的历史，土豆推广者们特别提到此点。苏格兰商人帕特里克·科尔克霍恩（Patrick Colquhoun）把他的伦敦施粥所的成功归因于它只供应常见食物，尤其是土豆。因此，它的食客"没有（对他）抱怨推出了一种他们不习惯的新食物"。他解释说，由于土豆"长期以来为许多劳动阶层所熟知"，所以"基本上没有什么偏见……需要对付"。[103] 倡导者们解释说，即使某个地方没人吃土豆，它在欧洲别处也非常普遍。例如，在帕蒙蒂埃写给法国读者的许多关于土豆的论著中，他详细介绍了土豆作为一种普通食物，是如何在荷兰、佛兰德斯、洛林、阿尔萨斯、爱尔兰和兰开夏郡得到种植和食用的。在启蒙运动之前，土豆是劳动人民种植和食用的日常食物，这段历史，为它在18世纪被视为国家富强的根源奠定了基础。

结论

土豆在近代早期的地位变化，表明了日常饮食习惯在政治意义上的转变。在17世纪，土豆的营养特性曾惹恼了官员和政治理论家，其中包括威廉·配第。他抱怨说，土豆使得爱尔兰人每天只工作两小时就能过得很好。"他们何必工作呢？土豆能满足他们，一个人的劳动可以喂饱四十个人。"他恼怒地写道。[104] 到18世纪，这种恼人的多产，却成为对渴望增加国家财富和权力的"开明政治家"的吸引力。国家的福祉在一定程度上取决于个人的健康，这一共识，使劳动人口的饮食习惯具备了政治重要性。

作家们明确表述了土豆、人口和政治经济之间的这些关系。在西班牙政府支持下发表的一篇论著中，爱尔兰土豆热衷者亨利（或恩里克，Enrique）·道尔阐明了土豆与经济繁荣之间的关系：

> 随着人口的增长、人类的繁衍，不仅有必要改善土壤、增加耕地面积，而且还有必要利用适合土质、适合普通人食用以维持生计的其他块根作物、植物和蔬菜的优势，以在任何时候都能以适中价格保持贸易平衡。作家们一致认为，土豆提供了这种必要的帮助，不仅因为它产量大、价格低，还因为它有益健康。[105]

他这篇不断重印的论著，到1804年已经增加到250多页，道尔在书中强调，尽管富人乐于吃土豆，但它的功用，根本上在于它作为劳动人民粮食的潜力。他强调，土豆营养丰富，吃土豆的人"健康而强壮"。像其他许多人一样，他把健壮的爱尔兰农民作为土豆有益健康、能维持生计的证据。而且，由于农民们食用土豆，爱尔兰才能够出口数百万磅的小麦，使地主和国库受益。[106] 因此，至少从国家和大地主的角度来看，以土豆为生的劳动者推动了农业和商业的成功。

土豆作为士兵食物的潜力，也证明了土豆与国家强大的关系。道尔还注意到，土豆在马德里的军队中很受欢迎。它还有更多的吸引力：它能够促进哺乳期母亲泌乳，并且是合适的母乳替代品。道尔总结说，西班牙的经济和政治福祉不仅取决于政府，还取决于推动穷人消费土豆的"爱国者的热忱和警觉"。[107] 道尔的论著概括了人民个体健康与国家整体健康、稳定及其商业之间的新关系。与17世纪的政治理论家不同的是，道尔等18世纪的土豆赞颂者，将土豆的营养性看成是对国家的巨大有利条件。

*

这些在18世纪形成的普通人日常饮食习惯与国家福祉的关系，出现在各种独立的、不相关的行为与更大的非个人力量之间的关系引起持续关注的时刻。微小物理粒子的随机运动，会自己组成有意义的序列吗？个人的迫切希望和渴求，能否与更大的国家的福祉相统一？贸易和商业如何安排，才能带来最大的益处？哲学家、数学家、植物学家等许多人，都被这种"自组织"（self-organization）的过程深深吸引。在这个过程中，秩序从无数看似无序的不协调事件中出现。[108] 本书下一章将追溯18世纪政治哲学家开始感知的饮食与经济自组织之间的关系。引导市场的这一只看不见的手，是否也会引导个人的饮食偏好？到18世纪末，认为个人饮食行为影响国家富强的信念，与通过市场开放实现经济自组织的新观念融合在一起。这种融合，引发了成功的饮食自组织的美好愿景，在这种愿景中，导致个人幸福的事物，同时也被证明是有益于整个社会的。

注释

1. Legrand d'Aussy, *Histoire de la vie privée des Français*, I: 112.

2. Varenne de Béost, *La cuisine des pauvres*, 2(first quote); *Journal historique & litteraire* 156(1780), 128–129; Buchan, *Observations Concerning the Diet of the Common People*, 43 (second quote).

3. Toussaint-Samat, *A History of Food*, 646.

4. Virey, 'De la vie et des ouvrages d'Antoine-Augustin Parmentier', 60–61. 也可参见 Parmentier, *Observations on Such Nutritive Vegetables as May be Substituted in the Place of Ordinary Food*, 8–9.

5. Thomopoulos, *The History of Greece*, 71.

6. 例如，一位芬兰同事解释说，他曾经学到过，芬兰牧师阿克塞尔·劳雷尔

（Axel Laurell）"通过在他的田里派夜班警卫的方式，把土豆传播给了教区居民。他命令这个警卫必须让教区居民晚上从地里偷土豆。劳雷尔认为，如果教区居民认为土豆是一种被'禁止'或受'保护'的果实，他们就会尝试自己种植土豆"。Joonas Tammela, 16 Oct. 2017, personal communication.

7. Parker, *Global Crisis*.

8. Riley, *Population Thought in the Age of the Demographic Revolution*; Tomaselli, 'Moral Philosophy and Population Questions in Eighteenth-Century Europe'; Cole, *The Power of Large Numbers*; Rusnock, *Vital Accounts*, 95 (quote, my emphasis); Charbit, *The Classical Foundations of Population Thought*; Mc Cormick, 'Population: Modes of Seventeenth-Century Demographic Thought'.

9. Ward, *Proyecto económico*, 70.

10. Ward, *Proyecto económico*, 70; Hutchison, 'Swedish Population Thought in the Eighteenth Century', 85–86.

11. Riley, *Population Thought in the Age of the Demographic Revolution*; Tomaselli, 'Moral Philosophy and Population Questions in Eighteenth-Century Europe'; Charbit, *The Classical Foundations of Population Thought*, 124–131.

12. G. Warde, 'An Idea for the Relief of the Poor', Bradfield, 15 Feb. 1795, *Annals of Agriculture* 24 (1795), 370. 为此，他所依赖的权威不亚于亚当·斯密，斯密本人曾断言"任何国家繁荣的最决定性标志是其居民数量的增加"，Smith, *An Inquiry into the Nature and Causes of the Wealth of Nations*, I.viii.23。

13. Mill, 'Colony'.

14. Foucault, *Security, Territory, Population*. 也可参见 Dean, *The Constitution of Poverty*。

15. Moheau, *Recherches et considérations sur la population de la France*, 250.

16. Ward, *Proyecto económico*, 196 (my emphasis).

17. Ward, *Obra pía*; Ward, *Proyecto económico*, 'Al lector', v (quote).

18. Ward, *Proyecto económico*, 58.

19. Ward, *Proyecto económico*, 58 (my emphasis). 也可参见 Argumossa y Gandara, *Erudicción política*; Rodríguez de Campomanes, *Discurso sobre el fomento de la industria popular*, 136; Campillo y Cosío, *Nuevo sistema de gobierno económico*, 261; Cabarrús, 'Discurso sobre la libertad de comercio'; Uriz, *Causas prácticas de la muerte de los niños expósitos*。Tavárez Simó, 'La invención de un imperio comercial hispano' 认为 *Nuevo sistema de gobierno económico* 的作者应为 Melchor Rafael de Macanaz。

20. Hutchison, 'Swedish Population Thought in the Eighteenth Century', 84.

21. Parmentier, *Les pommes de terre, considérées relativement à la santé & à l'économie*, 133.

22. Lagrange, 'Essai d'arithmétique politique sur les premiers besoins de l'intérieur

de la république', 1796, *Oeuvres*, VII: 578 (quote); Simmons, *Vital Minimum*, 231–232.

23. Dalrymple, *The Poor Man's Friend*, 21 (appendix).

24. Becher, *Politische Discurs*, 205 (quote, my emphasis); Petyt, *Britannia Languens*, section7; Tribe, *Governing Economy*, 85; Tribe, 'Cameralism and the Sciences of the State', 527.

25. Melon, *Essai politique sur le commerce*, 12. 或参见 Béguillet, *Traité des subsistances et des grains*, I: 252; and Lavoisier, 'Résultats extraits d'un oeuvrage intitulé de la richesse territoriale du royaume de France', 1784, *Oeuvres*, VI: 422。

26. Hanway, *A Candid Historical Account of the Hospital*, 10.

27. Girdler, *Observations on the Pernicious Consequences of Forestalling*, 53 (quote), 88.

28. Benavídez, *Secretos de chirurgia*, 26–27 (quote); Huarte de San Juan, *Examen de Ingenios*, 21–22; Lemnius, *The Touchstone of Complexions*, 25–31; Moffet, *Health's Improvement*, 59; Earle, *The Body of the Conquistador*; Gentilcore, *Food and Health in Early Modern Europe*.

29. Fox, 'Food, Drink and Social Distinction in Early Modern England'.

30. Smith, *An Inquiry into the Nature and Causes of the Wealth of Nations*, I.viii.45.

31. Fielding, 'An Enquiry into the Causes of the Late Increase of Robbers', 90. 另可参见 Lémery, *Traité des aliments*, I: xxix–xxx; Des-Essartz, *Traité de l'éducation corporelle des enfants*; Ballexserd, *Dissertation sur l'éducation physique des enfans*, 29; Bonells, *Perjuicios que acarrean al género humano y al estado las madres que rehusan criar a sus hijos*; Dalrymple, *The Poor Man's Friend*, 21; Girdler, *Observations on the Pernicious Consequences of Forestalling*, 53, 88; Uriz, *Causas prácticas de la muerte de los niños expósitos*, I: 85; Arteta, *Disertacion sobre la muchdumbre de niños que mueren en la infancia*, I: 16。

32. Des-Essartz, *Traité de l'éducation corporelle des enfants*, vi (first quote); Guimarães Sá, 'Circulation of Children in Eighteenth-Century Portugal', 29 (second quote).

33. Uriz, *Causas prácticas de la muerte de los niños expósitos*, I: 85 (quote), II: 127–130, 157.

34. 可参见 Bonells, *Perjuicios que acarrean al género humano*; Uriz, *Causas prácticas de la muerte de los niños expósitos*, I: 85 (quote), II: 127–130, 157; Arteta, *Disertación sobre la muchdumbre de niños que mueren en la infancia*, I: 16; Morel, 'Théories et pratiques de l'allaitement en France au XVIIIe siècle', esp. 403–404; Sherwood, *Poverty in Eighteenth-Century Spain*; Andrews, *Philanthropy and Police*; Guimarães Sá, 'Circulation of Children in Eighteenth-Century Portugal', 29 (quote); Morel, 'Children', I: 247。

35. Turnbull, *The Naval Surgeon*, 23–46; Admiral Philip Patton, 'Strictures on Naval

Discipline and the Conduct of a Ship of War', c. 1807, *Shipboard Life and Organisation*, ed. Lavery, 629 (quote); Macdonald, *British Navy's Victualling Board*; Knight and Wilcox, *Sustaining the Fleet*; Macdonald, *Feeding Nelson's Navy*.

36. Buchet, *The British Navy*, 125.

37. González, *Tratado de las enfermedades de la gente de mar*.

38. Lémery, *Traité des aliments*, I: xxx; Lorry, *Essai sur les alimens*, II: 216; Poissonnier Desperrières, *Traité des maladies des gens de mer*, 11 (first quote); Poissonnier Desperrières, *Mémoire sur les avantages qu'il y aurait a changer absolument la nourriture des gens de mer*, 7 (second quote); Lebeschu de la Bastays, *L'Ami des navigateurs*, esp. iii, 60–87.

39. Post, 'Nutritional Status and Mortality in Eighteenth-Century Europe'; Gallego, *El motín de Esquilache, América y Europa*.

40. 关于这一点，参见 Stedman Jones, *An End to Poverty?*

41. Bignami, *Le patate*, 15, 4, respectively.

42. Henry, *The Complete English Farmer*, 275–276. 或可参见 Smith, *An Inquiry into the Nature and Causes of the Wealth of Nations*, I.xi.b.41; Parmentier, *Les pommes de terre, considérées relativement à la santé & à l'économie*, 195; Parmentier, *Traité sur la culture et les usages des pommes de terre, de la patate, et du topinambour*, 17–18; Doyle, *Tratado sobre el cultivo, uso y utilidades de las patatas* (1797), 24–25; *Feuille de cultivateur*, 10 Feb. 1799。

43. Bentham, 'Panopticon; or, the Inspection-House: Containing the Idea of a New Principle of Construction applicable to any sort of establishment, in which persons of any description are to be kept under inspection ... written in the year 1787', *The Works of Jeremy Bentham*, IV: 155; and Good, *Dissertation*, 77 (quote).

44. Ferrières, 'Le cas de la pomme de terre dans le Midi'.

45. Engel, *Traité*, 45–46.

46. Cadet de Vaux, *L'Ami de l'économie*, 59.

47. *Essay on Modern Luxuries*, 7.

48. Hanway, *A Candid Historical Account of the Hospital*, 11 (quote). 对酒精饮料的抱怨，参见 Brennan, *Public Drinking and Popular Culture in Eighteenth-Century Paris*; Kümin, *Drinking Matters*, 153; Warner, *Craze*。

49. [Lobb], *Primitive Cookery*; [Colquhoun], *An Account of a Meat and Soup Charity*, 19.

50. Eden, *The State of the Poor*, I: 491–492. 也可参见 Dean, *The Constitution of Poverty*, 50。

51. Buchan, *Domestic Medicine*, 46.

52. Buchan, *Domestic Medicine*, 67.

53. Buchan, *Observations Concerning the Diet of the Common People*, 7.

54. Buchan, *Observations Concerning the Diet of the Common People*, 31; Richard Pearson, 'Analysis of the Potatoe-Root', *Report of the Committee of the Board of Agriculture*, Board of Agriculture, 79–86.

55. Buchan, *Observations Concerning the Diet of the Common People*, 31–32. 关于土豆菜园，参见 Burchardt, 'Land and the Laborer'。

56. Board of Agriculture Minute Books, SR RASE A/I-II, B/I (fols. 31–34 slugs), B/II, B/VI–VII, B/X, B/XIII (fol. 266 earless sheep), MRELSC; 'Plan for Establishing a Board of Agriculture and Internal Improvement, Originally Printed in May 1793', *Communications to the Board of Agriculture*, Board of Agriculture; and Mitchison, 'The Old Board of Agriculture'.

57. 试验性供应的一批干土豆的确被一艘军舰送到了植物学湾（Botany Bay），"这样制作的块茎对海员的健康非常有益"。Board of Agriculture Minutes, 23 May 1797, SR RASE B/XIII, fol. 199, MRELSC.

58. 土豆委员会的会议记录现藏于 SR RASE B/X, MRELSC。也可参见 Board of Agriculture, *Account of the Experiments tried by the Board of Agriculture in the Composition of Various Sorts of Bread*。

59. 'Sir John Sinclair's Address to the Board of Agriculture, on Tuesday, the Twenty-fourthof May, 1796', *Communications to the Board of Agriculture*, Board of Agriculture, lxiv.

60. Board of Agriculture, *Report of the Committee of the Board of Agriculture*, vii (quote);'Substance of Sir John Sinclair's Address to the Board of Agriculture, on Tuesday the14th July, 1795', Communications to the Board of Agriculture, Board of Agriculture, lx–lxiii.

61. Board of Agriculture, *Report of the Committee of the Board of Agriculture*, 74.

62. Board of Agriculture, *Report of the Committee of the Board of Agriculture*, 70 (quote); Board of Agriculture, *Hints Respecting the Culture and the Use of Potatoes*. 发土豆面的难点在于，土豆本身不含足够的麸质，所以很难膨胀。

63. Board of Agriculture, Report of the Committee of the Board of Agriculture, 73. 伦敦园艺学会会长赞成以下观点：由于土豆为相当一部分英国居民提供了食物，因此"其栽培的每一项改进都成为国家的重要目标"，Knight, 'On Raising New and Early Varieties of the Potatoe', I: 57。法国初期土豆推广者一篇文章的译者观察到，鼓励土豆种植"是国家最重要的目标"，Parmentier, *Observations on Such Nutritive Vegetables as May be Substituted in the Place of Ordinary Food*, vii。

64. Phillips, *The New World of Words*, entry on 'potatoes'.

65. 可参见 Switzer, *The Practical Kitchen Gardiner*, 217–219, 378; Miller, *The Gardeners*

Dictionary, II: lycopersicon entry; Adam, *Practical Essays on Agriculture*, essay 7, II: 1 (quote); Henriette Charlotte Gräfin von Itzenplitz, treatise on potatoes, Kew, 18 Aug. 1793, LRO, DDSc 9/47; *Annals of Agriculture* 24 (1795), *passim*; Aikin, *A Description of the Country from Thirty to Forty Miles Round Manchester*, 45–46, 362; Board of Agriculture, *Communications to the Board of Agriculture*, Board of Agriculture, 37–45, 93; Overton, *Agricultural Revolution in England*, 102; Horrell and Oxley, 'Hasty Pudding versus Tasty Bread'; Zylberberg, 'Fuel Prices, Regional Diets and Cooking Habits'。

66. *The Gentleman's Magazine and Historical Chronicle* 34 (1764), 599; *Extractos de las Juntas Generales celebradas por la Real Sociedad Bascongada de los Amigos del País en la ciudad de Vitoria por julio de* 1786, 28; Doyle, *Tratado sobre el cultivo, uso y utilidades de las patatas*(1797). 其他国家层面的指令，参见 *Instrucção sobre a cultura das batatas*; *El Excmo. Sr. Secretario de Estado y del Despacho de Hacienda*; Wiegelmann, *Alltags-und Festspeisen in Mitteleuropa*, 83n.40; Spary, *Feeding France*; Jones, *Provincial Development in Russia*, 154–155; Smith and Christian, *Bread and Salt*, 199–200; and Kisbán, 'The Beginnings of Potato Cultivation in Transylvania and Hungary: Government Policyand Spontaneous Process'。

67. Parmentier, *Manière de Faire le Pain de Pommes de Terre sans Mélange de Farine*; Parmentier, *Les pommes de terre, considérées relativement à la santé & à l'économie*; Baldini, *De' Pomi di terra ragionamento*; Parmentier, *Mémoire couronné le 25 aout 1784*; Parmentier, *Traité sur la culture et les usages des pommes de terre, de la patate, et du topinambour*; Berg, 'Die Kartoffelund die Rübe'; Talve, 'The Potato in Finnish Food Economy'; Gentilcore, *Italy and the Potato: A History*.

68. John Howard, Cardington, Bedfordshire, 17 Feb. 1769, RSA, RSA/PR/GE/110/26/70and 71; *Københavnske Nye Tidende*, 1758, in *Norges landbrukshistorie*, ed. Kåre. 阿蒙德·佩德森（Amund Pedersen）提供了挪威语翻译。

69. Laurell, *Lyhykäinen kirjoitus potatesten eli maan-päronain wiljelemisestä*; Wilse, *Physisk, oeconomisk og statistisk Beskrivelse*, 255–256; 'Carta del cura del Linares sobre el cultivo y aprovechamiento de las patatas', SAA 2 (1797), 222–227, 245–252, 261–269, 277–282; 'Carta del cura de Linares', SAA, 26 Dec. 1799, 401–402; 'Carta del cura de Linares', *SAA*, 13 Aug.1801, 97–106; Simonen, *Raivaajia ja rakentajia*, 33–36; Drake, *Population and Society in Norway*, 54–55. 阿蒙德·佩德森和尤西佩卡·卢克宁（Jussipekka Luukkonen）解读了挪威语与芬兰语材料。

70. Campini, *Saggi d' Agricoltura*, 388–393.

71. Spary, *Feeding France*, 167–186.

72. *Cuisinière républicaine*; More, *The Cottage Cook*.

73. 在例如 Miller, *The Gardeners Dictionary*; Rudenschöld, *Almanach*; Halldórsson,

Korte Beretninger om nogle Forsög til Landvæsenets; Zanon, *Della coltivazione, e dell'uso delle patate*; Hammer, *Afhandling om Patatos*; Engel, *Traité*; Henry, *The Complete English Farmer*; *Beskrifning om jordpärons plantering*; Hertzberg, *Underretning for Bønder*; Baldini, *De' Pomi di terra ragionamento*; Occhiolini, *Memorie sopra il meraviglioso frutto americano*; Adam, *Practical Essays on Agriculture* II; Parmentier, *Traité sur la culture et les usages des pommes de terre, de la patate, et du topinambour*; Doyle, *Tratado sobre el cultivo, uso y utilidades de las patatas* (1797); Larumbe, *Epítome cristiano de agricultura*; *Metodo facile, e sperimentato per coltivare le patate*; Gómez de Ortega, *Elementos teóricos-prácticos de agricultura*; Estéban Boutalou, 'Memoria sobre las patatas', *SAA* 19 (1806); Riera Climent and Riera Palmero, 'Los alimentos americanos en los Extractos de la Bascongada' 中都对农业技术作了解说。1797年，亚瑟·杨给他的《农业年鉴》(*Annals of Agriculture*) 中关于土豆的文章编制了索引，这份索引很好地展示了此类著作所涉及的主体范围——从肥料到收获时间。*Annals of Agriculture* 29(1797), 38–62.

74. 上面列出的大部分作品都阐述了储存问题。也可参见 *Angående potaters förvarande*; M. de Bullion, 'Moyen de conserver les Pommes-de-terre', *Mémoires d'agriculture, d'économie rurale et domestique* (1789), 183–186; Board of Agriculture, *Hints Respecting the Culture and the Use of Potatoes*; *Annals of Agriculture* 24 (1795), 64–71; Amoretti, *Della coltivazione delle patate*, 40–44; 'Modo de conservar las patatas muchos años', *SAA* 17 (1805), 392–393; Virey, 'Pomme-de-terre, ou papas des Américas', 542ff.; Popplow, 'Economizing Agricultural Resources in the German Economic Enlightenment', 277; Spary, *Feeding France*。

75. Engel, *Traité*, 68; *Bibliothèque physico-économique, instructive et amusante* (1787); *Mémoires d'agriculture, d'économie rurale et domestique* (1789), 34; Adam, *Practical Essays on Agriculture*, II: 1–37; Parmentier, *Traité sur la culture et les usages des pommes de terre, de la patate, et du topinambour*; Billingsley, 'On the Culture of Potatoes, and feeding Hogs withthem'; John Boys, 'Experiments on Fattening Hogs', *Annals of Agriculture* 29 (1797), 150–159; Doyle, *Tratado sobre la cría y propagación de pastos y ganados*; *SAA* 19 (1806), 249; Drake, *Population and Society in Norway*, 55.

76. Varenne de Béost, *La cuisine des pauvres*; Parmentier, *Les pommes de terre, considérées relativement à la santé & à l'économie*; 'Potatoes, by a Lancashire Man', *Annals of Agriculture* 24 (1795), 568–575; 'Modo de cocer las patatas en Irlanda', *SAA* 17 (1805), 256; and Guðmundsdóttir, 'Viðreisn garðræktar á síðari hluta 18. Aldar'. 惠普娜·罗伯茨多蒂尔（Hrefna Róbertsdóttir）翻译了冰岛语材料。农业委员会写信给他们的一位记者："人们对各种煮土豆的方法赞不绝口，但似乎还没有进行足够准确的实验来确定哪种方法最好。如果你能够决定这个问题并形成一个比较实验，请

告知我们结果。" Board of Agriculture Minutes, 8Mar. 1796, SR RASE B/XIII, fol. 71, MRELSC.

77. Skytte, 'Ron at utaf potatoes brånna brånnavin', 231–232; *Magazzino toscana* 22:1 (1775), 114; Parmentier, *Les pommes de terre, considérées relativement à la santé & à l'économie*, 32, 176–180; *Encyclopédie méthodique: Arts et métiers mécaniques* 2 (1783), 196; James Anderson, 'Of Ardent Spirits Afforded by Potatoes', *Letters and Papers on Agriculture* 4 (1788), 43–52; Board of Agriculture Committee on Potatoes, 28 Jan. 1795, SR RASE B/X, fol. 5, MRELSC; Keith, *A General View of the Agriculture of Aberdeenshire*, 266–268; Alonso de Herrera, *Agricultura general*, II: 253; Kisbán, 'The Beginnings of Potato Cultivation in Transylvania and Hungary', 180–181; Koerner, *Linnaeus*, 132, 149.

78. James Stonhouse, 'Expedients for alleviating the Distress occasioned by the present Dearness of Corn', Northampton, *Universal Magazine* 21 (1757), 270; Bolotov, 'O delaniiiz tartofelia muki'; Parmentier, *Manière de Faire le Pain de Pommes de Terre; Extractos de las Juntas Generales celebradas por la Real Sociedad Bascongada de los Amigos del País en la villa de Vergara por setiembre de 1779*; Baldini, *De' Pomi di terra ragionamento*, 23–27; William Augustus Howard, Grays Inn Great Square, 14 July 1795, TNA, HO 42/35, fols. 157–158; Board of Agriculture Minutes of Committee on Potatoes, Committee on the Scarcity of Grain, 23 Aug. 1794–27 Nov. 1795, SR RASE B/X, fols. 1–45, MRELSC; Board of Agriculture, *Account of the Experiments tried by the Board of Agriculture in the Composition of Various Sorts of Bread*; Doyle, *Tratado sobre el cultivo, uso y utilidades de las patatas* (1797); 'Carta del cura del Linares sobre el cultivo y aprovechamiento de las patatas', *SAA* 2 (1797), 222–227, 245–252, 261–269, 277–282; 'Report from Saxon Electoral Society of Agriculture on the Cultivation of Potatoes', *Communications to the Board of Agriculture*, Board of Agriculture, 295; Kisbán, 'The Beginnings of Potato Cultivation in Transylvania and Hungary'; Koerner, *Linnaeus*, 149; Popplow, 'Economizing Agricultural Resources in the German Economic Enlightenment', 277; Hiler, 'La pomme de terre révolutionnaire'; Spary, *Feeding France*. 科勒姆·莱基（Colum Leckey）为我翻译了俄语内容。

79. Doyle, *Tratado sobre el cultivo, uso y utilidades de las patatas* (1797), 8, 24, 31; Uriz, *Causas prácticas de la muerte de los niños expósitos*, II: 127.

80. Lavoisier, *Oeuvres*, II: 815; Grimaux, *Lavoisier*, 165; *SAA* 3 (1798), 61; Larriba, 'Un intento de reforma agraria'.

81. *SAA* 11 (1802), 316.

82. Doyle, *Tratado sobre el cultivo, uso y utilidades de las patatas* (1797).

83. Engel, *Traité*, 7–8, 12, 35, 42, 56, 60, 68.

84. Mr. Tschiffeli, 'An Experiment to Make Potatoes Thrive without Dung', *Foreign Essays on Agriculture and Arts*.

85. Skytte, 'Ron at utaf potatoes brånna brånnavin', 231–232; *L'Avantcoureur*, 18 Nov. 1771; *Magazzino toscana* 22:1 (1775), 114; Keith, *A General View of the Agriculture of Aberdeenshire*, 266–268; Henriette Charlotte Gräfin von Itzenplitz, treatise on potatoes, Kew, 18 Aug.1793, LRO, DDSc 9/47; George Way on spirit from potatoe apples, 21 Dec. 1797, RSA/PR/MC/105/10/396; Koerner, *Linnaeus*, 149.

86. Miodunka 'L'essor de la culture de la pomme de terre au sud de la Pologne'. 也可参见 Parmentier, *Mémoire couronné le 25 aout 1784*, 5; Adam, *Practical Essays on Agriculture*, II:20–21。

87. *SAA* 18 (1805), 4 and 11 July 1805; 19 (1806), 20 Feb. and 6 Mar. 1806; Gómez de Ortega, *Elementos teóricos-prácticos de agricultura*, II: 137; Estéban Boutalou, 'Memoria sobre laspatatas', *SAA* 19 (1806); and Alonso de Herrera, *Agricultura general*, II: 248 (quote).

88. Cadet de Vaux, *L'Ami de l'économie*, 58–59.

89. *Journal historique et politique des principaux événemens des différentes cours de l'Europe* 1, 10 Jan. 1779, 69.

90. Thompson, 'Of Food, and Particularly of Feeding the Poor', *Essays, Political, Economical and Philosophical*, I: 277.

91. Colum Leckey, personal communication, 3 Nov. 2015, 引用了圣彼得堡俄罗斯国家历史档案馆的资料（Russian State Historical Archive, St Petersburg），fond 91, opis' 1, delo 35, ll. 191–196; Khodnev, *Istoriia Imperatorskago Vol'nago Ekonomicheskago Obshchestva*, 382。

92. Stapelbroek and Marjanen, eds., *The Rise of Economic Societies in the Eighteenth Century*.

93. *Annalen der Braunschweig Luneburgischen Churlande; Verhandlungen und Schriften*, 72; Henriette Charlotte Gräfin von Itzenplitz, treatise on potatoes, Kew, 18 Aug. 1793, LRO, DDSc 9/47; 'Report from Saxon Electoral Society of Agriculture on the Cultivation of Potatoes', *Communications to the Board of Agriculture*, Board of Agriculture, 295–300; Eichler, 'Die Leipziger Ökonomische Sozietät', 369; Lowood, *Patriotism, Profit and the Promotion of Science*, 159–160; Popplow, 'Economizing Agricultural Resources in the German Economic Enlightenment', 277–283.

94. Talve, 'The Potato in Finnish Food Economy'. See also Kreuger, 'Mediating Progressin the Provinces', 50.

95. *Journal historique et politique de Genève*, 1 Dec. 1779, 497–498; *London Review of English and Foreign Literature* 11 (1780), 325; *Letters and Papers on Agriculture* 1

(1783), 32–33, 242–262, and 3 (1786), 292–293; *Monthly Magazine and British Register for 1797* 3 (1797) no. 13; Mackenzie, *Prize Essays and Transactions of the Highland Society of Scotland*; *Correio mercantil e economico de Portugal*, 24 Apr. 1798; White, *Natural History of Selborne*, 210.

96. Piqueras Haba, 'La difusión de la patata en España', 83–84.

97. *Extractos de las juntas generales celebradas por la Real Sociedad Bascongada de los Amigos del País en la ciudad de Vitoria por julio de 1786*, 28; Riera Climent and Riera Palmero, 'Los alimentos americanos en los Extractos de la Bascongada'. 关于此类翻译的重要性, 参见 Usoz, 'Political Economy and the Creation of the Public Sphere'. 也可参见 'Explicación de la voz batata para incluir en un diccionario de la lengua', *Memorial literario, instructivo y curioso de la corte de Madrid*, Nov. 1790, no. 121, 365; *SAA* 7 (1800), 13 Mar. 1800, 173; *Junta Pública de la Real Sociedad Económica de Amigos del País de Valencia*; *SAA* 11 (1802), 173; Piqueras Haba, 'La difusión de la patata en España'。

98. Kisbán, 'The Beginnings of Potato Cultivation in Transylvania and Hungary', 181. 关于18世纪农艺的更广阔图景, 参见 Blum, *The End of the Old Order*, 247–304。

99. Board of Agriculture, *Report of the Committee of the Board of Agriculture*, 73; Tribe, *Land, Labour and Economic Discourse*; Tribe, *Governing Economy*; Tribe, *Strategies of Economic Order*; Tribe, 'Cameralism and the Sciences of the State', 525–526; Phillips, *Acolytes of Nature*; Spary, *Feeding France*; Serrano, 'Making *Oeconomic* People'.

100. Joseph Dombey to André Thovin, Lima, 1 Dec. 1778, *Joseph Dombey*, ed. Hamy, 42; Tabáres de Ulloa, *Observaciones prácticas sobre el cacahuete*; Targioni-Tozzetti, *Cenni storici sulla introduzione di varie piante*, 46; Koerner, *Linnaeus*; Drayton, *Nature's Government*; Spary and White, 'Food of Paradise'; Bleichmar, *Visible Empire*, 123–148; Zilberstein, 'Inured to Empire'; Earle, 'Food, Colonialism and the Quantum of Happiness'.

101. Smith, *An Inquiry into the Nature and Causes of the Wealth of Nations*, I.xi.39; *SAA* 19(1806), 20, 27 Mar. 1806; Popplow, 'Economizing Agricultural Resources in the German Economic Enlightenment'.

102. 18世纪的营养（nourishment）概念, 参见 Earle, 'The Political Economy of Nutrition'。

103. [Colquhoun], *An Account of a Meat and Soup Charity*, 10, 15. 或可参见 Parmentier, *Les pommes de terre, considérées relativement à la santé*; Parmentier, *Traité sur la culture et les usages des pommes de terre, de la patate, et du topinambour*; Young, *The Question of Scarcity Plainly Stated*, 65。

104. Petty, 'The Political Anatomy of Ireland', 1672, *Tracts*, 366.

105. Doyle, *Tratado sobre el cultivo, uso y utilidades de las patatas* (1797), 4–5 (quote), 78–79, 85.

106. Doyle, *Tratado sobre el cultivo, uso y utilidades de las patatas* (1797), 24–25, 33, 81.

107. Doyle, *Instrucción*, 26–27; Doyle, *Tratado sobre el cultivo, uso y utilidades de las patatas*(1797), 6 (quote), 8, 24, 31; Doyle, *Tratado sobre el cultivo, uso y utilidades de las patatas o papas, corregido y considerablemente aumentado* (1804), 7.

108. Sheehan and Wahrman, *Invisible Hands*.

第三章　自由市场的土豆

> 我们实际谈论幸福时，不能不考虑它是指谁的幸福。[1]
> ——乔治·格罗特（Grote）《功利主义哲学考察》
> （*An Examination of the Utilitarian Philosophy*）

协调个人自由与公共利益，是长期困扰着哲学家和政治思想家的挑战。这个问题是当今大多数政治理论的中心问题，如约翰·罗尔斯（John Rawls）①的正义理论，以及查尔斯·泰勒（Charles Taylor）②对原子论的批判，后者称"社会的愿景，在某种意义上由个人构成，以主要实现个人的目标"。[2] 这也是当今政治家们迫切关注的现实问题。对食品、饮料的监管不断导致的冲突清楚地表明，将公众健康置于个人自主权之上的尝试，虽然会得到褒扬，但也同样招致批评。用谷歌查询"保姆式

---

① 约翰·罗尔斯（1921—2002），美国政治哲学家、伦理学家，普林斯顿大学哲学博士、哈佛大学教授，著名论著有《正义论》《政治自由主义》《作为公平的正义：正义新论》《万民法》等，是20世纪70年代西方新自然法学派的主要代表之一，同时也是20世纪最伟大的哲学家之一。——译注
② 查尔斯·泰勒（1931—　），当代著名的思想家、政治活动家，1961年在牛津大学取得哲学博士学位，现任加拿大麦吉尔大学教授。最著名的作品是《自我的来源》。——译注

国家"（nanny state）一词，会发现有成千上万的网站在谴责政府通过监管影响公众健康的工作。例如，"保姆式国家指数"（Nanny State Index）就是根据对烟、酒和电子烟的征税程度，对欧洲国家从"最自由"到"最不自由"进行排名。2017年，世界卫生组织认定芬兰是欧洲"最不自由"的国家，远远超过（或取决于你的观点，落后于）排名第二的英国。捷克共和国被认为是最自由的，因为它对烈酒和香烟的税率很低。[3] "老大哥，让开点！"记者大卫·海萨尼（David Harsanyi）在一本论述"食品法西斯主义者、禁酒主义者、自高自大的道德家和其他愚蠢的官僚正在把美国变成一个儿童国家"的书中如此敦促道。[4] 这些评论家以及与他们观点相同的选民们认为，如果在让人们随意吃喝和限制他们的选择范围之间必须作出选择，那么自由民主的原则是让他们自行选择。反对者们反驳说，其结果是对每个人都更糟。例如，在颇具影响力的《新英格兰医学杂志》（New England Journal of Medicine）上发表的一篇文章，为了支持对含糖饮料征税，列举了美国人减少糖摄入量将带来的经济和公共健康益处。报告的作者认为，"不断上升的医疗保健费用和不良饮食所致疾病的负担，造成了对解决方案的迫切需要，从而证明政府收回成本的权利是合理的"。[5] 当涉及个人饮食习惯时，平衡自主权和公共利益之间的矛盾绝非易事。

但如果一开始就没有冲突呢？如果最健康、最有益的食物也是最美味的呢？这正是土豆倡导者在18世纪所宣称的。正如政治经济新纪律的制定者们所认为的那样，允许个人管好自己的经济事务，与更大的整体经济福祉最终是没有冲突的。因此，土豆倡导者们坚持认为，建立强大而繁荣的国家所需的土豆，正是穷人会自发选择食用的。所需要的，只是一场教育运动和增加供应。土豆倡导者因此发行小册子，有时甚至是整本烹饪书，对土豆等健康食物的最佳食用方法进行"提示"。[6] 许多向收获或消费最多作物的劳动者提供奖品和奖金的做法，同样也是为了使个人有条件——而不是强迫他们——为自己的利益而行动。对于许多热心人士而言，土豆提供了一个具体的例子，它证明了运转良好的经济的所谓发动机——个人利益，如何化解个人权利与公共利益之间的所有紧张关系。这代表了自由放任主义和市场支持者所倡导的一切。

由于美味，土豆达到了这一目的。因此，土豆给个人带来的好处，据说已经胜过了它们消灭饥饿的能力。它的倡导者坚持认为，土豆能带来幸福。这一点非常重要，因为按照启蒙主义的治国理论，让人民幸福是国家的基本职责。因此，土豆使人们幸福，既是个人吃土豆的动机，也是国家履行上述义务的途径。土豆带来幸福的潜力具有政治意义。

把对推荐食物的消费看作与个人利益相关之事，通过这一点，18世纪的这类争论，为如今强调个人饮食习惯应在知情选择的框架内理解的主张奠定了基础。我们作为"选择者"，应该根据什么对自己最好的自我评估来决定吃什么。2018年，饼干制造商麦维他（McVitie）被指通过营销含糖零食而助长了肥胖和不健康，面对这一批评，麦维他辩解称，它提供了大量关于产品的信息，使消费者能够作出"知情的零食选择"。[7] 知情选择正是18世纪的土豆倡导者们宣称要进行鼓励的。

18世纪提出的观点，也奠定了一种信念的基础，即试图通过更直接的政府干预来塑造我们的饮食，侵犯了我们作为个体支配自己事务的权利。如果注意到与18世纪的土豆推广同时期流传的经济论述，我们能够更清楚地看到，从一开始，个人饮食习惯是国家关注点的概念，是如何与个人对自己的（我们今天所说的）饮食选择要有多大程度的作用和责任的争论纠缠在一起的。

本章首先回顾重土豆的饮食给个人带来的益处和幸福。它们被认为足以激励个人吃土豆，而这种动机独立于吃更多土豆会带来更大的政治和经济利益的想法。接下来，本章把对个人饮食利益的关注与政治经济学的新语言关联起来，后者同样将个人对幸福的追求视为国家安康的基石。它们的关系，反映了饮食根本的、但在18世纪政治经济讨论中很大程度上未被认识到的重要性。值得一提的是，很多关于自由贸易的讨论，重点并未放在总体的贸易上，而是具体地放在欧洲的主要粮食——谷物的贸易上。对贸易规则自由化的提议，对人民饮食习惯的兴趣，以及对选择的说辞，在18世纪的个人主义与有效治国之道的新模式中，并不是简单地呈平行关系，而是内在地相互关联。

选择土豆

18世纪的土豆支持者坚持认为，土豆在增强国家实力、增加国家财富的同时，对个人也有许多好处。首先，它可以终结饥饿和贫困。如苏格兰农业文献所言，有了土豆，"穷人永远不会缺乏食物"。苏格兰医师、土豆菜园的倡导者威廉·巴肯描绘了多吃土豆带来的幸福情景："人们将会生生不息，而贫穷，仅在挥霍无度者当中才会出现。"[8]（我们将在稍后谈论挥霍无度者的情况。）不仅如此，土豆还会带来快乐。18世纪对土豆的讨论一致认为，土豆在减少饥饿的同时也会促进幸福。进步地主、土豆推广者彼得罗·马利亚·贝格美认为，土豆不仅会让博洛尼亚变得更富有，而且也会带给它更多的幸福，这样他的家乡就会成为"全意大利最富有、最幸福的地区之一"。安东尼奥·詹农（Antonio Zanon）是乌迪内（Udine）实用农业学会（the Society for Practical Agriculture）的创始人之一，撰写了许多关于经济和农业问题的著作，他同一时期在威尼托所写的文章中，也认为土豆是"幸福和富裕"的源泉。[9] 弗雷德里克·伊登在18世纪90年代的著作中，总结了对土豆的这种乐观看法。他写道，土豆种植的普及"或许是人类的快乐得到延伸的有力例证"。[10] 个人和整个国家都将受益于这种土豆幸福效应。一本法国烹饪书写道，如果穷人多吃土豆等营养丰富的蔬菜，他们自身就会更健康，同时，他们更具活力的身体也会更好地为国家的荣耀和繁荣作贡献："他们多么安逸！国家多么幸福啊！"[11]

土豆使人幸福，因为吃土豆是一种乐趣，至少土豆倡导者是这样认为的。按法国土豆推广者安托万·奥古斯丁·帕蒙蒂埃的说法，只需看看喜欢土豆的小孩的自然而然的倾向就足以证明这一点。它也表明，土豆是多么适合人类的消化系统，而且，土豆还把幸福从满足的食用者传递给了满意的旁观者。另一位土豆热衷者想象着，对一位穷爸爸而言，能给孩子提供如此完美地符合他们饮食需求的食物，该多么令人高兴啊。支持者把以土豆为主的食谱说得十分美味甚至奢华。[12] 法国作家称赞这种"美味、多产、健康的块根"是仁慈的神存在的证据。欧洲各地的推广者解释说，那些最顽固的反对者——通常是固执己见的劳动者——在被说服进行品尝之后都改变了自己的看法。[13] 理性的考虑加上感官的愉悦，会引导每个人，尤其是那些贫穷的劳动者，以自己的意愿去接受土豆。将这些催生幸福的个人选择结合起来，就会让国家更强大、更成功。自我利益和国家利益和谐共存。

这些概念出现在政治经济学原理正在发展的时候，并非巧合。倡导者们坚持认为，确保商业繁荣的最佳方式，是让人们关心自己的福祉。正如亚当·斯密所指出的那样，如果个体能够做到这一点，那么无数利己主义行为的汇聚，将会组成一个运转良好的经济体系。1776年，他撰文解释说，尽管个人通常只追求自身的经济利益，但这常会导致总体人群变得"最幸福、最舒适"。[14] 事实上，允许个人追求自己的利益，比直接干预更有可能产生这样的结果。在斯密看来，"通过追求自己的利益,(一个人)往往能比他真正愿意促进社会利益之时更有效地促进社会的利益"。[15] 因此,个体对个人利益的追求是成功的经济的基础,而成功的经济又会给所有人带来更大的幸福。这些思想在 17 世纪晚期开始得到表达,同时,理论家们开始认为,享受营养饮食的人口是国家富强的组成部分。这两个概念是同时出现的。

穷人仅因为饥饿而吃土豆,对这一点的强调,反映了一个客观存在的事实:土豆在更昂贵的谷物之外提供了一种选择,至少在大量种植土豆的欧洲那些地区如此。要断言土豆是一种美味的选择,还意味着更多:人们选择吃土豆,不是因为他们饿了,而是因为土豆是幸福和安慰的来源。按照土豆倡导者的观点,吃土豆不但是个人促进幸福的一种愉快的方式,而且也是对国家的福祉作出贡献的一种愉快的方式。总体而言,普通人很少被认为够格来评判更重要的国家事务。正如西班牙剧作家尼古拉斯·费尔南德斯·德莫拉廷(Nicolás Fernández de Moratín)所言,"无知的人应该知道,他们没有资格对自己不理解的事情作出决定,而在那些研究过这些事情的人说话时,他们应该保持最深的沉默"。[16] 然而,应该相信,即使是这样的人,也能确定自己是否喜欢某种特定的食物。味觉享乐应该属于一个"无知的人"够格进行判断的范围,至少对土豆来说是这样。这样就会使国家更强大、更富有,至少土豆倡导者是这么认为的。"他们多么安逸!国家多么幸福啊!"这种饮食就相当于斯密的"看不见的手"。

政治经济学、个人利益和粮食供应

将个人利益与更广泛的国家福祉相结合,这个挑战困扰着 18 世纪的经济作家。17 世纪的一些评论家已经开始提出,如果个体的自私行为能够刺激商业或鼓励新奢侈品的发明，那么这些行为应该是有益的。伯纳德·曼德维尔（Bernard Mandeville）①在《蜜蜂的寓言》（*Fable of the Bees*）（1714年）中，以引人入胜的押韵诗句来呈现这个论点。他写道，奢侈"亦在支配上百万穷苦之士，可恶的骄傲则主宰着更多人"。还有人提出，个人野心和贪婪的潜在破坏性力量将相互制约，创造一种不稳定的平衡，最终对政体有利。例如，意大利哲学家詹巴蒂斯塔·维柯（Giambattista Vico）②1744年撰文说，"出于人们各自追求个人利益的激情"，神圣的天意创造了一种文明秩序，使所有人都能在社会中共同生活。法国哲学家孟德斯鸠（Montesquieu）1748年说："事实证明，每个人都在为大众福利作贡献，同时都想着自己是在为自己的利益而工作。"在整个18世纪，这些主要起源于政治领域的思想，越来越多地被用于分析政治和经济事务之间的相互关系。例如，孟德斯鸠就认为，推动国际贸易的对经济利益的渴望，在减少政治冲突上起到了作用。[17]

　　如何恰如其分地将私人经济收益与公共利益结合起来，是在18世纪被反复提出的难题。国家对贸易的管制引起了特别的争议。越来越多的作者认为，试图通过关税等手段来控制贸易，干扰了个人追求利润带来的有益影响。这种干预不仅无益，而且可能注定要失败。米拉波侯爵

---

① 伯纳德·曼德维尔（1670—1733），英国哲学家、古典经济学家。曼德维尔作为18世纪道德哲学的主要贡献者之一，向人们揭示了在强制性的行为限制中，个人追求自身利益的行为可能会推进整个社会的福利。在曼德维尔看来，如果从道德的角度看，受自利驱策的商业社会是应该受到谴责的，但如果想以"公共精神"为基础而建立起一种充满美德的繁荣社会，那纯粹是一种"浪漫的奇想"。这就是著名的"曼德维尔悖论"。曼德维尔的观点——"私人恶德即公众利益"成了人类经济活动和经济实践中无法挣脱的噩梦。——译注

② 詹巴蒂斯塔·维柯（1668—1744），意大利政治哲学家、修辞学家、历史学家和法理学家。他为古老风俗辩护，批判了现代理性主义，并以巨著《新科学》闻名于世。维柯作为近代历史哲学之祖和精神科学原理奠基者的地位逐渐为人们所认识。——译注

维克多·里克蒂（Victor Riqueti）特别清楚地表达了这一观点——他解释说，一位君王的作用，不是"领导人们的艺术"，而是：

> 通过遵守自然秩序以及构成自然法律和经济秩序的物理法则，来保障他们的安全和生存的艺术。通过这种艺术，国家的，特别是每个人的生存才能得到保证；这个目标实现了，人的行为就决定了，而每个人都领导他自己。[18]

里克蒂是重农主义的追随者，这种经济模式是由弗朗索瓦·魁奈（François Quesnay）①提出的。魁奈是蓬帕杜夫人（Madame de Pompadour）②的医生，也是极具影响力的解除经济管制的倡导者。重农主义提出了以两种看法为前提解释经济交换的理论，一种看法是人类受到对个人利益的内在渴求的支配，另一种是经济运行遵循某些基本的自然法则。这些法则可以被忽略，但不能被推翻。重农主义者认为，国家的职责是使其政策与自然秩序保持一致，让私人利益不受阻碍地运行，从而确保所有人的利益。因此，君主的个人利益，应该引导他们允许有资格的个人不受规则的阻碍而追求自己的经济野心。[19]

关于市场监管优点的争论主要围绕着粮食贸易展开。[20] 重农主义者强调，经济增长仅仅来自农业，他们认为，对谷类贸易的诸多限制，是

---

① 弗朗索瓦·魁奈（1694—1774），古典政治经济学奠基人之一，法国重农学派的创始人和重要代表。他早年研究医学和哲学，后转到经济学，并在各领域都有卓越成果。尤其在经济学方面，他有许多著作和一系列重要的观点学说。著作有《经济表》《谷物论》《人口论》《赋税论》等。——译注

② 蓬帕杜夫人（1721—1764），又译蓬巴杜夫人，全名让娜-安托瓦妮特·普瓦松，蓬帕杜女侯爵（Jeanne-Antoinette Poisson，Marquise de Pompadour），法国国王路易十五的著名王室情妇、社交名媛。——译注

背离统治所有交换的自然法则而带来恶果的明显例子。长期以来，调控粮食供应，特别是谷物市场，一直被认为是政府的一项道德义务。为了防止粮食短缺，谷物的流通和价格通常是受到控制的，"囤粮"或投机性购买谷物通常是被禁止的。市政府官员还会调查市场上出售的粮食的质量和价格。这个长期存在的价格、运输和质量控制体系的前提，是国家被认为有责任避免疾病和饥荒，这个体系还反映出对粮食短缺的政治后果的担忧。魁奈等新政治经济学的倡导者坚持认为，这种方法会适得其反，而且很可能会引发它旨在防止的粮食短缺，因为它鼓励农民只生产他们能够确定卖掉的粮食数量。重农主义者主张让市场来决定价格和供应问题。他们希望，这将确保更稳定的价格，并从长远看，能保证更可靠的供应。总而言之，他们相信，确保健康的经济和人口的政治约束，最好通过允许个人对商业利益更自由的支配来实现。

重农主义者关于支持农业和改革税收体系的重要性的观念，影响了许多地方的政策，尽管令重农主义者烦恼的是，他们的建议通常只得到零散的实施。许多君主，如瑞典的古斯塔夫三世（Gustavus Ⅲ）、巴登大公卡尔·弗雷德里希（Charles Frederick），都尝试实施受重农主义启发的法律，有时还直接与重要的"经济学家"（économistes，重农主义者以此自称）通信。魁奈和他的追随者在法国18世纪六七十年代路易十五时期取得了相当大的影响力，他们利用这个时期来实现他们的中心目标，即解除对粮食贸易的管制。对粮食的国内流动和海外出口的控制在1763年和1764年首次得到放松，并于1774年再次得到放松，结果都造成了灾难性的影响，于是两次都被迅速恢复。卡洛斯三世1765年在西班牙也推出了这样的法律，在面临高物价和公众抗议后又予以撤销。意大利的几个城邦也做过类似的尝试，但同样不成功。[21] 由此可见，这些自由放任经济的早期实验，明确关注的不是普通贸易，而是粮食贸易。对这种做法的批评，如那不勒斯人费迪南多·加利亚尼神父（abbé Ferdinando Galiani）颇具影响力的《粮食贸易对话》(*Dialogues on the Grain Trade*)，同样也把重点放在谷物自由贸易的缺点上。粮食是政治经济学辩论的中心，正如它也是治国新思想的中心。

18世纪研究经济问题的作家们认识到，管理商业与培养健康幸福人口的任务是密不可分的。比如，魁奈就把自由贸易、粮食、幸福和国家富裕之间的关系阐释得非常清楚。在《百科全书》(*Encyclopédie*)一篇有影响力的关于农民的文章中，他坚持认为，如果他的想法得到实现，法国的居民将会更加富裕，他们增加的物力会"为他们提供更好的食物、满足他们的需求、给予他们幸福，会增加人口，（并）提高土地所有者和国家的收入"。[22] 在他看来，所需要的仅仅是政府停止干预粮食贸易，并对整个税收体系做少许调整。简而言之，对组织经济流通的最佳方式的兴趣，是由确保充足粮食供应的特定需求，以及充足粮食供应能够——尤其是在穷人中——传播幸福的信念所强力激发的。确保两者兼顾的最佳方式，是允许个体追求自身的利益，或也可以说，是人们越来越坚称这一点。

对个人幸福之合理追求的同样强调，导致了对劳动人民饮食习惯的讨论，他们对食物的偏好已成为这样一个政治问题。两者的核心都是一样的：一个以个人对幸福和私利的追求为导向的体系，是否可能被证明是将一个政体中不同的成员统一为一个平稳运行整体的最有效方式。劳动者可能缺乏专门知识来判断他们自己的经济利益，也必定缺乏干预国家事务的权力，但可以信赖的是，他们能辨别饭菜是否美味。至少，这是许多土豆热衷者的看法。

幸福之汤

　　这一点，在关于土豆汤的讨论中最清楚不过了。18 世纪充斥着"经济汤"的食谱。这些汤的基础是淀粉，通常来源于土豆，再配上少量的肉和一些辛辣的调味料。炖菜和浓汤长期以来一直是平民的主食，而且在早几个世纪里经常出现在济贫院等机构的菜单中。[23] 在 18 世纪，随着劳动阶层的饮食具备了越来越多的政治意义，炖汤不仅成为人们常吃的食物，而且也成为公众讨论的主题。烹饪书向家庭慈善家介绍如何烹饪分发给穷人的汤，厨师把详细的做法抄到自己的菜谱中。有一种汤是将牛肉、米饭、土豆、胡萝卜、芜菁、韭菜、芹菜、百里香和欧芹混在一起，用 13 加仑的水炖煮，一位女士将它抄在笔记本上，称之为"以合理的成本烹制的有益健康的肉汤，用来喂养这个国家的穷人"。[24] 报纸也刊登食谱，记者们讨论炖汤的优点并提出改进建议。各国政府散发传单，介绍如何烹饪这种汤供个人消费或大量分发。[25] 爱国的个人和机构建立施粥所，为应该得到救济的穷人提供慈善食物，并为那些能够发明"最健康、营养最丰富且花费不超过每加仑 5 便士的汤"供他和他的家人食用的劳动者提供额外津贴。[26] 廉价汤也已经进入了公共领域。（参见"供 25 名士兵吃的杂烩汤"食谱。）

　　18 世纪的最后十年，精英们对施粥所的兴趣尤其浓厚，当时战争、人口增长和歉收导致了严重的粮食短缺。自 18 世纪 50 年代开始，不断增长的人口给谷物供应带来了越来越大的压力，它们一直供不应求。例如，在英格兰，人均小麦产量从 1750 年的 5.9 蒲式耳下降到 1800 年的 4.6 蒲式耳。厄尔尼诺效应造成的气候波动也可能导致情况进一步恶化。[27] 粮食价格大幅上涨。18 世纪 90 年代的革命战争调动了数百万人，更是

供 25 名士兵吃的杂烩汤

在 18 世纪，富有的欧洲人对养活了普通人上千年的淀粉汤、炖菜开始产生兴趣。这类营养丰富的食物将有助于培养充满活力的劳动人口，至少他们希望如此。这道简单的土豆牛肉汤食谱，是 18 世纪末米德尔塞克斯郡海斯（Hayes，Middlesex）一座庄园的住户抄写在笔记本上的。他注意到，这道汤只需花费 2 便士就能供 25 名士兵食用。这道食谱后来有一个不含大葱的版本，还有一个被描述为"以合理的成本烹制的有益健康的肉汤，用来喂养这个国家的穷人"的版本。

供 25 名士兵吃的 2 便士杂烩汤

4 加仑水

2 夸脱剥好的豌豆

6 磅牛腩

6 颗芹菜头

6 根大葱、6 个洋葱

欧芹和百里香

6 磅土豆

胡椒和盐

面包

将劳动力从农业中抽离，同时为了补养军队，小麦等主要作物的需求更为增长。不断上涨的粮食价格和蔓延的革命情绪引起了普遍的不满，让整个欧洲大陆的政府感到恐慌。英国议会委员会警告说，"雅各宾派的报纸"充斥着"因面包的高价格而对政府的最粗野的谩骂，他们想煽动暴徒发动骚乱，因为他们觉得，负责公共事务的人有严重失误，而且没有采取必要的预防措施来避免这场严重的灾难"。[28] 慈善施粥所的建立，被看成是确保社会和政治安宁的真正贡献。1800 年，在兰开夏郡克里塞罗（Clitheroe）镇，古文物研究者、助理牧师托马斯·威尔逊（Thomas Wilson）报告说，"大多数较大的城镇都开设了施粥所，优质的营养汤以每夸脱 1 便士的价格分发。这当然是为穷人提供帮助的一种合适的方式"。威尔逊抱怨说，受益于这种慷慨的人通常都是忘恩负义的。他们抗议说，提供的汤不符合他们的要求，而且他们花不起从分发点把汤拿回家的时间。威尔逊总结道，这就是"与下层人民打交道的困难所在，他们不能作出正确的判断，也不相信自己是错的"。[29]

威尔逊认为，这些汤并不总会得到因其巨大的营养价值而应该得到的热情追捧，而他并不是唯一持有这种观点的人。研究英国工人阶级的历史学家弗雷德里克·伊登在 1797 年悲哀地写道，英格兰南部各郡的穷人对廉价汤有着几乎无法克服的厌恶。他报告说，"他们共同的抗议是，这些又稀又淡的东西没有营养：我们'不能像猪一样，被喂食切碎的土豆！'"[30] 在一些城市，慈善施粥所吸引了大批顾客——据报道，日内瓦的施粥所每天供应 1200 碗汤，每 20 个居民就有一碗——但很明显，即使在真正短缺的时期，人们对这些汤的接受程度也各不相同。[31]

尽管如此，整个欧洲却异口同声地咬定，这些汤非但没有遭到鄙弃，反而被其目标消费者津津有味地享用了。医师（同时也是一位非圣公会的牧师）西奥菲勒斯·洛布（Theophilus Lobb）在他 1767 年的节俭食谱集中介绍说，他所描述的任何一种炖菜和浓汤都"非常营养、健康、口感宜人"。同时代的一份法国菜谱也明确表示，"我们的农民发现这道汤非常棒"。1800 年，农学家亚瑟·杨在下议院委员会上发表演讲时，首次介绍了如何制作一种内有精瘦肉的实验汤，然后他说，这样

"做出的汤最棒,穷人都非常喜欢"。[32] 这种断言并不是对廉价汤整体受欢迎程度的公正评价。对于许多消费者来说,土豆汤既不能缓解他们的饥饿,也不能使他们快乐,伊登和克里塞罗的助理牧师所记录的抱怨和不安,可证明这一点。坚持认为这些汤是一种快乐的来源,是在试图掩盖实情,同时也是为了坚持一种信念,即个人利益会将个人饮食习惯组织成为对国家更大目标的支持。所谓的土豆汤能带来的幸福,是更大政治愿景的一部分。

国家的责任是让国民幸福,这一点在启蒙运动时期几乎是老生常谈。沙特吕侯爵(Marquis de Chastellux)在论幸福的专著中说,"不可否认的是,或至少我有理由相信的是,在这个开明的世纪,任何政府的首要目标都是使它的人民幸福,这是一个公认的真理"。[33] 欧洲各地的作家投入大量精力讨论幸福的本质和源泉,用历史学家达林·麦马翁(Darrin McMahon)①的话来说,他们在世上追求的是"本世纪最伟大的目标"。人们普遍肯定,促进公共幸福符合,或者构成了国家的最高目标。哲学家和官员们都一致认为,无论"人民的幸福"是直接得益于官方的政策,还是间接得益于明智治国之道支持下的经济全面改善,如果一个国家不追求这个目标,它就无法获得成功。普鲁士腓特烈大帝坚持认为,君王的主要目标,必须是提升他所统治的人民的"快乐和幸福"。整个欧洲大陆也发出了相同的和声。[34] 因此,在理论上,劳动人民的幸福、他们身体的强健,是与国家富强直接相关的。

一个国家应该如何实现这一核心目标,是一个被无休止讨论的话题。有一种将个人幸福与大众幸福结合起来的途径,那就是食物。如

---

① 达林·麦马翁(1963—),历史学家、作家、公共演说家。现任达特茅斯学院历史教授。1997年在耶鲁大学获得博士学位,后来曾在哥伦比亚大学、耶鲁大学和纽约大学担任客座教授。著有《幸福的历史》(2006)等。——译注

马尔萨斯所见,饥饿的人当然很可能不幸福,而不幸福的人就容易对政治现状不满。但食物增加个人和公众幸福的能力,大大超越了食物短缺导致不幸福的负面潜力。充足的营养食物会直接增加幸福,而幸福的主体,用米歇尔·福柯的话说,"正是国家的力量"。³⁵ 正因如此,农业委员会将其增加英国粮食供应的工作描述为"为(英国)未来的繁荣和幸福奠定基础"。³⁶ 心满意足的食客,在理论上被认为是国家迈向强大、安全的过程中的一环。

将营养食物、治国良道和幸福组合在一起的力量,反映在18世纪对它进行量化的不懈努力上。从越来越多地引入数据、配料和方法介绍的烹饪书,到森林的管理,量化,这一"现代思想的典型形式",迅速在18世纪许多生活领域传播开来。量化为分析世界提供了一种权威的新语言,它用数学的方法来展示幸福的组成部分,强调它们之间相互联系的重要性,并增加了精确度。³⁷ 创造一种"幸福微积分"将这些因素联系起来的努力并不少见。爱尔兰哲学家弗兰西斯·哈奇森(Francis Hutcheson)① 提出了一个复杂的算术公式,用以衡量美德、邪恶和幸福之间的关系,用一系列方程予以表示。年轻的西班牙军官特奥多罗·文图拉·甘达拉(Theodoro Ventura de Argumossa y Gandara)在一篇政治经济学论著中提出,一位政治家造就的幸福人数,乘以他为实现这个目标所克服的障碍,就构成了他的荣耀及名望的"算术表达式"。沙特吕侯爵根据这种量化精神创建了一个"幸福指数",这个指数证明,拥有丰富健康食物之地区的居民是最幸福的。³⁸ 从数学和哲学的角度而言,丰富的食物会带来幸福,而幸福又导致国家更强大。从土豆汤中产生的

---

① 弗兰西斯·哈奇森(1694—1746),出生于爱尔兰,是18世纪苏格兰启蒙运动的奠基人,苏格兰哲学之父,其著作涉及伦理学、形而上学、逻辑学、美学。哈奇森曾经为亚当·斯密授课,从斯密的著作《道德情操论》中可以看出哈奇森带给他的思想烙印。著名作品有《论美》《论道德的善与恶》等。——译注

幸福，是与政治直接相关的。

18世纪最有名的汤，是本杰明·汤普森（Benjamin Thompson）发明的。汤普森于18世纪50年代出生在马萨诸塞，于美国独立战争期间离开了北美。他当时是反对独立的，因此离开也是恰当之举。为英国军队工作一段时间后，他成为巴伐利亚选帝侯卡尔·西奥多（Karl Theodor）的顾问。汤普森的工作内容十分广泛。在慕尼黑，汤普森重组了巴伐利亚军队，并建立了他所谓的"工业之家"，这是一种为乞丐和穷人设立的收容营。正是看到了他的这些努力，卡尔·西奥多才授予他拉姆福德伯爵（Count Rumford）的头衔。拉姆福德和他的雇主们一直持续关注的问题，是把巴伐利亚的穷人变成具有生产力的军人公民。不过，拉姆福德的兴趣远远超出了救济穷人，扩展到大炮设计、热力性质的实验、烟囱改进、士兵制服等领域。他是一位著名的公众人物，他的种种活跃行为，成为简·奥斯汀1818年《诺桑觉寺》（*Northanger Abbey*）中人物提尔尼将军（General Tilney）的灵感来源。[39]（参见图10）

在拉姆福德将穷人变为有用的巴伐利亚人的计划中，食物是核心元素。他解释说，经营慕尼黑济贫院的经验，为他提供了充足的机会来鉴别"最便宜、最美味、最有营养的食物"，它是一种"将珍珠麦、豌豆、土豆、小麦面包丁、醋、盐和水混在一起"煮三个小时的汤，土豆是这道食谱的主材，每一份配干豌豆的大麦需要两份土豆。[40]拉姆福德详细计算了煮汤的成本，但他明确表示，简单的节约不是他唯一的关注点，美味也是汤所必需的。他引用希波克拉底的话，坚持认为"凡是愉悦味觉的必有营养"。这就是他的汤需要烤面包丁的原因。他解释说，烤面包丁需要长时间咀嚼，咀嚼因产生唾液而有助于消化，但对拉姆福德而言，它的重要性超越了它在消化过程中的作用。咀嚼还能增加幸福感，因为它"延长了享受进食的时间，这的确是一件非常重要的事情，而且

图 10　拉姆福德伯爵在他特别设计的壁炉前的漫画。拉姆福德伯爵本杰明·汤普森，是一位多产的发明家、著名的穷人土豆汤的推广者。图中，他在根据热力学改进的炉子前为他的屁股取暖。

到目前为止还没有得到足够的重视"。站在道德高地上，拉姆福德坚持认为，大多数人都不认同穷人有权享有幸福的观点，但他却非常认同。他虔诚地提出："大多数人的享受，并不多得足以放弃增加更多享受的尝试。"[41] 他的配上烤面包丁的土豆汤，会使得慕尼黑最可怜的乞丐也高兴起来。

很明显，加入烤面包丁是一种节省汤的方法（拉姆福德认为 20 盎司汤就足够吃一顿），人们无从知晓慕尼黑的乞丐们到底多享受他的创造，但拉姆福德自己却坚定不移。他重申，从这种汤中获得乐趣，是它的实用性最精华所在，而让他的汤成为成功的美食，土豆是功不可没的。他声称，由于慕尼黑的穷人一开始对土豆怀有敌意，因此他最初是"偷偷地"把土豆加到汤里"……以防被发现"。然而，土豆改进了他的汤，"穷人……证明了他们对这种改进的认可……普遍而响亮，他们现在慢慢变得非常喜欢吃土豆，以至于没有土豆他们都不易感到满意"。当然，他指出，所有人都认识到降低成本的重要性，但这不应该以牺牲吃的乐趣为代价，即使是在最穷困潦倒的人当中。幸运的是，拉姆福德报告说，土豆大大降低了他这种汤的成本，所以实际上不需要作出这样的让步。[42]

1800年在巴黎出版、与法国政府关系密切的期刊《哲学、文学和政治年代》（*The Philosophical, Literary and Political Decade*）中，有一篇文章优美地描写了拉姆福德的汤所带来的美食愉悦，这篇文章的作者是帕蒙蒂埃的同事，同为土豆推广者的化学家安托万-亚历克西斯·卡代·德沃克斯。文章讲述了卡代·德沃克斯和一位爱好美食的朋友之间的一次对话。卡代·德沃克斯的朋友描述了他前一天晚上吃的一顿奢华的晚餐，包括酥皮馅饼、一些排骨、一大只松鸡配松露、一条鲭鱼、一些"美味的"小点心和一整瓶沃尔奈葡萄酒（Volney）。不出所料，这顿大餐之后是消化不良和痛风带来的惨状。卡代·德沃克斯和他的朋友一致认为，提供这种食物的那些新流行的"餐馆"简直徒有虚名，因为它们非但不能带来健康，反而还会严重损害健康。[43]

卡代·德沃克斯建议他们去一处风格不同的地方，一家真正的餐馆，在那里他们能吃到的饭菜，既美味又能真正地使人恢复元气。第二天，这两位巴黎人出现在美尔街16号，他们看到一块欢迎牌上写着："拉姆福德汤。一碗汤包满意，中午至下午两点。"两位朋友走了进去并坐下来用餐。汤由在施粥所当志愿者的那些衣着干净、仁慈大度的女士们端上来。喝汤时，卡代·德沃克斯喜好美食的朋友赞美它说，"不是好，而是*极好*"。在这顿经济而美味的晚餐中，两位朋友讨论了如何改变汤的配料以避免单调乏味，并一致认为，把这样的汤供给士兵，取代那些令人讨厌的食物，会是大有裨益的。他们希望农村的人多种土豆、大麦等健康作物，用来制作美味的汤，同时也会减少他们对面包的消耗。这样也会使得法国能出口更多的粮食，增加其财富。晚餐结束时，他们对拉姆福德伯爵赞不绝口，他烹制的美味汤既能满足美食家的要求，又能促进法国的经济繁荣、增强军事力量。卡代·德沃克斯的故事，从讲述拉姆福德的汤给人带来的味觉享受，顺畅地转移到法国的更加繁荣昌盛。个人私利和更大的利益，在一碗汤里相遇。

拉姆福德的著作和他的风味酸汤风靡了整个欧洲。他那篇关于喂饱穷人的文章（包括题为《吃的快乐，以及增加快乐的方法》的章节）被广泛翻译和转载。从拿破仑到英国农业委员会主席等许多人，都对他的食谱赞不绝口，还称赞他发明的专利炉，因为它缩短了煮汤的时间。在18世纪90年代的战争年代，欧洲许多城市都出现了受拉姆福德启发的慈善施粥所，拉姆福德自己于1796年在伦敦孤儿院开办了一家。到1800年，仅在首都就有近50家这样的机构。在慕尼黑以外的其他德国城市，以及瑞士、意大利、瑞典、西班牙和法国，更多的机构也被开办起来，1802年的时候，巴黎已有超过20家。日内瓦综合医院的主任在一本法国杂志上读到拉姆福德的革新发明，于是他亲赴慕尼黑去考察汤的分发情况。他一回来，就监督修建了一家由公众捐款资助的大型施粥所。[44]

这种汤在西班牙引起了乐善好施的精英和政府的兴趣。一些爱国组织翻译了拉姆福德著作的摘要，而由国家资助的《教区牧师农业艺术周刊》则定期更新自己的拉姆福德风味汤试做。[45] 西班牙多个城市建起了慈善施粥所。在巴伦西亚，当地的经济学会受到媒体报道的启发，也决定建立自己的机构。他们仔细比较了拉姆福德的信徒们在欧洲不同地区的报告，并安排两名成员照抄拉姆福德的方法。拉姆福德最初的食谱很快就因被认为过于平淡无味而遭到摒弃，随后学会着手对其进行了调整，以适应西班牙人的口味。学会介绍说，这是通过多种方式改变食谱、增加"吃的乐趣"来实现的。在学会看来，快乐和营养一样重要。该学会在该市圣纳西索监狱的囚犯当中测试了他们的实验配方。由于他们和拉姆福德一样重视人们喝汤的愉悦体验，学会很高兴地报告说，他们的第三种改进配方获得了成功。这一版被证明是最便宜也是最受欢迎的，里面含土豆、大麦、黄豆、洋葱、油、盐、薄荷和辣椒。以该学会的话说，"按照大多数品尝过它的人的说法，它最符合当地的口味"。据说它受到囚犯们的喜爱，他们每天都想喝。[46]（参见图11）

		Rs.	ms.	vn.
Harina de Cebada.	30 onz.	1	12	
Abichuelas.	30 onz.		31	
Patatas.	6 lib.	1	14	
Pan.	30 onz.	1	9	
Sal.	8 onz.		6	
Aceyte.	8 onz.		30	
Cebollas 3.			2	
Hierba buena y pimiento picante			2	
Agua. 50 libras.				
Leña. 24 libras.		1		
		7	4	

图 11 巴伦西亚经济学会第三版汤食谱。1800年的整个秋天，巴伦西亚的经济学会都一直在考虑土豆汤的理想食谱。经过一些实验，获得经验的成员们决定采用一种将6磅土豆与较少量的大麦面粉、黄豆、干面包和调料混在一起的版本。他们认为，这样的汤既便宜又美味。他们把它分发给当地监狱的囚犯。

1803 年，马德里经济学会为了应对当年小麦的歉收，决定分发拉姆福德汤，因此也进行了类似的调查。学会的主席是富埃尔特·希贾尔侯爵（Marquisde Fuerte Híjar），他成立了一个委员会，整个秋天每天都开会，以仔细考虑分发点的最佳位置、汤的理想配方等。最早的德国食谱再次被摒弃，因为它完全不适合西班牙人的口味。主席解释说，无论多么饥饿，都没人会从一顿不适合当地口味的食物中获得"愉悦、幸福和满足"。经过一番试验，学会定下了一份食谱：六份土豆配一份干黄豆，外加少量的洋葱、大蒜、小茴香、甜辣椒粉、油、醋和盐。先将调味料放在油里煎炒、碾碎，然后将之前煮好的土豆和黄豆加醋一起加入锅中，一份浓浓的土豆汤便大功告成，每份仅需区区 7 个铜板（Maravedí）。食谱的变种包括用豌豆或大麦面粉来代替黄豆，还可用像瑞士甜菜那样的蔬菜，也可加入土豆丸子。和巴伦西亚学会一样，马德里的小组在大量的穷人当中测试食谱，并根据他们的反应进行调整。最终，学会信心十足地认为，他们的汤"满足了西班牙人的口味"。[47]

类似的增加愉悦感的改动在整个欧洲都在发生。法国慈善施粥所根据当地口味对汤进行了调整，不用西班牙常用的小茴香、红辣椒粉和橄榄油来替代拉姆福德的醋，而是改用欧芹、百里香、月桂叶和另一种烤面包丁。据法国报纸报道，拉姆福德亲自考察了巴黎的施粥所，并赞同这些改变，因为它们改善了汤的味道。在纽夏特（Neufchâtel），土豆却被完全省去了，"因为穷人更喜欢大米、大麦、豌豆和意大利面团"。的里雅斯特（Trieste）的汤里加入了当地的特色调料香蒜酱。[48] 1802 年的一份建立施粥所的指南，总结了愉悦对于慈善机构取得成功的中心作用。"最重要的事情，"它提到，"是确保汤的经济性，确保它美味又确保它有益健康。"[49]

在一项关于 18 世纪法国食物的重要调查中，历史学家艾玛·斯帕里（Emma Spary）提出，汤的推广者们认为，在给穷人提供食物时，"习惯或偏好等标准可以在考虑营养时被排除在外"。[50] 施粥所当然不是给喜好美食的人准备的，它们也没有给服务对象提供多少机会来行使真正的饮食自主权。抱怨的人会被谴责为忘恩负义。尽管如此，习惯和偏好远不是不相干的，相反还被认为是一种汤能否成功的关键。汤的推广者们煞费苦心地描述穷人接受者对他们调制的汤可能的认可。在描述中，除开成本计算和限制穷人获得喝汤权的技巧之外，享用汤的乐趣占据了突出的位置。就像引导囚犯为巴伦西亚汤点赞一样，对18世纪60年代向饥饿的巴黎人供应的经济汤的描述，不仅包括重农主义者米拉波侯爵等富人的肯定，还有对穷人自己也能愉快地食用这些土豆汤的反复强调。甚至连巴黎圣罗克教区养的弃婴也被拿来证明土豆汤的愉悦性。这些婴儿的味觉天性不会撒谎，据说和平时吃的食物相比，他们更喜欢土豆米汤。[51]

对愉悦的这种坚持，揭示了这些方案固有的政治因素。幸福一词强调的是个人的选择，而不是体制性的贫穷。伦敦慈善施粥所的一位组织者坚持说，任何事都无法"超出城市各处勤劳的穷人前来的渴望"。他们认为，这不仅证明了短暂的困苦，而且也证明了"劳动阶级的习惯已经发生了有利的变化"。[52] 法国一本同样关注自愿消费（事实上也承认这些汤并不总是受欢迎）的手册也坚持认为，在任何情况下都不应该强迫穷人喝这些汤：他们必须自愿选择这样做。[53] 选择，而不是必需，被认为是消费的主要驱动力。正如文学评论家桑德拉·谢尔曼（Sandra Sherman）所说，"问题"在于，这种选择是"一面镜子"。[54] 实际上，18世纪90年代许多施粥所的出现，反映出经济环境的不断恶化给一些欧洲国家的粮食供应带来的压力。比如，在18世纪晚期的英格兰，许多劳动人民的饮食完全不够支撑他们所承担的高强度体力劳动。在奥尔德姆（Oldham），如手工织布机织工威廉·罗伯顿（William Rowbottom）在1800年的日记中所写，高粮价使穷人陷入了"极其糟糕的境地"。很少有人能买得起"比大麦面包、大麦粥更好的东西"，甚至土豆都贵得让"穷人买不起"。结果，他们"处于非常虚弱的状态"。[55]

然而，汤的推广者们却依然认为，慈善施粥之所以成功，贫穷并不是唯一的，甚至都不是主要的原因。而是如谢尔曼所指出的那样，据说它反映了"穷人内在的、积极主动的对汤的优点的发现"。推动者们坚称，穷人想要喝土豆汤。[56]在饮食中，正如在经济学中一样，个人利益被想象成一种将个人幸福与大众幸福结合起来的有效的发动机，它消除了穷人的个人福祉与要求穷人享受土豆汤的经济及政治秩序之间的紧张关系。

亚当·斯密与土豆

或许亚当·斯密不可避免地应该特别推荐土豆。倍受推崇的斯密的《国富论》（1776），详细阐述了让合格的个人在经济事务中行使自主权所带来的公共利益。在有力地进行论证的两卷中，他通篇反复地解释：

> 个人利益与对资财的追求，通常会让他们将资本投在对社会最有利的行业上……根本不用法律的干涉，仅仅就是个人的利益和对资财的追求，就会引导人们把社会上的资本尽可能地按照对社会利害关系最合理的比例分配给国内所有不同的行业。[57]

他强调，它的发生"根本不用法律的干涉"；《国富论》的很多篇幅都在仔细剖析此类干预的有害影响。将奖励、垄断等的扭曲作用消除，会让每个人"完全按照自己的方法追求自己的利益"。结果，统治者们将被解除不可能履行的"监督、指导私人产业使之最有利于社会"的义务。[58] 由于每个人总体上都比他人更能判断自己的利益，其结果将会是一个更富有、更幸福的国家的诞生。

对于让个人去追求自身利益的好处，斯密的信心并非没有止境。大型海外贸易公司等垄断组织，以牺牲其他所有人的经济利益为代价，推进了自身的经济利益，而其缺陷在于当初允许此类机构存在的立法。斯密也并不认为每个人都能平等地认识到自己的真正利益。和其他许多人一样，他认为土地所有者通常会因为轻松的生活而不会把注意力放在经济事务上，因此他们很难判断哪些可能有利于他们的产业。[59] 占人口大多数的劳动者则缺乏有关经济事务的"必要信息"，而且即使提供了这些信息，他们也不具备理解这些信息所需的时间、教育和习惯。因此，他们在促进自身经济利益方面处于不利地位。尤为不幸的是，劳动者的福祉与社会的福祉是"紧密相连"的。斯密对"完全的无知和愚蠢"使他们的分析能力变得迟钝而感到遗憾，并主张扩大小学教育（大部分由私人资助）来解决这个问题。[60]

《国富论》尤其明确地表示，对个人利益的合理追求，是为整个政体的利益而组织贸易的一项有效战略。斯密认为，这种方法会使所有人更安逸、更舒适、更安全，因此有助于促进幸福，他认为幸福是根植于人性中的一种渴望。[61] 斯密对自由市场的理论建立在这样一个信念的基础上：只有当人们感到满意时，经济成功才有可能。他坚持认为，如果一个社会中绝大部分成员贫穷而又悲惨，那么这个社会就不可能繁荣幸福。[62] 他坚持认为，经济增长为大多数人提供了幸福而舒适的环境。增长需要大量令人愉快且营养丰富的粮食供应，而这就是土豆所能提供的。土豆不仅比小麦多产——据斯密计算，种植土豆的土地产生的营养，是种小麦的土地的三倍——而且土豆更容易耕种，至关重要的是，它是一种"宜人的、有益健康的食物"。正如他所指出的，英国"最强壮的男子和最美丽的女子"都以土豆为生。"没有食物能比它更确凿地证明其营养价值，或特别适合人类。"他总结说。[63]

斯密将个体从更多地吃土豆中获得的个人利益，与经济的更加繁荣联系在一起。如果种植土豆，耕地将养活更多的人口，"劳动者常以土豆为食"，他们的产出就会有更多盈余，给自己、地主和整体经济都带来益处。[64] 同巴黎的慈善汤推广者威廉·巴肯等许多土豆倡导者一样，斯密也认为，如果人们选择多吃土豆，其好处将会落在每个人身上。多消费土豆，在增加个人幸福的同时，也会增加国家的财富。一部英国百科全书的编者曾说，没有任何庄稼"能像土豆那样对人类的繁荣产生如此重大的影响"。[65] 其结果，将会是无处不在的幸福的增加，而它又将有助于建立更富强的国家。因此，劳动人民的饮食直接关系到整个国家的幸福和福祉。

斯密的政治经济学和农业改进模型的基础是社会组织。为了与对社会组织更广泛的理解保持一致，他并不建议人们必须种植和食用土豆，如同他也不赞成限制个人的贸易权一样。试图控制公众的饮食，与试图控制市场一样是一种误导。信奉自由市场模式的作家们，将迫使公众消费不喜欢食物的尝试描述为"独裁"，这绝非偶然。[66] 下议院的一个委员会断言，试图"用法律改变社区大部分人的食物"的想法，是"极其令人反感的"。[67] 因此，解决英国18世纪后期粮食短缺问题的这些建议，涉及的是信息和自我改善，而不是强制。谢尔曼写道，"没人'强迫'穷人改变"。[68]

与自身利益和选择的这类讨论格格不入的，是贫穷劳动者的饮食习惯在很大程度上被强制约束的现实。对食物没有太多的控制权，几乎就是贫穷的定义。由于缺钱缺时间，加上慈善机构和济贫系统提供的食物种类有限，他们每天的饮食选择都受到限制。在关于选择的话题背后，强迫的可能和贫困的现实都时隐时现。圣纳西索监狱试吃巴伦西亚学会实验配方的囚犯们，除了一号、二号、三号汤之外，也没有更多的选择余地。喝慈善汤很难被看成是一种主动的选择，但经济自组织新模式却鼓励政治作家恰恰这样来看待它。

土豆、幸福和治国之道在政治经济学语言中紧密相连。18世纪对土豆的坚持不懈的推广，成为治国新模式的一部分，这种模式的前提是：个人不服从政府的直接命令，而是服从支配人类和经济行为的自然法则。这也是亚当·斯密花了几页篇幅来颂扬它的原因。到18世纪末，如历史学家基思·特赖布（Keith Tribe）所言，"财富、自由、需求和幸福（即满足感）已经结成了一条意义链"。[69] 如何确保人们吃到最有益的食物的讨论，与政治经济学新学科更广泛的原则并行，而政治经济学转而又反映了粮食、治国之道、商业和人口管理之间的相互关系。

结论

哲学家伊曼努尔·康德在1784年的论著《什么是启蒙》中，列出了他认为能判断不成熟、蒙昧存在的三种行为。它们是：毫无疑问地接受书本上读到的东西，允许宗教权威决定个人的道德准则，允许医生决定个人的饮食。[70] 放弃饮食自主权就是放弃一部分自由。在饮食问题上，正如在良心问题上一样，康德认为，任何外部权威，当然也包括国家，都不应干涉个人决定自己最佳行动方针的权利。选择食物的能力，与独立思考信仰问题的决心一样，是智力成熟的一种要素。

康德的道德哲学建立在个人作为道德自治主体的形象上，与新政治经济学热烈讨论的原则在许多方面都存在分歧，后者把人口看作在本质上是被社会引导的消费者。在亚当·斯密看来，消费的欲望深深植根于人类的天性中，迫使个人努力获取能让他们在同伴眼中得到尊重的商品。[71] 康德所言的思想自由、开明的个人，并没有受到同一种冲动的驱使，但康德的自主观同样把选择的自由作为成熟人性的一个必要条件。他坚持认为，自由，意味着"不受他人选择的约束"，它是"因人性而属于每个人的唯一原始权利"。[72] 无论是对于呈现判断力和自我克制的自由而自主的个人的哲学观点，还是作为消费者群体的社会的经济模式，在个人选择自己行为的基本权利方面，都把吃作为最可理解的行为。决定自己的饮食构成了启蒙运动对个人自由定义的一部分。

在对于18世纪英格兰自由经济崛起的经典分析中，历史学家乔伊斯·阿普尔比（Joyce Appleby）[①] 关注的是一种内在的矛盾，它试图将人类设定为既能按自己的意愿行动并对此负责，也受到经济上自我改善的内在欲望驱使的角色。经济学作家们发现，要将这些已经互相冲突的对人性的理解，与普遍认为穷人缺乏经济理性的观点调和起来，是尤其困难的。[73] 劳动人民能否成为哲学上成熟的个人，或者理智的消费者？政治经济学家们对这两种可能性都表示怀疑。然而，当穷人面对美味可口的食物时，他们对穷人作出明智选择的能力却不那么怀疑了。在饮食方面，穷人劳动者既可以是康德式的自主个体，也可以是斯密式的消费主

[①] 乔伊斯·阿普尔比（1929—2016），当代最有成就的历史学家之一。毕业于斯坦福大学，1966年获博士学位，美国历史学会和美国历史学家组织前会长。曾任教于加州大学洛杉矶分校，其著作曾获2009年的小阿瑟·M. 施莱辛格奖。主要作品有《无情的革命：资本主义的历史》《历史想象中的自由主义与新的社会秩序：1790年代的杰弗逊式看法》《17世纪英国的经济思想与意识形态》等。——译注

体。饮食提供了一个空间，即使是没有受过良好教育的人，也能在此享受哲学家塞缪尔·弗莱施哈克尔（Samuel Fleischacker）所说的在赋予生命意义的微小、具体的活动中行使判断而产生的"适当的乐趣"。[74]

把土豆微不足道的历史与18世纪经济思想更大的轮廓相互参照，就有助于解释"我们吃什么既是自己的事，也是国家的事"这个观念的根源。今天，支持正确饮食最有说服力的论据通常是：一方面它是我们个人的自身利益，而另一方面，如果我们都照顾自己的福祉，那么社会作为一个整体也就会更好。18世纪所解释的为什么劳动人民应该吃必需的食物以确保他们对国家的功用，也是基于同样的观念。土豆倡导者强调，增加土豆消费会带来政治和经济上的好处，而同时也坚持认为，土豆是一种美味的健康营养来源，人们根本不考虑给政府带来了多少好处也可以愉快地食用。正如米拉波侯爵所言："一个秩序良好的社会的一切魅力，在于每个人都在为他人工作，同时又相信自己只是在为自己工作。"[75] 在18世纪政治经济学的新语言中，饮食、幸福和治国之道是密不可分的。

进行治理需要对当地人口的饮食习惯给予一些关注，这种观念一直伴随着18世纪晚期、19世纪在巴达维亚（Batavia）、加尔各答（Calcutta）和博特尼湾（Botahy Bay）开设商店的官僚和殖民者。土豆也在四处奔波，有时是被装在精心设计的用于运输珍贵植物标本的箱子里，有时是被塞进无名海员的裤兜里、储物柜里。第四章跟随着土豆，以及对普通人饮食习惯的争论，一路来到印度、中国等地。

注释

1. Grote, *An Examination of the Utilitarian Philosophy*, 85; Tribe, 'Henry Sidgwick',

921.

2. Taylor, 'Atomism', *Philosophical Papers*, II: 187.

3. 保姆式国家指数。由商人和自由市场主义者查尔斯·科赫（Charles Koch）资助建立的自由意志主义的卡托研究所（Cato Institute）同样认为，"在过去30年里，政府扩张的一个更令人不安的趋势是法律、法规和有约束力的法院裁决的总和，它们组成了'保姆式国家'。这些法律法规代表了政府最傲慢的一面"。Cato Institute, 'The Nanny State'.

4. Harsanyi, *Nanny State*.

5. Brownell et al., 'The Public Health and Economic Benefits of Taxing Sugar-Sweetened Beverages'.

6. 有关的提示和鼓励，参见 *Underrättelse om potatoës*; *Gentleman's Magazine and Historical Chronicle* 34 (1764), 599; Varenne de Béost, *La cuisine des pauvres*; Board of Agriculture, *Hints Respecting the Culture and the Use of Potatoes*; More, *The Cottage Cook*; Buchan, *Observations Concerning the Diet of the Common People*, 7; Doyle, *Tratado sobre el cultivo, uso y utilidades de las patatas* (1797); Wirilander, *Savon Historia*, 647–650; Drake, *Population and Society in Norway*, 55; Jones, *Provincial Development in Russia*, 154–155; Kisbán, 'The Beginnings of Potato Cultivation in Transylvania and Hungary'; Koerner, *Linnaeus*, 120, 132, 149; Gentilcore, *Italy and the Potato*; Spary, *Feeding France*, 167–186。尤西佩卡·卢克宁翻译了芬兰语材料。

7. Coveney, *Food, Morals and Meaning*; Sarah Boseley, '"Ultra-processed" Products Now Half of all UK Family Food Purchases', *Guardian*, 3 Feb. 2018.

8. Marquis de Langle, *Voyage de Figaro en Espagne*, 60; *Joseph Dombey*, ed. Hamy, xxxiv; Adam, *Practical Essays on Agriculture*, II: 20 (quote); Parmentier, *Traité sur la culture et les usages des pommes de terre, de la patate, et du topinambour*, 26; 'Substance of Sir John Sinclair's Address to the Board of Agriculture, on Tuesday the 14th July, 1795', *Communications to the Board of Agriculture*, Board of Agriculture, lxii; Buchan, *Observations Concerning the Diet of the Common People*, 31; Doyle, *Tratado sobre el cultivo, uso y utilidades de las patatas* (1797), 27; *Feuille de cultivateur*, 10 Feb. 1799; *SAA* 6 (1799), 21 Nov. 1799; Dubroca, *Conversaciones de un padre con sus hijos*, II: 74–75; Alletz, *Agronomía*, 166; Virey, 'Pomme-de-terre, ou papas des Américas', 526, 531–533, 548; Alonso de Herrera, *Agricultura general*, II: 248.

9. Zanon, *Della coltivazione, e dell'uso delle patate*, 9; Bignami, *Le patate*, 4.

10. Eden, *The State of the Poor*, I: 505.

11. Varenne de Béost, *La cuisine des pauvres*, 11; Spary, *Feeding France*, 181.

12. James Stonhouse, 'Expedients for alleviating the Distress occasioned by the present Dearness of Corn', Northampton, *Universal Magazine* 21 (1757), 269; *Bibliothèque*

physico-économique, instructive et amusante (1787), I: 201–202; Parmentier, *Les pommes de terre, considérées relativement à la santé & à l'économie*, 184–185; 'Lettre d'une fermiere de ... surdifferens moyens de nourrir les pauvres', *Bibliothèque physico-économique, instructive et amusant* (1790), I: 245; Good, *Dissertation*, 79; Delle Piane, *De' pomi di terra*, 1; [Colquhoun], *An Account of a Meat and Soup Charity*, 10–15; Cadet de Vaux et al., *Recueil de rapports*, 30; Cadet de Vaux, *L'Ami de l'économie*, 56.

13. *Magazzino toscana* 22:1 (1775), 109; Marquis de Langle, *Voyage de Figaro en Espagne*, 60; Thompson, 'Of Food, and Particularly of Feeding the Poor', *Essays, Political, Economical and Philosophical*, I: 206–207. 真正顽固的人被描述为在美食上十分盲目，以至于他们的意见无关紧要，Carta de San Lucar de Barrameda, *SAA* 3 (1798), 272 (26 Apr. 1798)。

14. Smith, *An Inquiry into the Nature and Causes of the Wealth of Nations*, I.viii.43 (quote), IV.ii.4.

15. Smith, *An Inquiry into the Nature and Causes of the Wealth of Nations*, IV.ii.9.

16. Covarrubias, *En busca del hombre útil*, 26–27.

17. Mandeville, *The Fable of the Bees*, 9; Vico, *The New Science*, 62; Montesquieu, 'De l'Espritdes loix', 1748, *Oeuvres*, III: 132; Hirschman, *The Passions and the Interests*, 10, 17; Hont, *Jealousy of Trade*; Berg and Eger, 'Introduction', 2.

18. Riquetti, *Philosophie rurale*, I: xlii–xliii; and Tribe, 'Continental Political Economy', 160.

19. Hirschman, *The Passions and the Interests*; Foucault, *Security, Territory, Population*; Vardi, *The Physiocrats and the World of the Enlightenment*; Sheehan and Wahrman, *Invisible Hands*, 249–257.

20. Persson, *Grain Markets in Europe*, 1–22.

21. Kaplan, *Bread, Politics and Political Economy*; Miller, *Mastering the Market*; Gallego, *El motín de Esquilache*; Hochstrasser, 'Physiocracy and the Politics of Laissez-Faire'.

22. Quesnay, 'Fermiers'.

23. Camporesi, *The Magic Harvest*; Jütte, *Poverty and Deviance in Early Modern Europe*, 72–78.

24. Manuscript Recipe book of W. Walker, Hayes Middlesex, c.1798–1826, 'Recipes for curesand cookery, c. 1802–1826', SL, AR297, 85–86. 也可参见 Manuscript Recipe Book, LRO, DDB acc 6685 Box 179 b. 34, third folder, small notebook; English manuscript receiptbook, McGL, Doncaster Recipes Collection, MSG 1230; Mason, *The Lady's Assistant*, 189; Varenne de Béost, *La cuisine des pauvres*; More, *The Cottage Cook*; *Ensayos de comidas económicas á la Rumford*; Hunter, *Culina Famulatrix Medicinae*, 37;

Housekeeper's Receipt Book, 59.

25. *Appendix to the Scots Magazine* (1740); also James Stonhouse, 'Expedients for alleviating the Distress occasioned by the present Dearness of Corn', Northampton, *Universal Magazine* 21 (1757); Young, *Farmer's Letters to the People of England*, 193; *Bibliothèque physico-économique, instructive et amusant* (1790), 238–249; *English Review* 24 (1794), 116; *Annals of Agriculture* 24 (1795); *Gentleman's Magazine* 65:1 (1795), 15, 393; *Weekly Entertainer*, 4 Jan. 1796, 15; *Correo mercantil de España y sus Indias*, 10 Aug. 1801, 508; *SAA* 11 (1802), 184; *Memorial literario, ó, Biblioteca periódica de ciencias y artes* 3 (1802), 35, 90–100; Cadet de Vaux, *L'Ami de l'économie*.

26. Zanon, *Della coltivazione, e dell'uso delle patate; Annals of Agriculture* 24 (1795); [Colquhoun], *An Account of a Meat and Soup Charity*; Doyle, *Tratado sobre el cultivo, uso y utilidades de las patatas* (1797), 68; Good, *Dissertation; European Magazine and London Review* (June 1800), 427 (quote); Cadet de Vaux et al., *Recueil de rapports*; Amoretti, *Della coltivazione delle patate*; *Junta Pública de la Real Sociedad Económica de Amigos del País de Valencia*, 36–37; *Ensayos de comidas económicas á la Rumford*; Demerson, 'La distribución de sopas económicas'; Redlich, 'Science and Charity'; Wells, *Wretched Faces*; Gonnella, 'L'assistenza pubblica a Trieste'; Valles Garrido, 'La distribución de sopas económicas del Conde Rumford'; Spary, *Feeding France*, esp. 32–34.

27. Grove, 'The Great El Niño of 1789–93'; Muldrew, *Food, Energy and the Creation of Industriousness*, 322.

28. 'High price of bread', Board of Agriculture, SR RASE B/XIII, MRELSC, fol. 2.

29. Thomas Wilson, Questionnaire for Clitheroe, Lancashire, 8 Nov. 1800, TNA H.O. 42/35, fol. 188.

30. Eden, *The State of the Poor*, I: 533. Wells, *Wretched Faces* 清楚地说明了对汤的厌恶。

31. Redlich, 'Science and Charity'. 当时该城人口约 2.4 万, 'Genève(commune)', *Dictionnaire Historique de la Suisse*。

32. [Lobb], *Primitive Cookery*, 4(first quote); Duhamel du Monceau, *Moyens de conserver la santé aux equipages des vaisseaux*, 153–154; and 'Report respecting bread, corn, &c. &c', 10Feb. 1800, 38 (second quote), *House of Commons Sessional Papers of the Eighteenth Century*.

33. Marquis de Chastellux, *De la félicité publique*, I: 15.

34. Tribe, *Strategies of Economic Order*; Bruni and Porta, '*Economia civile* and *pubblica felicità* in the Italian Enlightenment'; Wahnbaeck, *Luxury and Public Happiness*; McMahon, *Happiness*, 200 (first quote); Beales, 'Philosophical Kingship and Enlightened Despotism', 513 (second quote); Tribe, 'Cameralism and the Sciences of the State',

528–529; Paquette, *Enlightenment, Governance, and Reform*, 56–92; Usoz, 'Political Economyand the Creation of the Public Sphere'; Earle, 'Food, Colonialism and the Quantum of Happiness'.

35. Foucault, *Security, Territory, Population*, 327 (quote), 338–339.

36. 'Substance of Sir John Sinclair's Address to the Board of Agriculture, on Tuesday the 14th July, 1795', *Communications to the Board of Agriculture*, Board of Agriculture, lxiii.

37. 关于量化精神，参见 Foucault, *Discipline and Punish*; Cohen, *A Calculating People*, 110–112; Frangsmyr et al., eds., *The Quantifying Spirit in the Eighteenth Century*, 2 (quote).

38. Hutcheson, *Inquiry into the Original of Our Ideas of Beauty and Virtue*, 163–178; Argumossa y Gandara, *Erudicción política*, 390; Marquis de Chastellux, *De la félicité publique*, II: 97–144; Bentham, *An Introduction to the Principles of Morals and Legislation*, 26–27 (for 'felicific calculus'); McMahon, *Happiness*, 205–222.

39. Sokolow, 'Count Rumford and Late Enlightenment Science, Technology and Reform', 71.

40. Thompson, 'Of Food, and Particularly of Feeding the Poor', *Essays, Political, Economical and Philosophical*, I: 192 (emphasis as in original).

41. Thompson, 'Of Food, and Particularly of Feeding the Poor', *Essays, Political, Economical and Philosophical*, I: 193–195, 202, 210–211. 关于唾液，参见 Lémery, *Traité des aliments*, I: lxix–lxxii; Manetti, *Delle specie diverse di frumento e di pane siccome della panizzazione*, 55–60; Clericuzio, 'Chemical and Mechanical Theories of Digestion in Early Modern Medicine'.

42. Thompson, 'Of Food, and Particularly of Feeding the Poor', *Essays, Political, Economical and Philosophical*, I: 206–207 (quote), 256–257.

43. Cadet de Vaux, 'Variétés', 367–368 (quotes; emphasis as in original). 关于餐馆，参见 Spang, *The Invention of the Restaurant*。

44. 《大英图书馆》(*Bibliothèque britannique*) 广泛报道了拉姆福德的作品，例如，可参见 'Morale Politique: An Account of an Establishment, &c. ... par le Comte Rumford', *Bibliothèque britannique* 2: *Littérature* (1796), 137–182; and Letter of Abraham Joly, Geneva, 25 Nov. 1797, *Bibliothèque britannique, ou Recueil extrait des ouvrages anglaise périodiques* 6: *Sciences et Arts* (1797), 300–301。也可参见 *Ausfuhrlicher Unterricht zur Bereitung der rumfortschen Spaarsuppen*, 17; Bonaparte, 'Note Dictated ata Ministerial Council', Paris, 11 Mar. 1812, *Letters of Napoleon*, No. 239; Redlich, 'Science and Charity'; Jaffe, '"Noticia de la vida y obras del Conde de Rumford"'; Gentilcore, *Italy and the Potato*, 3–4。

45. *SAA* 7 (1800), 54–64, 71–80, 119–128, 132–138, 148–160, 184–188, 340–349,

393–400; 8 (1800), 120–128, 271–272, 369–376; 10 (1801), 54–58, 63–64; and 11 (1802), 284–287; [Thompson], *Ensayos, políticos, económicos y fisosóficos*; *Memorial literario, ó, Biblioteca periódica de ciencias y artes* 3 (1802), 35, 90–100; Abad-Zardoya, 'Arquitectos en los fogones', 657.

46. *Junta Pública de la Real Sociedad Económica de Amigos del País de Valencia*, 60.

47. *Ensayos de comidas económicas*, 15 (second quote); Demerson, 'La distribución', 123 (first quote).

48. *SAA* 10 (1801), 64 (quote); Cadet de Vaux et al., *Recueil de rapports*, 193, 199; *Memorial literario, ó, Biblioteca periódica de ciencias y artes* 3 (1802), 98; Gonnella, 'L'assistenza pubblica', 1595–1596. 那不勒斯的版本和西班牙的一样包含辣椒，都灵的则加入了栗子，*Correo mercantil de España y sus Indias*, 9 Feb. 1801, 90; Gentilcore, *Italy and the Potato*, 3–4。也可参见 [Colquhoun], *An Account of a Meat and Soup Charity*, 10。

49. *Memorial literario, ó, Biblioteca periódica de ciencias y artes* 3 (1802), 35.

50. Spary, *Feeding France*, 32.

51. Varenne de Béost, *La cuisine des pauvres*, 24–34.

52. *General Report of the Committee of Subscribers*, 9; Sherman, *Imagining Poverty*.

53. Cadet de Vaux et al., *Recueil de rapports*, 46.

54. Sherman, *Imagining Poverty*, 28–29.

55. Shammas, 'The Eighteenth-Century English Diet and Economic Change'; Wells, *Wretched Faces*, 63, 66 (quotes); Muldrew, *Food, Energy and the Creation of Industriousness*; Meredith and Oxley, 'Food and Fodder'. 当然，存在着显著的区域差异，Zylberberg, 'Fuel Prices, Regional Diets and Cooking Habits'。

56. Sherman, *Imagining Poverty*, 182–183, 192 (quote).

57. Smith, *An Inquiry into the Nature and Causes of the Wealth of Nations*, IV.vii.88.

58. Smith, *An Inquiry into the Nature and Causes of the Wealth of Nations*, IV.ix.51.

59. 格拉斯哥（Glaswegian）学者约翰·米勒（John Millar）坚称，这位有地绅士"经常不顾自己和其他人的利益"，Hirschman, *The Passions and the Interests*, 122。

60. Smith, *An Inquiry into the Nature and Causes of the Wealth of Nations*, I.xi.9 (conclusion)(first quote), V.i.50 (third quote), V.i.53, V.i.61 (second quote); Fiori, 'Individuals and Self-Interest in Adam Smith's Wealth of Nations'; and Mehta, 'Self-Interest and Other Interests'.

61. Smith, *The Theory of Moral Sentiments*, I.i.1.1. 关于18世纪对"自我组织"的理解的更广泛图景，参见 Sheehan and Wahrman, *Invisible Hands*。

62. Smith, *An Inquiry into the Nature and Causes of the Wealth of Nations*, I.viii.36.

63. Smith, An Inquiry into the Nature and Causes of the Wealth of Nations, I.xi.39. 令

人惊讶的是，这些强壮的男人、美丽的女人竟然是爱尔兰的煤炭搬运工和妓女。

64. Smith, *An Inquiry into the Nature and Causes of the Wealth of Nations*, I.xi.39. 参见 Spary, *Feeding France*; Kaplan, *The Stakes of Regulation*，它们清晰地讨论了18世纪晚期法国的类似情况。

65. Alcedo, *The Geographical and Historical Dictionary of America and the West Indies*, III: 198.

66. Turton, *An Address to the Good Sense and Candour of the People*, 85–86.

67. 'Report respecting bread, corn, &c. &c', 10 Feb. 1800, 38 (second quote), *House of Commons Sessional Papers of the Eighteenth Century*, 6.

68. Sherman, *Imagining Poverty*, 38–39, 155 (quote).

69. Tribe, *Strategies of Economic Order*, 12.

70. Kant, 'What is Enlightenment?' 也可参见 Kant, 'What Does it Mean to Orient Oneself in Thinking?'

71. 斯密认为，贸易冲动是"人类本性中的一种倾向……是所有人共有的"。他还认为，人们被一种同样深刻的渴望所驱使，希望得到同伴们的爱和尊重。正如他所解释的，"我们追求财富，避免贫穷，主要是出于对人类情感的这种关注"。Smith, *An Inquiry into the Nature and Causes of the Wealth of Nations*, I.ii.1–2 (first quote); and Smith, *The Theory of Moral Sentiments*, I.iii.2.1 (second quote). 也可参见 Hont, *Jealousy of Trade*。

72. Kant, *Metaphysics of Morals*, 30 (quote); Bielefeldt, 'Autonomy and Republicanism'; Fleischacker, *A Third Concept of Liberty*.

73. Appleby, 'Ideology and Theory', 515.

74. Fleischacker, *A Third Concept of Liberty*.

75. Riquetti, *Philosophie rurale*, I: 138; Tribe, 'Continental Political Economy', 160.

第四章　全球的土豆

是谁把土豆带到了波斯？19世纪早期，两位英国外交官为谁占有这个成就的功劳而争吵不休。哈福德·琼斯爵士（Sir Harford Jones）和约翰·马尔科姆爵士（Sir John Malcolm）同时都担任着在德黑兰的波斯国王法特赫-阿里沙（Shah Fath-Ali）的宫廷代表，在混乱的外交环境中，两人成为死敌。在他们的诸多分歧中，有一条就是他们各自声称是自己将土豆引进到该地区。两人都认为推广土豆是一项慈善事业，并争先恐后地表示对此负责。马尔科姆自豪地说，他曾煞费苦心地在伊朗各地传播关于这种"贵重蔬菜"的知识。在他的指导下，有三十多袋土豆种得到分发，他还安排发行了一本有关最佳耕作方法的专著。而让他大为光火的是，有报道称法国外交使团早在一个世纪前就已经把土豆带到了伊朗。他在日记中写道："我希望获得将土豆引进波斯的好名声。"当得知据称由法国人引进的"土豆"其实是一种不相干（且淡而无味）的冒牌货时，他甚为安心。他的好名声得到了保证。然而，琼斯却狠狠地驳斥了马尔科姆的说法。他说，"已说了很多遍了……把土豆引入波斯是居民的功劳"，然后坚持认为所有荣誉都不属于马尔科姆。"早在约翰·马尔科姆爵士来波斯之前，我就把这种植物的块茎送给了几个波斯人。"他在自己的回忆录中很是肯定。[1]

土豆，随着现代早期的贸易扩张和殖民征服走向全球，如今已成为全世界最常见的作物之一。但将它变为全球主食的过程，不仅反映了这些力量，同时也反映了它在途中遇到的各种不同的境况。在波斯，土豆于19世纪80年代已被当地农民与菠菜、大麦等更传统的作物一起"大量种植"。现在的伊朗人把它放在沙拉里吃，有时还会用土豆在米饭上面加一层脆皮。[2]不过，伊朗的厨子是否是因哈福德·琼斯和约翰·马尔科姆的活动而接受了土豆，却无法得到确定。在18世纪中期，土豆已经开始种植在哈格岛上，这座岛屿是波斯海岸边的荷兰贸易站。两位自我宣传的外交官，在烹饪方面的影响实际上是很有限的。[3]（参见波斯土豆脆皮饭食谱）

尽管他们对波斯菜系的贡献无法确定，但他们希望把引入土豆的功劳归于自己的原因是十分清楚的。从18世纪晚期开始，欧洲的外交官、传教士和殖民官员，就把在伊朗、印度、西印度群岛等具有战略利益的地区推动淀粉类主食消费的工作，当作他们施善的切实证据。在18世纪的欧洲，土豆推广被视为增加贫穷劳动者幸福感的一种方式，同样，土豆在世界范围内的传播，也成为欧洲人无私行善的极其意识形态化的一种叙事。哈福德·琼斯和约翰·马尔科姆等人公开称赞土豆是一种有益的食物，它能赶走饥饿的威胁，鼓励健康的农业实践。因此，对它的接受，是当地整体文明水平的标志。相比之下，不接受土豆则展示出一种可悲的顽固态度。安第斯土豆，已经成为彰显欧洲优越性的证据。

这样的叙事清楚地表明，通过将对推荐的饮食方式的接受，与正式或非正式的帝国主义的所谓好处及其改善和进步的承诺相联系，实际或潜在的帝国臣民的农业及饮食习惯，就这样被塞进更广泛的殖民意识形态当中。我们不应该只看到这些叙事的表面价值。18世纪欧洲对土豆

波斯土豆脆皮饭

脆皮饭是伊朗经典菜肴。在烹饪过程中，将一些大米与黄油或酸奶混合，就可以制作出"锅巴"（tahdig）即金色的脆皮。（"波斯式"电饭锅有一个特殊的设置以实现这一点。）锅巴也可以用土豆等其他原料制成。至少从8世纪开始，大米就出现在波斯菜肴中。土豆作为后来加入的食物，创造了一种新菜式，也融入了已有的饮食方式。

波斯土豆脆皮饭

6人份

2杯香精米或香糙米

2汤勺海盐

4粒豆蔻，碾碎

1汤勺玫瑰水

6汤勺融化的黄油

1茶匙磨碎的藏红花，溶于2汤勺玫瑰水中

2个中等大小的土豆，切成1/4英寸薄片

挑净大米内的小渣，将大米放在大容器内洗净，倒入温水，用手缓缓搅动，然后将水沥干。反复5次直至水完全清澈。洗净的大米煮的时候会散发出一种令人愉悦的香气，而未洗净的大米则无。洗净后的大米已可用，但还需在加了1汤勺盐的8杯水中浸泡2—24小时。用盐分充足的水浸泡和煮大米，可使米粒变结实以经受长时间烹饪并防止米粒破裂。米粒会膨胀但不互相黏结，煮好之后就是被誉为"波斯美食珍珠"的松软米饭。

在5夸脱大的不粘锅内，将10杯水用高火煮沸。加入1汤勺盐、豆蔻和玫瑰水，将洗净沥干的大米倒入锅内。

> 精米用高火快煮 6—10 分钟（糙米 15 分钟），用木勺轻轻搅动 2 次，以防止米粒粘锅。咬一下米粒，如果感觉松软且米饭已膨胀至锅顶，则已煮好。将米饭倒入细筛的滤器中，用 2—3 杯凉水淋洗。
>
> 在搅拌锅内将 4 汤勺黄油、2 汤勺水、几滴藏红花水混在一起搅拌，并在锅内抹匀，将土豆片铺在锅底。将米饭在土豆上薄薄铺一层，用手压实，再将剩余的米饭在上面堆成金字塔形状，盖上盖子用中火煮 10 分钟以制作出金黄色脆皮。
>
> 将剩余的 2 汤勺融化的黄油与 1/4 杯水混合，淋在米饭金字塔上，再将剩余的藏红花水浇在上面，将锅盖用干净的洗碗巾缠紧，以防止蒸汽泄漏。用低火煮 70 分钟。
>
> 将锅从热源上端开，置于潮湿的表面，如垫有湿洗碗巾的烤盘上，冷却 5 分钟，勿揭开盖子。这样做是为了让脆皮起离锅底。
>
> 开盖后将浅盘扣紧在锅口，将二者同时翻转，米饭倒入浅盘内。米饭呈金黄色脆皮蛋糕状。与薯条一起食用。

的大力推广，并不是土豆进入欧洲饮食的主要原因，同样，土豆在全球传播的功劳，也不能简单地归于帝国代表们的推动行为。全球化的力量和当地情况总是结合在一起的。本章概述了土豆全球传播的一些复杂因素，展示了这种新食物进入当地饮食的不同方式，以及它的重要性如何随着时间的推移而变化。土豆的意义从来就不是一成不变的，即使在它的南美故乡也是如此。

在毛泽东之后的共产主义中国，土豆重要性的变化，尤为清晰地证明了现代粮食面貌同时存在的本地性和全球性。土豆于 17 世纪传入中国，长期以来一直是偏远贫困地区农村居民的重要粮食来源，而直到最近，中国政府才注意到它的存在。不过，在过去的几十年里，它已经成为中国粮食安全计划的一个明确组成部分，中国现在是世界上最大的土豆出产国。土豆地位的转变，反映了市场改革、农村供给战略以及历史学家马克·维斯罗基（Mark Swislocki）所说的"营养治国"（旨在促进国力和经济成功的粮食政策的制定）的综合影响。[4]市场经济、日常饮食和诞生于18世纪的现代国家之间的密切关系，继续影响着当代对健康饮食和国家安全的讨论。

面包果与善行

托马斯·杰斐逊在刚独立的美国就任总统后不久，就起草了一份备忘录，他在其中评价了自己付出许多心血，为建立这个国家所作出的贡献。杰斐逊的《公共服务声明》列出了他为结束跨大西洋奴隶贸易所付出的努力、对宗教自由的支持，当然还包括脱离英国的《独立宣言》。他还将鼓励商业化种植水稻包括在内。杰斐逊为他对旱稻的支持感到自豪，因为"对任何一个国家最大的贡献，就是在其栽培中添加一种有用的庄稼"。当有用的庄稼是做面包的粮食或其他主食时，这样的成就就更值得称赞了。增加淀粉类粮食的供应是一项公共服务，可与政治人物可能取得的任何成就相媲美，至少杰斐逊是这样认为的。（杰斐逊并没有考虑到谷物将一种特别残酷的劳动制度强加给种植它的奴隶的事实，他也没有赞扬西非人，其技术知识对新世界的水稻栽培至关重要。）[5]在杰斐逊看来，水稻种植的传播，证明了像他这样的白人的善意和成就。

杰斐逊坚定地认为，传播含淀粉的新型粮食是一种重大的人道主义服务。他的同时代人也认同这一信念。一本极受欢迎的西印度群岛历史书的作者布莱恩·爱德华兹（Bryan Edwards）（他还在牙买加拥有几座种植园，共有大约1500名奴隶劳动力）也赞同道：

> 在一生的所有劳动中，如果有一种追求比其他追求更充满仁爱，更能增加世上人们的舒适，甚至能通过增加他们的生活资料来增加他们的数量，那必定是通过把对人类有益的天然产物从地球的一处移植到另一处的方式，向海外传播丰富作物的追求。

爱德华兹对英国将面包果引进加勒比地区的工作特别热心。1793年，威廉·布莱（William Bligh）船长成功地将面包果从太平洋运至西印度群岛，按照爱德华兹的评价，它标志着"英属西印度群岛历史上的一个重要时代！"[6] 面包果被引入加勒比海的历史，为殖民官员同时在其他地方传播土豆的工作提供了背景，因此，那一个更广为人知的故事，值得花一点篇幅来讲述。

布莱船长是在他的第二次尝试下，才完成了这一历史壮举。在英国政府的资助下，布莱远航至塔希提岛（Tahiti），此行的目的是搜集一种被称为"rima"或"soccus"，或英国人所称的面包果树（breadfruit）的样本。[7] 本来的打算是将这种植物引进牙买加。布莱的第一次航行，因1789年"邦蒂号"（Bounty）上的船员哗变而中断，当时他的船员控制了运输船。面包果货物受到的优越对待所引起的愤怒，是引发叛乱的重要原因。几百棵面包果树占用了船上的全部空间和水源，叛乱船员的第一个行动，就是把树苗扔到船外。布莱和几名随从被扔到一艘小船上随波漂流，船上只有很少的补给和导航设备。布莱以顽强的决心，成功地指挥这艘敞篷船返回了英国。返回之后，他产生了强烈的复仇欲望。在威廉·皮特（William Pitt）政府资助第二次面包果树远征后，他如愿以偿。这次远征，布莱完成了将面包果树运送到英属加勒比殖民地的任务,同时也找到并逮捕了活着的叛乱船员。[8]

布莱的成功得到了大西洋两岸观察家的称赞。评论家们一致认为,在返回英国的戏剧性旅途中,他表现出了非凡的毅力和专业知识,但运送面包果树,才是伟大得多的成就。这种新植物被誉为营养粮食的无尽来源,被引入加勒比海后将会改变当地的饮食习惯。按照英国和加勒比地区的作家的观点,面包果树和它的果实,"对于西印度群岛具有极大的重要性"。[9]面包果的营养价值,是它引起关注的主要原因。

事实上,英国政府决定资助不是一次而是两次耗资巨大的越洋探险,将塔希提的面包果树带到西印度群岛,与其美洲殖民地的实际粮食需求并没有多大关系。面包果如它的名字所暗示的那样,被植物学家认为是一种类似面包的淀粉类食物。可是,谁会吃这种面包的替代品呢?爱德华兹这样的种植园主,对面包果树的到来报以热烈的掌声,但他们并不想用这种异国舶来物装点自己的餐桌。岛上的奴隶劳动者也不需要额外的淀粉来源。当时的作家也承认,西印度群岛已经有丰富的淀粉类蔬菜供应,包括红薯、木薯和芋头。在殖民地种植园里劳动的奴隶,肯定会从粮食定量供应的增加、额外的蛋白质来源中受益,但他们最不需要的就是淀粉。西印度群岛的种植园主一开始通常也不关心他们的奴隶劳动力对饮食的需求。大部分耕地都被用来种植利润丰厚的出口糖料作物,而没有用来为劳动者种植粮食,奴隶们往往是自己想办法解决粮食问题。[10]出于这个原因,在1793年面包果树被引进之后的几年里,种植园主对它在饮食方面的潜力兴趣不大。1806年,圣文森特植物园的负责人说:"事实上,种植园主对它占用自己的土地很是反感,因为他们认为它是甘蔗田的入侵者,他们不喜欢甘蔗以外的任何东西。"[11]尽管种植园主们对这种树不以为然,但那位负责人还是称赞乔治三世对这件事的支持"孕育着仁爱之心",因为它同时增加了"舒适和维持生活的手段……（以及）人的幸福和人的数量"。[12] 至于有没有人吃这种无与伦比的食物，那倒不是重点。

面包果树，更多的是欧洲对其殖民地承担责任的表现和启蒙运动的象征，而不是解决供应困境的方法。帝国对面包果的兴趣来源于这些关联，而不是源于对加勒比地区粮食供应的实际关切。事实上，在资助"面包果树航行"的同一年，英国政府明确驳回了种植园主因加勒比殖民地正遭受粮食短缺而提出的要求。[13] 英国政府资助从塔希提岛运送数千棵面包果树，并不是为了解决它认为并不存在的生存危机。相反，这件事被塑造成英国利他主义的典范。布莱坚持认为，"如果一个人在一生中种十棵（面包果树），这事可能一个小时内就完成"，那么他就即刻履行了"对他自己和未来几代人的责任"。[14] 据大家说，这一营养方面的慷慨行为的受益者都拒绝这种新食物，这只是表明了他们的忘恩负义，却丝毫无损于这份使命所取得的成就。[15]（据一位种植园主说，奴隶们"冷漠地"看待面包果。[16]）不管对饮食的实际影响如何，营养丰富的面包果象征着人们的善行。

18世纪后期，这种引进植物方面的善行，在整个大英帝国各处都在重演。在"邦蒂号"远征的同一年，加尔各答植物园的创建者罗伯特·基德（Robert Kyd），也在鼓励将西米椰树从马来半岛引进印度。[17]（西米是从一种热带树的树茎中提取的淀粉类物质，可用于制作各种面包，以及像木薯粥一样的粥。）在考虑基德的提议时，英国皇家学会主席、英国著名植物学家约瑟夫·班克斯（Joseph Banks）① 很高兴地权衡了它的多重优势。班克斯认为，引进西米椰树将会为印度人提供非常有

---

① 约瑟夫·班克斯（1743—1820），英国植物学家、探险家和自然学家。1768—1771年随同詹姆斯·库克作环球考察旅行，首先提出澳大利亚有袋类哺乳动物比胎盘哺乳动物更原始的说法，还发现了许多植物新品种。1778年起任英国皇家学会会长，直至1820年去世。——译注

益健康的食物。因此，基德的计划为"政府的善意"提供了实实在在的证据。最终，得到好处而心怀感激的印度人会"敬畏他们英国征服者的名声"。更重要的是，印度人会感谢英国人把他们"从上天给他们带来的最严重的灾难——饥荒"中解救出来。[18] 班克斯从引进西米椰树到英国人的仁慈，再到缓解饥荒，最后到印度人的感激之情的简单递进，充分地反映了与推广淀粉粮食有关的各种思想。

这类计划更关心的是使特定的治理形式合法化，而不是减少饥饿。事实上，英国在印度防止饥荒的成绩极其糟糕。在班克斯洋洋自得地评论西米之前的十年里，孟加拉经历了一场毁灭性的饥荒，哪怕不是由英国的政策导致，至少也因它而加剧。这次饥荒中死亡的人数或许达1000万。在整个殖民统治时期，造成上百万人死亡的可怕饥荒继续折磨着印度人民。事实证明，英国的政策完全不足以缓解饥荒，而且在许多情况下还直接导致了饥荒。[19] 班克斯自以为是地认为，饥荒是由天意的神秘运转造成的，这种说法无比狡猾虚伪。更过分的是，基德的植物园占用了从印度农民那里没收的土地，为了给新物种让路，他们的粮食作物被铲除。西米椰树的引进更可能恶化了当地人的饮食状况。[20] 西米的倡导者并不认为这与他们声称改善印度人福祉的说法相矛盾。和面包果一样，西米椰树的重要性并不在于它对解决粮食短缺问题的实际贡献，而是在于它能让人感觉到它代表着英国对殖民地臣民的关心。在英属印度鼓励土豆种植，也同样反映了类似的信念和做法。食物的隐喻力量，积极地塑造了政治论调和殖民政策。

土豆在印度的栽培

土豆早期在印度次大陆传播的年代顺序并不比它在欧洲更清晰。当

16世纪中期欧洲人第一次在安第斯山脉遇到土豆时，葡萄牙的商船也在定期驶往印度港口，并在印度西海岸的科钦（Cochin）建立了一处殖民地。因此，辣椒、土豆和红薯等食物被从美洲运到印度的机会是十分充足的。不太清楚它们何时传入印度，但到16世纪30年代，马拉巴尔海岸（Malabar coast）已有辣椒生长，而且它很快在印度南部的菜肴中确立了地位。在随后的几十年里，腰果等美洲新奇食物也进入了当地的饮食领域。至于新世界的块茎植物，16世纪晚期欧洲一些零星的记录，提到了一种消费很广的"batatas"——在加勒比的词汇中常指的是红薯，但很可能它们实际上是山药——在印度已经早被食用。[21] 17世纪早期服务于东印度公司的专职教士爱德华·特里（Edward Terry）报告说，1615年，他在拉贾斯坦（Rajasthan）莫卧儿王朝的大臣阿萨夫汗（Asaf Khan）举办的宴会上，看到了"精心烹制的土豆"。它们究竟是山药、土豆还是红薯，还是不能确定。但是，由于1590年之前，加勒比的菠萝已经出现在莫卧儿皇帝自己的食品柜内和餐桌上，因此阿萨夫汗的土豆肯定是上述美洲块茎中的一种。[22] 第一个明确的证据出现在17世纪70年代，当时一位东印度公司的外科医生发现，在印度西海岸的卡纳塔克（Karnataka），人们经常食用这类东西。在那时，土豆在这位外科医生的家乡伦敦已经很常见，因此我们确定他的鉴定是可信的。[23] 不管土豆何时到达印度，它的传播者，无疑是那些驾驶船舶将美洲的银矿和甘蔗种植园与欧亚大陆相连的船员，以及从16世纪中期开始定居在港口城市果阿（Goa）和吉大港（Chittagong）的葡萄牙商人。因此，"新世界"块茎最先出现在印度西海岸也就不足为奇了。

18世纪英国对印度的殖民统治，导致英国作家们为这些侵略行为进行辩护，虽然他们明明知道印度是主权国家。对印度君主和印度文化的合法性进行污名化，是这类做法的一贯特点。作家们将欧洲人到来之前印度的商业和社会状况与当时的状况进行对比，他们发现有了很大的改善，他们还列举了英国"温和而理性"统治的优势。²⁴农业创新在其中占有突出地位。殖民主义思想家坚持认为，当地的农业形式比欧洲低劣，而且，坚持采用低劣农业方式的人已放弃了他们的自治权。正如约翰·洛克（John Locke）①在他的《政府论》第二篇（Second Treatise of Government）中所指出的那样，虽然上帝将世界给予人类所共有，"但不能假设上帝的意图是要使世界永远归公共所有而不加以耕植。他是把世界给予勤劳和有理性的人们利用的"。从这个角度而言，主权与某些形式的具有经济效益的农业是相关联的。因此，勤劳理性的英国人完全有理由夺取控制权。再说，据说印度人自身也从这个过程中获益，因为他们低效的农业技术被欧洲改进的技术所取代。这种"改进"的概念提供了一个框架，以展示殖民主义不仅对殖民者有利，而且对被殖民者也有利。它意味着，如后殖民时期的批评家爱德华·萨义德（Edward Said）②所说，"某些领土和人民需要并恳求被统治"。²⁵

在农业方面，印度是"可怜而无知的"，而英国在"改良和指导手段"方面是异常先进的，1820年的一份报刊如此说道。因此，这份报刊总结说，幸运的是，上天把改良印度农业的任务交给了英国，由此将"舒适和幸福"传播给了印度人民。促进"当地居民的利益和幸福"新

---

① 约翰·洛克（1632—1704），英国哲学家、医生，被广泛认为是最有影响力的启蒙思想家和"自由主义"之父，英国最早的经验主义者之一。主要著作有《政府论》《论宽容》《人类理解论》等。——译注
② 爱德华·萨义德（1935—2003），著名文学理论家与批评家，是巴勒斯坦立国运动的活跃分子，也是东方主义的提出者。他出生于巴勒斯坦，毕业于美国普林斯顿大学，取得学士学位后又在哈佛大学获得硕士和博士学位。之后多年他在哥伦比亚大学担任英语和比较文学教授，也曾执教于约翰霍普金斯大学、哈佛大学和耶鲁大学。代表著作有《世界·文本·批评家》《东方主义》《文化与帝国主义》《知识分子论》等。——译注

近被纳入了东印度公司的特许状。显然，传播英国的农业技术是履行这一义务的一种方式。[26] 毫不意外的是，除了西米椰树和面包果树之外，土豆也成为了这项慈善改革计划的一部分。18世纪后期驻扎在孟加拉的陆军专职教士威廉·坦南特（William Tennant），是诸多土豆热衷者之一。18世纪90年代，坦南特写了许多随笔，后来以《印度的娱乐》（Indian Recreations）为题发表。它除对印度教的非理性和印度地主的残忍进行批评之外，还长篇累牍地抱怨传统的农业实践。针对这些弊端，坦南特列出了印度人从殖民主义中获得的好处。其中最突出的是英国农业技术，尤其是像土豆这样的"英国"作物的引进。

　　坦南特认为土豆对印度人的福祉有巨大贡献。在他看来，单凭它对穷人的功用，就使得英国在印度的穷人当中推广食用土豆的工作，已足以弥补早期英国殖民者在美洲殖民时期可能犯下的所有罪行。在地球这一侧的殖民改良，由此便弥补了早先对地球另一侧不那么仁慈的入侵。土豆是某种帝国镇痛膏，能够减轻所有挥之不去的罪恶感。坦南特列举了这种植物的优点，并建议给予印度农民种植土豆的小块私人土地。这样他们就能够用土豆取代孟加拉的主食大米。坦南特认为，大米在当地粮食中的中心地位，是该地区易受饥荒影响的原因。坦南特强调，如果印度人多吃土豆，少吃大米，他们不仅可以从饥荒的痛苦中解脱出来，还可以减轻政府在如20年前水稻歉收时照顾他们那样的繁重义务。把接受土豆和消除饥荒联系起来，坦南特由此暗示，如果孟加拉人再次遭受18世纪70年代的致命饥荒，那只能怪他们自己。土豆，如西米和面包果一样，免去了英国对未来饥荒的责任，同时也展示了英国的粮食和农业传统的优越性。[27]

　　印度农业和园艺学会（The Agricultural and Horticultural Society of India）成立于1820年，目的是"全面改善印度的农业状况"，它同样深信，其鼓励种植土豆的工作，展示了英国对其印度臣民的善意。[28] 学会不断尝试在孟加拉推广土豆种植。它的创办章程清楚地表明，一个"农业学会所追求的目标，是*引进新型的、有用的作物*"，既包括粮食作物，也包括具有商业利益的作物。学会的第一任会长坚持认为，一些殖民者对土豆的推广，已经展现了这种抱负的功绩。他还认为，"如果许多人都朝同一个方向艰辛追求，他们共同努力，将会带来更多的成果！"[29] 为此，学会创建了一座适应性菜园，学会成员们在园内种植土豆，以及烟草、苹果、柑橘、油桃、番荔枝和牛油果。它还从欧洲进口土豆种并分发给成员，让他们试验不同肥料和土壤对"这种贵重蔬菜"产量的影响。[30]

这种绅士般慢条斯理的做法并不能满足该学会的抱负，它希望印度人也能接受这些蔬菜。学会使用与欧洲相同的方法，为"本土农民"种植土豆、豌豆、花椰菜等受欢迎的作物提供奖励。虽然这些激励措施在一定程度上是为了弥补市场上家用"优质农产品"供应"严重缺乏"的问题，但它们更大的目标是促使印度人不仅种植，而且食用土豆和花椰菜。因此，学会就印度不同宗教社区接受土豆的情况进行了调查，在发现某些村庄已经接受土豆时，他们就感到十分高兴。[31] 学会想要鼓励的是一群温顺、感恩的吃土豆的印度人。它把这项事业作为一场艰苦的斗争，对抗的是"当地人对涉及自身利益之事的无知或错误想法"，但总体而言它确信，它正在努力提高印度农村人口的幸福和福祉。对学会而言，这些鼓励土豆消费的工作，生动地证明了殖民者的善意，同时也证明了"本土农民"对明显有利于自身之事的顽固拒绝。[32]

印度农民的所谓落后，是学会反复强调的内容。"目前印度的农业状况，"第一任会长坚称，"比两百年前的英国还要原始。"田地和菜园杂乱无章，看得出对农艺毫不了解。他对印度农民的自我满足感到惋惜。他们是"习俗的奴隶"，缺乏好奇心，也没有合适的器具，而且难以理解地并不为自己的落后而烦恼。他认为，印度人生活在"对目前的悲惨境地愚蠢地感到满足的状态中"。[33] 因此，学会的成员们不能确定，他们对土豆及合适园艺技术的辛勤鼓励，是否能得到印度农民的热情欢迎，但他们毫不怀疑的是，他们的行为展现了他们对增加"普遍幸福"的真实承诺，并且因此也使英国的统治合法化。[34]

对土豆带来幸福的颂扬，忽略了英国统治对印度自给自足能力的实际影响，而这种影响在很大程度上是破坏性的。出口农业抽走了基本的主食，殖民政府疏于维护莫卧儿帝国复杂的灌溉系统。几乎没有证据表明印度人的福祉能胜过宗主国的发财致富。在灾难性的饥荒期间，政府一如既往地勤勉收税。任何关于有责任为印度人民提供英国济贫法所规定的最低生活保障的建议，都遭到殖民地官员的强烈反对。[35] 在印度，推广土豆是将饥饿和饥荒归罪于印度人的一种方法，而不是切实可行的提高粮食安全的一项计划。

学会推广土豆的实际影响，无论如何都是有限的，并且，在他们作出努力之前，土豆就已经进入了当地的饮食。18 世纪末，受英国驻军和殖民地居民需求的刺激，商业土豆种植在加尔各答和西德干（Deccan）地区发展起来。到 19 世纪初，土豆已成为当地烹饪的常见食材，被人们"贪婪而大量地"食用。在整个 19 世纪，土豆出现在印度次大陆许多地方的厨房里，出现在烹饪书籍和医疗手册上，并在某些地区成为"本土常见烹饪方式"的一部分。它的生产和销售是如何组织的，哪些阶级或种姓参与其中，这些都有待进一步研究，但如今，土豆已成为数百万印度人的日常食物，而印度已成为主要的土豆出产国。[36] 尽管殖民主义理论家努力把土豆和英国所谓的优越性联系在一起，但它还是挣脱束缚，发出了自己在当地的声音。

至 20 世纪早期，反殖民主义作家自己也开始推广土豆了，他们把推广土豆作为振兴印度人口更大计划的一部分。民族主义者也在担心，他们的同胞太多的素食会削弱他们的体质，使他们更难抵抗英国的统治。圣雄甘地（Mahatma Gandhi）记得自己小时候在学校操场上唱的歌："看那威猛的英国人，统治着弱小的印度人，因为他们吃着肉，他们身高5腕尺。"（5腕尺约为2.3米，或7.5英尺。）他的好友谢赫·梅赫塔布（Sheikh Mehtab）也认为，"我们的人民弱小，因为我们不吃肉。英国人能统治我们，是因为他们吃肉"。民族主义者根据最新的营养科学，对印度饮食中蛋白质的明显缺乏，以及白米中营养成分的有限都予以了重视。多食用土豆，为这种欠缺的饮食提供了另一种选择。当欧洲的营养学家谴责土豆的营养不如小麦面包时，班恩思瓦·辛哈（Baneswar Singha）等印度营养学家却认为土豆优于大米。因此，土豆成为通过增强印度人的体质来加强民族主义运动的更大计划的一部分。在推广土豆的同时还开展了一系列活动，以证明素食主义不是印度教的教义，并鼓励健身、棍棒格斗等。[37] 被帝国主义者视为英国优越性象征的土豆，变成了结束殖民统治的工具。

土豆也开始在宗教仪式中发挥作用。虔诚的印度教教徒实行的定期斋戒（upawasa）要求禁食大米或小麦等食物，但不禁止食用土豆、木薯、西米和其他一些曾经属于异邦的食物。这类淀粉作物如今成为禁食期间的基本膳食。[38] 近代早期的殖民主义和欧洲贸易的扩张把土豆带到了印度，但仅凭这些力量，并不能决定土豆如今在印度文化中所扮演的角色。

以土豆汤结束一天

土豆在世界其他各处的传播，依赖的也是同样的两股力量相结合的方式，一股是殖民和商业力量，为这种植物提供了到达新目的地的途径；另一股是当地境况，为接受它做好了准备。近代早期殖民主义极大地加速了粮食在全球的传播。历史学家阿尔弗雷德·克罗斯比（Alfred Crosby）①在论述他所称的"哥伦布大交换"（Columbian exchange）的开创性著作中明确指出，欧洲在美洲的殖民主义，与它所开启的全球动植物转移是不可分割的。正如克罗斯比所示，殖民者在旅行时，带着他们的庄稼和牲畜（以及疾病），并带走了他们不熟悉的动植物。这些新的植物改变了世界各地的饮食习惯，西红柿来到了意大利，辣椒来到了印度，牛来到了美洲，还有无数其他的新口味、新风味来到世界各地的人们面前。欧洲人并非这些转变的唯一推动者。当西非人自愿或被迫航行到美洲时，同样也带着许多食物。[39] 食物从世界的一处向另一处运动的驱动力，一方面是想复制家乡熟悉的和保证健康的食物的愿望，另一方面是通过种植甘蔗和咖啡等商业农作物赚钱的渴望。植物学的发展对这些商业冒险的成功起到了重要作用。适应性菜园、农艺技术的系统调查、远距离运输植物所需的海上基础设施也都作出了贡献。自然历史、经济植物学和试图在殖民地空间重建欧洲景观、作物景观和食物景观的"生态帝国主义"，共同构成了欧洲殖民主义这个庞然大机器。[40]

这些力量把土豆带到了世界各地，正如它们把甘蔗运到美洲，把茶叶从中国运到印度一样。在葡萄牙商人和水手把土豆带到西印度后，英国19世纪的殖民基础设施进一步推动了土豆在印度次大陆的传播。在孟加拉，商业土豆贸易沿着连接不同村庄的殖民铁路线发展起来。[41] 在澳大利亚，土豆是英国建立驻领殖民地的大计划的组成部分，并且它有效

---

① 阿尔弗雷德·克罗斯比（1931—2018），美国历史学家。主要研究环境史和全球史，被视为环境史的奠基人之一，致力于探索人类的历史进程。重要作品有《哥伦布大交换》等。——译注

地发挥了作用。18世纪80年代，约瑟夫·班克斯向最早的英国殖民地提供了土豆种、醋栗木、烟草和咖啡苗，这是他建立自给自足、具有商业价值的殖民地前哨计划的一部分。1788年，运面包果树的布莱同样在塔斯马尼亚岛（Tasmania）种植了土豆。到1817年，这个岛向澳大利亚的殖民者出口了近400吨土豆。[42] "园艺殖民"在把土豆带到印度洋的法国殖民地的过程中也扮演了相同的角色。[43] 土豆的传播，同样也得益于在非正式帝国主义下蓬勃发展的商业企业。波士顿阿特金斯糖业公司（Atkins Sugar Company of Boston）引以为荣的是，他们20世纪初在与哈佛大学联合建立的植物园里，种植了古巴的第一批土豆。[44]

殖民主义同样推动了土豆在北美东海岸的传播。犹他在1.2万年前就有野生土豆被食用，因此很可能在欧洲人到来之前，土豆就已经来到了大西洋沿岸。虽然如此，早期定居的欧洲人并没有注意到它们。最早的书面记载认为，它们可能是通过在加勒比海的殖民地而从欧洲传入的。[45] 在印度的英国殖民者视土豆为一种优秀的欧洲文化，同样，北美洲的殖民者也将南美洲的土豆视为他们自己的一种饮食传承。1685年，在新成为殖民地的宾夕法尼亚，有一位居民报告说，他种植了一种"爱尔兰土豆"，他希望这种土豆"明年可大幅提高产量以用于移植"。对不列颠群岛的土豆种植早已熟悉的苏格兰-爱尔兰长老会教徒，早在18世纪初就在新罕布什尔种植了土豆。英格兰、苏格兰和爱尔兰的殖民者可能是第一批种植土豆的人，但土豆很快就传播到了这些社区以外的地方。特拉华的瑞典移民种植土豆供个人消费并出售，加拿大的法裔阿卡迪亚人（Acadia）和北卡罗来纳的摩拉维亚人（Moravian）也是如此做法，纽约北部的豪德诺索尼（Haudenosaunee）（易洛魁）农民同样如此。有旅行者报告说，在芬格湖（Finger Lakes）地区塞内卡族（Senaca）的土地上，土豆种植"数量巨大，质量不等"。到18世纪中叶，土豆已成为人们熟悉的商业作物,它们在大西洋沿岸被交易,并与其他商品一样受到监管。[46]

大量的证据表明,到18世纪,土豆已经成为东部沿海地区许多人的普通食物。在七年战争(1756—1763)期间,在阿迪朗达克山脉(Adirondacks)作战的一位新英格兰的造船工在他的日记中写道,"土豆汤"标志着他们"一天的结束"。在美国独立战争中,参战双方的士兵都把土豆作为口粮每周食用。年轻的女士们仔细地将土豆酵母和土豆面包的食谱,连同椰子布丁、杏仁奶酪蛋糕的做法,治疗百日咳的方法,以及"绝妙"的查尔斯顿夫人腌牛肉的食谱,一起抄进她们的食谱笔记中。1796年,一位自我标榜的"美国人"出版了这个新建国家的第一本印刷烹饪食谱,书中包括土豆填火鸡和玫瑰水香味土豆布丁的食谱。它的作者阿米莉亚·西蒙斯(Amelia Simmons)形容它们已得到"普遍食用"。1790年,本杰明·富兰克林的孙子受邀在宾夕法尼亚大学的毕业典礼上发表演讲,他选择了土豆作为主题。他称,土豆不仅是无与伦比的食物,而且还把"红白相间的美"赐予人们的脸庞。[47]如在英属印度一样,安第斯的土豆成为了关于欧洲所谓优越性的种族化叙事的一部分。

土豆在非洲各地的传播,同样也极大地归功于殖民者、殖民官员和传教士的活动。在斯瓦希里语(Swahili)中,土豆的意思大致可以翻译为"欧洲根"。殖民主义者认为,推广土豆有助于欧洲统治的合法化,这种信念一直持续到20世纪。在布基纳法索的法国官员要求当地人种植土豆等外国蔬菜,把它作为一项更大的改造计划的一部分,这项计划旨在用据称更先进的欧洲技术取代西非的种植方式。1918年,英属肯尼亚的本地人理事会(Local Native Council),作为自身"改良"计划的一部分,也同样对土豆种进行分发。[48]在肯尼亚,这些鼓励土豆种植的工作是在白人定居者占用土地、强制推行新耕作方式的背景下发生的,这样的做法颠覆了已有的性别规范,破坏了非洲农业维持的生态平衡。不管殖民地官员的说辞如何,这些做法几乎不能被形容为进步。这些欧洲"改良"计划的强制性从口述历史中清晰地显现出来。"让我告诉你吧",一位刚果老妪在2017年回忆道:

> 我出生时是比利时人在掌权。农学家会来给人丈量一块田,他会丈量几公里,或许是15公里,然后说:"你把它犁出来种土豆。"你如果不完成就会被抓起来。[49]

和在印度一样,殖民政府反复地实施加重饥荒的政策,而同时,他们却将随后的粮食短缺归咎于"本地人的无知"和落后的农业实践。[50]

在任何地方,对土豆和其他"欧洲"蔬菜的接受,都被视为欧洲化的推动力和标志。1826年,《殖民时代和塔斯马尼亚广告报》(*Colonial Times and Tasmanian Advertiser*)认为,如果塔斯马尼亚人种植土豆和小麦,他们很快就会摒弃"四处漂泊的性格,养成些许的勤劳习惯,这是文明的第一步"。[51] 殖民时期的旅行者看到基督教皈依者的菜园里有土豆的踪影,他们就会赞不绝口。在他们看来,这些菜园往往比未皈依者打理的菜园更整洁、更漂亮。[52] 相反的是,非欧洲人的栽培习惯则揭示了在接受文明方面更大的失败。在伊朗,19世纪英国对卡扎尔宫廷的一项指控是:

> 土豆通常都矮小得可怜,这是园丁的懒惰造成的,他们把土豆种几乎就埋在地表面,而不费心去挖个洞。波斯盛产野生芦笋,但本土的植株比培植的植株低劣,正如霍屯督人(Hottentot)比不上

有教养的欧洲人一样。[53]

土豆和芦笋成为人的代表,他们的文明性在菜园里体现得淋漓尽致。

尽管这些报道可能有些泼皮和片面,但它们还是让我们得以一窥土豆进入当地饮食的方式。德黑兰的园艺师们显然在19世纪60年代就已开始种植土豆,并且已经发展出了自己的种植技术,不管这些技术在英国人看来有多不完善。在18世纪80年代欧洲水手种植土豆的新西兰,土豆很快进入了毛利人的农业中,既是粮食又是商品。作为粮食,它成为当地的主食红薯的补充。但它和红薯不同,红薯已经根植在先前存在的宗教规矩的禁忌中,它的耕种受到限制。而土豆缺乏宗教意义,因此可在更广泛的环境中种植。因此,土豆成为村庄农业中广受欢迎的新成员,并迅速成为一种重要粮食,在毛利人生活中经常出现的节日集会里被大量食用。它们也获得了经济上的重要性。它们和猪一起被用作货币,用以交易欧洲人的火枪和其他铁制品。欧洲人对土豆的渴求使得毛利人能够以此换取他们喜欢的商品。到19世纪早期,毛利农民一直在以商业规模种植土豆,专门用于国际贸易。欧洲人的探险航行和殖民行为将土豆引进了新西兰,而毛利农民将它带来的贸易和外交机会进行了积极的经营。[54](参见图12)

总之,土豆的世界之旅,毫无疑问是因欧洲在近代早期开展长途贸易和殖民计划能力的提高而开始的。然而,认为土豆只是由约翰·马尔科姆或布莱船长等个人"引进"的观点,几乎未能抓住土豆能融入某些环境,却不能融入另一些环境的复杂过程。这样的过程总是很具体的,这是土豆没以相同的速度在全球传播的原因。一种特定植物的植物学特征——它们对水的需求、产量、抗虫害能力、贮藏性以及其他性质——

图 12　毛利人在吃土豆。拜英国的帆船所赐,土豆到达了新西兰。但它们在毛利人社会中的角色却是由当地人决定的。毛利人很快将土豆种植发展成一种商业活动,使他们能够与途经的欧洲船只进行贸易。他们还将土豆融入了自己的公共宴乐文化。尽管土豆来自遥远的地方,但画中的男子沃特金尼(Watikini)并没有把土豆当成舶来品。

是植物转移史的重要组成部分。而新植物适应当地饮食方式和所有土地、社会组织和贸易体制的程度,也同样如此。19 世纪 10 年代,苏格兰传教士约翰·坎贝尔(John Campbell)在穿越西开普(Western Cape)的旅行中,惊讶地遇到了一个身穿欧洲服装的南非人。他指着一排果树、土豆、卷心菜和其他蔬菜自豪地宣称:"这座房子是我的!还有整个菜园!"[55] 也许这位自豪的菜园主是从当地的教会获得的土豆,但他已经把它们变成自己的了。

用土豆来让天平保持平衡

在秘鲁,与安第斯土豆有关的含义尤其多样化。启蒙运动的力量,同在欧洲一样,也塑造了秘鲁等西班牙殖民帝国前哨地区的政治文化。自 18 世纪下半叶起,植物园、天文台、矿业学校、小学教育改革、街道照明运动等现代事业也出现在整个南半球。有教养的学者们起草了改良主义论著，目的是在殖民社会的各个方面实现现代化；天主教牧师试图清除荒谬民俗中的宗教仪式和巴洛克式的奢侈。各地的殖民官员和当地的哲学家，都在从事一种微妙的平衡工作，既要颂扬人类理性的力量，又要保留甚至加固现有的社会、种族和经济阶层。拉丁美洲的启蒙运动与欧洲启蒙运动相比，既有许多相同点，也有其本身的鲜明特点。[56]

在秘鲁，一群大部分居住在相当于帝王首都的利马的富有知识分子，决定成立一个组织，以促进家乡的繁荣和"文明"，从而确立他们对启蒙价值观的认同。利马之友学术学会（The Academic Society of the Friends of Lima）于1790年成立，并在接下来的四年里定期举行会议。它受到欧洲、美洲其他地方成立的许多经济学会的启发，并和它们一样具有明确的爱国主张。它的主要活动是出版了一份报纸《秘鲁水星报》（*Peruvian Mercury*），它主要讨论文学和哲学、政治经济学、科学和商业发展、当地社会，尤其包括促进秘鲁经济发展的方法。其成员热情信奉实用主义和理性主义，并把自己当成启蒙运动的代言人。[57]

学会一直关注的问题之一，是反击欧洲对西班牙及其殖民地的批评。西班牙在欧洲启蒙哲学家当中声誉不佳，被普遍认为是一潭蒙昧而迷信的中世纪死水。西班牙的帝国冒险也没有获得肯定的评价，它的征服被广泛视为导致数百万美洲印第安人死亡的一系列屠杀行为。在1783年法国举办的、邀请投稿者探讨人类在西班牙对美洲的殖民统治中有何收获的征文比赛中，这种观点得到了有力的印证，舆论的一致意见是毫无收获。这种态度激怒了西班牙人和西班牙裔世界的殖民者，他们质疑某些批评的准确性，并提出了另一种说法，强调他们所认为的西班牙帝国主义带来的大量好处。在这样的辩护中，向土著社会传播福音占据了显著的位置，但辩护文章的作者并没有把自己局限在精神问题上。⁵⁸ 对秘鲁学会而言，土豆也成为一种有力的反证。

学会清楚地意识到土豆在 18 世纪欧洲的崇高声誉，其成员紧随欧洲人探讨人口、政治经济和农业发展之间的关系，也探讨这些观点是否适用于他们的家乡。受过教育的秘鲁人明确地肯定了土豆供养欧洲，从而促进人口增长和经济进步的贡献。西班牙的作家也作出了同样的举动，坚称西班牙通过土豆为这个不领情的世界提供了营养。他们强调说，是西班牙人把土豆带回了欧洲，而他们特别气愤的是，这一成就却被归功于沃尔特·罗利（Walter Raleigh）①和弗朗西斯·德雷克（Francis Drake）②。于这些作家而言，土豆证明了西班牙对世界福祉的贡献，从而证明了西班牙殖民大帝国的合法性。⁵⁹ 以土豆为基础的帝国宣传引发了反西班牙说客的愤怒，他们谴责这种以炫耀土豆来掩饰征服暴行的企图。发起了 1783 年征文比赛的里昂科学院（Scientific Academy of Lyon）谴责说："你们何以敢用胭脂虫（和）土豆来让天平保持平衡？"⁶⁰

而在同一时刻，受到启蒙观念影响的秘鲁人对吃土豆是极其热衷的。长期以来，土豆与玉米、藜麦等蔬菜，一直是安第斯山脉食物景观的重要组成部分，而移民们很快将这种不熟悉的块茎与土著居民联系起来。在欧洲最早的描述中曾提到，土豆是"印第安人吃的某种食物"。⁶¹ 年代史编者和旅行者一致强调，土豆在当地人的饮食中占据着中心地位，并详细描述了土豆的吃法和制作"*chuño*"的新奇方法。"*chuño*"是一种可以长期保存的冻土豆，是该地区的主食。殖民地作家认为，这些食物是"*印第安人的面包*"。⁶²

---

① 沃尔特·罗利（1552—1618），英国冒险家、历史学家、诗人。北卡殖民地的奠基者，美国北卡罗来纳州首府罗利即是以他命名。——译注
② 弗朗西斯·德雷克（1540—1596）是英国著名的私掠船船长、航海家，也是伊丽莎白时代的政治家。德雷克在 1577 年和 1580 年进行了两次环球航行。——译注

而且，殖民地作家们十分清楚的是，土豆在安第斯宗教中也很重要。在安第斯人的宇宙论中，人体、植物和宇宙深深交织在一起，土豆是由大地之母帕查玛玛（Pachamama）的女儿、土豆之母阿萨玛玛（Axamama）精心呵护的。在沦为殖民地之前，土豆通常作为祭品被献给神灵——既是礼物也是餐食。天主教对"偶像崇拜"的长期调查显示，在殖民统治建立之后，土豆仍是秘密宗教仪式的重要组成部分。这些调查记录了土豆、古柯叶、羊驼脂肪、豚鼠等与土著世界有紧密联系的物品。这些联系一直存在于社区记忆和持续的农业实践中。在殖民政权成立70多年后，安第斯年代史编者费利佩·瓜曼·波马·德阿亚拉（Felipe Guaman Poma de Ayala）在其著作中，完善地描述了前印加帝国的土豆种植、收获与宗教仪式之间的联系。从烹饪和宇宙哲学两者的角度而言，安第斯土豆都属于美洲印第安人世界。[63]

在秘鲁殖民时期的等级体系文化中，这几乎算不上声誉意识很强的移民对土豆作为一种食物的认可。一代代移民和他们的后代，都反复强调保持欧洲饮食的必要性，并详细说明了像印第安人一样吃东西所存在的风险。但无疑，安第斯殖民地的现实生活意味着，土豆实际上已经深深地渗透到了移民阶层的饮食方式中。就连在阿雷基帕（Arequipa）的圣特蕾莎和圣卡塔琳娜修道院养尊处优的修女们，也经常食用土豆，她们的账簿已经显露了这种状况。殖民地的饮食习惯不可避免的是现实和愿望之间的妥协。[64]然而，由于土豆与不受重视的土著世界的关联，它缺少了在欧洲所拥有的光芒。《水星报》关于土豆的讨论，反映了它与贫穷、落后的关联。在字里行间，"卑""微"是与土豆联系最紧密的两个字。[65]

更糟的是，一些秘鲁作家怀疑土豆实际上不利于健康。利马人何塞·曼纽尔·达瓦洛斯（José Manuel Davalos）获得了蒙彼利埃大学

（University of Montpellier）医学学位，在学位论文中，他将该市居民普遍患有的疾病归咎于他们过度食用当地食物，尤其是加了大量调料的猪肉、木薯和土豆。达瓦洛斯引用林奈的话，指出土豆"是一种真正的茄属植物，因此很容易判断它有问题。如果经常食用，即使很少的量也会带来害处[66]"。《水星报》同样发布了警告：旅客过度食用土豆，尤其是在食用辛辣食物和饮用烈酒的同时食用土豆，会造成致命后果。最好的情况下，旅客可能会消化不良，但更有可能发生的是致命的痢疾。[67] 秘鲁学会的创始人之一希波利托·乌纳努埃（Hipólito Unanue）医生认为，利马穷人吃的这些胀肚子的块根会在胃里发酸变质，所以最好不要吃。[68] 秘鲁的爱国学者团体不但不推广食用土豆，反而进行劝阻。在受到启蒙观念影响的秘鲁人的论述中，土豆的传播证明了西班牙给人类的贡献，但声名显赫的它，与美洲印第安人每天食用的平凡块茎却毫不相干。

当地农民的农业知识使安第斯山脉的土豆如此丰富，但他们却没有从 18 世纪对土豆的颂扬中得到任何好处。很少有欧洲和美洲土豆热衷者认为他们的土豆有终结饥饿、传播幸福的功劳，而土著居民却明白，土豆更常代表着他们自己的耻辱地位。1819 年，秘鲁城市阿雷基帕发生了一起有关乱扔垃圾的争吵，一名愤怒的居民对洛伦扎·埃斯库德罗（Lorenza Escudero）吼道，应该把她"像一袋土豆一样绑在羊驼身上"驮出城。他的嘲讽，巧妙地将这位土著女人与土著特有的食物联系在一起，这种侮辱深深地刺痛了她，令她在法庭上对此大加抱怨。[69] 土豆，与吃土豆的土著居民、驮土豆的羊驼一样，都属于卑微之物。而与此同时，土豆不仅仅是一种从属地位的象征，它也是一种重要的商业作物。土著和非土著的农民都在种植土豆，人们在安第斯山脉、太平洋沿岸交易土豆，它为那些有能力进行大规模贸易的人带来了可观的利润。[70]

土豆既是一种有价值的商品，又是日常食物的来源，也是低人一等的标志，同时又是西班牙对人类福祉作出贡献的证明。

土豆在秘鲁殖民时期所拥有的复杂含义，在塑造现代世界饮食的全球互动中颇为典型。帝国主义和贸易的全球进程，与新政治语言的出现，总是有着它们在当地的表现。这类当地的表现，绝不仅仅是别处发生的事件的回音，而且还具备自己的动力和内在逻辑。对地点加以考虑，有助于理解构成我们世界的这些多重的、有争议的经历。分析土豆在 18 世纪全球范围所具备的多种含义，为我们理解现代化所同时具备的流动与本地化的本质、饮食在现代化的成果中扮演的重要角色，都提供了一种具体的方法。[71]

欢乐的土豆家族

在中国，土豆一直被看成是穷人的粮食。它们可能是在 17 世纪，通过与在福建或菲律宾的葡萄牙商人的接触而进入中国的。不过，也有一些学者认为是经由西北省份陕西而被引入的。另一些美洲的食物，如花生、红薯和玉米，到达中国的时间甚至更早，也是从不同的途径而来。土豆引进后，似乎主要被贫穷山村的居民所食用。土豆吸引乡下人有几个原因。在 1750 年到 1850 年间，中国人口增加了一倍多。与此同时，丝绸和茶叶等非粮食作物的产量稳步增长，这些作物大量占用了种植主食所需的优质土地。农民被迫搬到缺乏充足灌溉的边缘土地上，土地面积往往太小，无法维持家庭生计。因此，能以更少的土地产出更多营养的新粮食是具备吸引力的。在中国的一些地区，红薯和玉米可以起到这种作用。而在一些北方省份，起到这种作用的则是土豆。19 世纪早期陕西有报告提到，"高山上生活的人以这种作物为他们的主食"。[72] 德国地质学家费迪南德·冯·里希特霍芬（Ferdinand von Richthofen）曾在19世纪六七十年代游历过中国北方，他发现富裕的人们羞于食用土豆。土豆使得人们能够在气候不适宜种植其他作物的山区定居，因此，它和玉米一样，对试图将部分土地改种棉花等经济作物谋生的贫穷农民特别具有吸引力。来自人口稠密地区、定居在北部省份陡峭的边缘土地上的移民，也认为土豆很有吸引力，尤其是，它只需要少数人，比如移民一家的成员，便可耕种。一位历史学家认为，它们与当地饮食的结合是"一场自下而上的革命"。[73]

在这段时间长河里，中国政府对土豆并没有表现出太大的兴趣。清王朝的统治从17世纪中期持续到1912年，它对粮食供应很关心。长期以来，"供养人民"一直是负责任的治国之道的核心内容。从公元前5世纪开始，儒家学者和官员就已经认识到政治稳定与充足的粮食供应之间的密切联系。不仅如此，中国政府还制定了令人印象深刻和有效的战略来防止饥荒。政府官员监控大米等谷物的供应情况，并鼓励建立覆盖广泛的粮仓网络以储备粮食，以便在粮食短缺时进行分发。他们开辟新的土地用于定居，资助大规模的灌溉工程，在饥荒时期分发救济物资。政府还努力传播推荐的农业技术知识，保证粮食供应。到了清代，这些历史久远的粮仓为数百万人口提供了生计。[74]

不过，清朝官员并未鼓励普通人改变日常饮食习惯，清朝的医师和政治家只关注富人过度奢侈饮食带来的危险。过度奢侈的饮食，削弱了他们认为确保中国成功最必不可少的精英阶层的活力和他们的身体。有效的治国要求皇帝和皇室能够摄入适当营养，从而为全国的贵族家庭树立榜样。它不需要监督农民、工匠或其他劳动人民的特定饮食习惯。这些人的饮食在政治上是无足轻重的。[75]

民族主义在20世纪起初几十年里的发展，伴随着国家在确保公共卫生方面角色的重新定义。全球营养科学的发展，在中国形成了关于经济、政治实力与劳动阶层饮食关系的争论。许多民族主义科学家和技术官员都在西方接受过教育，他们更强调，仅仅养活老百姓是不够的。劳动人民需要营养均衡、健康的饮食。这一点开始被看成经济成功的重要组成部分。因此，营养研究机构被建立、新的期刊问世，科学家和政治人物愈加坚信，中国要成为强国，必须极大地改变大众的饮食。由于这一时期，中国粮食供应的营养性实际上是在恶化，因此这些观点变得更加突出。大米机械化去皮降低了白米的价格，使城市人口能够比以前更多地消费这种高档但缺乏营养的粮食。在民国初期，由于新的政治、科学观念的出现，加之营养粮食的供应确实在恶化，穷人的饮食习惯成为治国必须面对的问题。[76]

　　1949年中国共产党开始治国，粮食营养与国家进步之间的关系更加紧密。历史学家西格丽德·施马尔泽（Sigrid Schmalzer，也译为舒喜乐）追溯了这个共产主义国家对现代农业方法的研究和传播所给予的长期的支持。现代农业方法被认为对经济发展至关重要。化学和生物虫害防治、施肥技术、水稻等主食高产杂交品种的培育等，都得到了政府的支持。雄心勃勃的项目旨在将受过教育的科学家（他们在技术上受过培训，但在意识形态上受到怀疑）与拥有实际专业知识的"老农"结合起来。在"文化大革命"期间，百万城市青年被下放到农村，他们也同样被寄予期望，参与发展新农业技术、使中国能够自给自足的全国性运动。[77]这些做法以20世纪式的对科学农业的关注，来灌输中国长期以来秉承的治国需"给予人民营养"的观念。第六章将对此进行更详细的探讨。

　　重点首先被放在提高谷物的产量上。在农村的试验推广工作以培育杂交水稻和高粱新品种为主。中国农学家对红薯和土豆的新品种进行了试验——到 20 世纪 60 年代，大约有 30 家研究机构专门研究土豆——但块茎作物并没有受到谷物那样的关注，几百年来，谷物都一直是地位最高、分布最广的粮食。[78] 由于水稻在研究和传播项目中的主导地位，依赖红薯等块茎作物的村民有时会感到不满，他们催促当地干部对这种地位低下的粮食进行研究。土豆种植区的农民尽量不理会国家要求他们种植更多谷物的指示。面对小麦的生产配额，村民们可能会继续种植土豆，但随后会将收获的土豆转化为他们认为等量的谷物。村干部会将土豆计入谷物产量向地方政府上报，以试图在继续种植土豆的同时也能完成配额。[79] 因而，土豆是当地的一种重要资源，但不是国家确保粮食安全总体战略的一部分。

相反的是，土豆等块茎作物在某种程度上让当地人得以避开政府的管辖。因"大跃进"期间的政策等而导致的 1959—1961 年的可怕饥荒中，当数百万人挨饿或死于疾病时，土豆却提供了活下去的机会，其原因，恰恰是它没受到政府官员的重视。尽管土豆在一些北方地区的饮食中占了很大一部分，但它们在统计上被归类为作为补充的蔬菜，而不是主食。饥荒期间，中国政府征收主要粮食，但不要蔬菜，因此也不要土豆。在北大荒地区的口述历史中，土豆在身心方面的重要性被口口相传。种植土豆的村庄没人死亡。来自这样一座村庄的一位冯姓农民回忆说，他和其他当地人之所以能挺过来，是因为"有很多土豆"。[80] 尽管幸存者心存感激，但土豆仍因饥荒的记忆而受到玷污。在今天的中国，"我吃着土豆长大"的意思是"我家里很穷"。经历过饥荒的人，对吃土豆至今心有余悸。[81]

这种情况对中国政府目前鼓励土豆消费的雄心构成了挑战。自 20 世纪 90 年代初以来，中国一直是世界上最大的农作物出产国，种植的作物占全球的 22% 左右，且土豆和红薯现在都被视为中国总体粮食安全计划的重要组成部分。土豆消费还被视为通过提供优质营养来提高"人口质量"的一种方式。2013年，农业部开始制定一项战略，将土豆从一种蔬菜，转变为像大米、小麦和玉米一样的主食。目前，中国有100多家研究所从事土豆育种研究，国家大力发展大规模的土豆农场，一方面供应新鲜的土豆，一方面也为更多的加工食品，如土豆粉等提供原料。[82]

中国政府在支持研究和鼓励商业化农业的同时，还出台了鼓励中国消费者多吃土豆的计划。工业化经济所特有的所谓营养转变，正在改变这个国家的饮食习惯。[83]与过去几十年相比，由于中国快速的城市化和富裕化，人们对肉类、乳制品、新鲜水果和大米的食用量增加了很多。目前，这些产品的生产规模都不足以满足新的需求。政府官员希望通过将对这类食物的消费转移到土豆上的方式，来满足全国的食物需求。到2015年，它已成为官方的政策。国务院总理李克强宣布了一项在全国范围内增加土豆消费的计划。农业部官员肯定土豆是"理想的现代食物"，是健康营养的来源，是环境成本更高的谷物的完美替代品。[84]为了实现将土豆转变为中国第四大主食的目标，官员们试图改变土豆作为低地位穷人食物的名声。人们不再把土豆与农村的饥饿联系在一起，取而代之的是一种将土豆与走向富裕的消费者联系在一起的新概念。

"快乐的土豆家族"活动完美地体现了这一雄心。在一则引人注目的广告中，两个被称为"土豆弟弟"和"土豆妹妹"的卡通土豆和一位衣着华丽、被称为"土豆姐姐"的女性在一起欢跃着，三人开心地笑着，对他们奇特的跨物种家庭感到泰然自若。（参见图13）

"土豆姐姐"是广告中的女子冯小燕的艺名。她自称是农民，来自山西，中国历来吃土豆的大省之一。她通过热情赞美土豆的宣传歌曲而成为全国明星。她和她的卡通弟弟妹妹一起出现，代表着把土豆与个人财富——而不是与顺从政府要求或与可怕的饥饿——相连的一种不断的努力。广告上的"土豪"指的是暴发户，发音和土豆很像。"和我们一起做土豪吧"，或许也可以理解成"和我们一起像土豆一样金光闪闪吧"。[85]

除了这样的广告，官方电视节目还介绍土豆食谱，鼓励公众讨论制作最佳的土豆菜肴。这一切都是将土豆重塑为一种健康生活方式之选择的努力。[86] 在联合国"国际土豆年"的2008年，中国出版了一本食谱，书中明确指出，吃土豆对健康有多种益处。作者介绍道：土豆，

> 营养成分齐全、维生素丰富、质地柔软，是理想的婴儿食品。有报道说新鲜的土豆泥可以外用治疗骨折。鲜榨土豆汁很有助于控制……便秘、胃溃疡、胃酸过多、十二指肠溃疡、鼻窦炎。土豆对癌症患者有益……它也被认为是一种抗衰老食品。[87]

图13 "快乐的土豆家族"。这则广告中的中国歌手冯小燕靠推广土豆获得职业的成功。中国政府计划到2020年将土豆产量翻一番。这个"快乐的土豆家族"提出土豆是走向富裕生活方式的一部分，希望鼓励消费者吃更多的土豆。

137 这一系列惊人的特性，都会给个人消费者带来益处。

并非巧合的是，土豆是个人幸福源泉的这种设计，与中国政府拥抱市场经济的做法不期而遇。1976年后，中国领导人开始重新调整经济，放宽中央计划控制，允许市场推动力发挥更大的作用。与此同时，政治话语也发生了变化，如今强调的是自身利益和个人财富的创造，而不是像以前那样只强调集体利益。尽管毛泽东时期对"为人民服务"的提倡并没有消失，但个人财富的增值和个人的自我发展已成为普遍现象。[88]同时期的土豆推广就是在这类价值观的框架内构建的。在今天的中国，正如18世纪晚期的欧洲一样，土豆被根本性地定义为保证国家粮食安全的一种手段，但它是作为个人健康和福祉的源泉而得到推广的。个人利益和公共利益，在饮食的理想自由市场状态中和谐运行。在今天的中国，土豆是以市场为基础的个人与国家观的组成部分。

结论

在全球范围内，土豆在过去500年的历史，是与殖民主义和海外贸易的力量密切相关的。它们将欧洲人和他们丰富的食物种类推向了世界各地。对于欧洲的外交官、传教士和官员而言，种植和食用土豆，是他们试图灌输的一大堆欧洲做法的一部分。在印度尼西亚，1824年出版的《英属印度及其属地之亚洲期刊及月刊》(*Asiatic Journal and Monthly Register for British India and its Dependencies*)报道说：以前不为人知的土豆

> 在斯坦福德·莱佛士爵士（Sir Stamford Raffles）治下于前几年引进苏门答腊岛，取得了可喜的成功。它们产量兴旺，不仅可

与其他地方的任何食物匹敌，而且现已成为居民饮食的重要组成部分。

在斯里兰卡中部，土豆的种植同样喜获成功。该刊物报道说，"也许这次实验中最令人高兴的是"当地人也开始种植土豆了，这是取得进步的真正迹象。[89] 土豆早已完全变成与欧洲人相关的东西，斯里兰卡人对它的接受，至少在殖民官员看来，体现了殖民主义的好处。

殖民主义、海外贸易虽然对土豆在全球的传播很重要，但它们，与莱佛士爵士一样，都不能解释土豆如何、为何能融入斯里兰卡等地的日常饮食。无论一种新引进的食物被多么热烈地推广，它本身都不足以刺激当地人接受它。历史学家威廉·贝纳特（William Beinart）和凯伦·米德尔顿（Karen Middleton）分析了影响不同植物物种全球发展轨迹的多种因素。[90] 除了植物学家和官员的活动外，每一种植物的植物学特征，还有普通人的行为，也能起到决定性的作用。乘坐火车穿越中西部的妇女、非洲的徒步者、被奴役的西非人以及其他许多人，在种子和植物在世界各地的传播中也起到了帮助作用。贝纳特和米德尔顿在研究了许多植物（包括玉米和刺梨）在全球的传播后，得出的结论是，这些进程很难被归纳，因为对于任何物种的接受，当地的境况都非常重要。

就土豆而言，决定它能被纳入特定膳食制度的因素，只能通过对当地境况的关注来解释。对于殖民和外交代表而言，种植和食用土豆是他们试图灌输欧洲实践的尝试之一。对于德黑兰的园丁、新西兰的毛利农民、孟加拉的村民和中国北方的农民而言，土豆却服务于其他的目的。在某些地区，土豆已扎下了深厚的根基，而在另一些地区，它们仍然只是大米等主食的补充。土地的可得性、土豆作为贸易商品的商业潜力，以及它逃避政府管制的能力，都影响着土豆在当地饮食中的渗透。

139　　　　　土豆的全球史也展示了殖民空间对于表达治理新理念的重要性。土豆在殖民地的起源，是西班牙将自己表现为一个现代、文明国家的一种努力。而土豆的推广也帮助在印度的官员保证了启蒙的目标，也就是他们声称的巩固英国殖民主义。改良语言深深渗入殖民话语，因此，当地人的农业和饮食习惯，既可以被解读为文明在不断发展，也可以被解读为野蛮在毫不妥协。国家层面对日常饮食习惯的关注，在欧洲和欧洲的殖民空间都有所显现，因为它们并不是彼此隔绝的区域。

中国政府将土豆重新定位为理想主食的工作，是18世纪欧洲历史的一种再现。在当时的欧洲，一种普通民众食用的乏味粮食，也同样引起了关心人口增长的政治家们的注意。中国与欧洲一样，人们对土豆的兴趣与市场经济的兴起不期而遇，而市场经济将消费者的选择视为个人能动性的重要展示。中国政府希望人们食用更多的土豆，但也希望人们是为了提高自己的福祉而作出这样的选择。如今，鼓励民众合理饮食，已成为人们期望现代政府所具备的一种功能，同样，人们也期望这种鼓励能够围绕个人利益的说法而构建。中国的土豆支持者强调更多的土豆消费能带来经济和健康方面的回报，这并非巧合，因为经济自由化也带来了对个人消费，包括食物消费的新态度。[91]

第五章探讨的是经济和日常饮食习惯之间的关系，并显示出土豆怎样为讨论资本主义带来的深刻社会、经济变革提供了一种途径。

注释

1. Kaye, *The Life and Correspondence of Major-General Sir John Malcolm*, II: 48 (first quote); Jones, *Account of the Transactions of his Majesty's Mission to the Court of Persia*, I: vii–viii (second quote); Binning, *Journal of Two Years' Travel in Persia Ceylon, Etc.*, II: 87–88.

2. Wills, *The Land of the Lion and the Sun*, 170 (quote), 174, 300; Matthee, 'Patterns of Food Consumption in Early Modern Iran'.

3. Hamilton, *A New Account of the East Indies*, I: 92; Wills, *The Land of the Lion and the Sun*, 174, 300; Matthee, 'Patterns of Food Consumption in Early Modern Iran'.

4. Swislocki, 'Nutritional Governmentality'.

5. Jefferson, 'Summary of Public Service', 已改为现代拼写。关于水稻，参见 Dusinberre, *Them Dark Days*; Carney, *Black Rice*; Eltis et al., 'Agency and Diaspora in Atlantic History'; Carney and Rosomoff, *In the Shadow of Slavery*; 'AHR Exchange: The Question of Black Rice'。

6. Edwards, *The History, Civil and Commercial, of the British Colonies in the West Indies*, I: 13–14.

7. 这种树结出的果实很大，烘烤或煮熟后会变得"又软又白，就像新烤的面包内部一样"，Ellis, *A Description of the Mangostan and the Bread-Fruit*, 11。

8. Bligh, *A Voyage to the South Sea*; Dening, *Mr Bligh's Bad Language*; Spary and White, 'Food of Paradise'.

9. Hinton East to Joseph Banks, Kingston, 19 July 1784, Banks, *The Indian and Pacific Correspondence of Sir Joseph Banks*, II: 62–63 (quote); Daniel Solander to John Ellis, London, 4 May 1776, Solander, *Daniel Solander*, 363–364; Ellis, *A Description of the Mangostan and the Bread-Fruit*, 11, 13.

10. West-India Planter, *Remarks on the Evidence Delivered on the Petition*; Parry, 'Plantations and Provision Grounds'; Marshall, 'Provision Ground and Plantation Labor in Four Windward Islands'; Tobin, *Colonizing Nature*.

11. Guilding, *An Account of the Botanic Garden*, 32. 也可参见 Alexander Anderson to Joseph Banks, St Vincent, 3 June 1793, James Wiles to Joseph Banks, Kingston, 14 Dec. 1793, and Alexander Anderson to Joseph Banks, St Vincent, 30 Mar. 1796, all in Banks, *The Indian and Pacific Correspondence of Sir Joseph Banks*, IV: 132, 181, 371–372; Thomas Dancer to Samuel More, Bath, Jamaica, 20 July 1794, RSA, PR/MC/104/10/240。

12. Guilding, *An Account of the Botanic Garden*, 12 (quote); De Loughrey, 'Globalizing the Routes of Breadfruit'.

13. Sheridan, 'The Crisis of Slave Subsistence'.

14. Bligh, *A Voyage to the South Sea*, 12.

15. Collins, *Practical Rules for the Management and Medical Treatment of the Negro Slaves*, 111–112; Sheridan, 'Captain Bligh, the Breadfruit and the Botanic Gardens of Jamaica'; Earle, 'Food, Colonialism and the Quantum of Happiness'.

16. Stewart, *An Account of Jamaica and its Inhabitants*, 98.

17. Proceedings of the governor general … relative to the establishment of a botanical

garden in Calcutta, 1786, BL, IOR H/Misc/799; Robert Kyd to the Court of Directors, 1786, Banks, *The Indian and Pacific Correspondence of Sir Joseph Banks*, II: 113–116; Mackay, *In the Wake of Cook*, 177; Drayton, *Nature's Government*, 118–120.

18. Joseph Banks to Henry Dundas, 1787, Banks, *The Indian and Pacific Correspondence of Sir Joseph Banks*, II: 205.

19. Arnold, 'Hunger in the Garden of Plenty'; Davis, *Late Victorian Holocausts*.

20. Proceedings of the governor general and council relative to the establishment of a botanical garden in Calcutta, 1786, BL, IOR H/Misc/799, 59–169; *Transactions of the Agricultural and Horticultural Society of India*, I: 8, 31, 130, 220, 225.

21. Linschoten, *The Voyage of John Huyghen van Linschoten*, II: 42, 279; Collingham, *Curry*, 52–73.

22. Terry, *A Voyage to East-India*, 195–197; Mazumdar, 'The Impact of New World Food Crops'.

23. Fryer, *A New Account of East India and Persia*, II: 76; Watt, *Dictionary of the Economic Products of India*, VI: 266; Raj, *Relocating Modern Science*, 42–43.

24. Tennant, 'On the Culture of the Potatoe', 1797, *Indian Recreations*, I: iv.

25. Locke, *Second Treatise of Government*, 20 (first quote); Said, *Culture and Imperialism*, 8 (second quote, emphasis as in original); Drayton, *Nature's Government*, 229. 关于"改进"的话语，参见 Drayton, *Nature's Government*; Arnold, 'Agriculture and "Improvement" in Early Colonial India'。

26. 'On the Agriculture of India', *Friend of India* 1 (1820), 42, 46; Porter, *Religion versus Empire?*, 74.

27. Tennant, *Indian Recreations*, I: 45–51.

28. Temple, 'The Agri-Horticultural Society of India', 341 (quote), 342–350.

29. William Carey, 'Prospectus of an Agricultural and Horticultural Society in India', 1820, *Transactions of the Agricultural and Horticultural Society of India*, I: 211–221 (216, 219 quotes, emphasis as in original).

30. *Transactions of the Agricultural and Horticultural Society of India*, II: 23, 27, 36, 81, 175 (quote), 253, 264, 265.

31. *Transactions of the Agricultural and Horticultural Society of India*, I: 21, 235 (quote), 237–238, II: 30ff., 81, 253, 264, 265; Temple, 'The Agri-Horticultural Society of India', 356–358.

32. *Transactions of the Agricultural and Horticultural Society of India*, I: 8; Temple, 'The Agri-Horticultural Society of India', 354–358; Ray, *Culinary Culture in Colonial India*. 关于善意，参见 Tobin, *Colonizing Nature*。

33. William Carey, Prospectus of an Agricultural and Horticultural Society in

India, 1820, *Transactions of the Agricultural and Horticultural Society of India*, I: 214; 'Introductory discourse, delivered by the president, 21 Sept. 1824', *Transactions of the Agricultural and Horticultural Society of India*, I: 3, 8 (quotes).

34. Proceedings of the governor general and council relative to the establishment of a botanical garden in Calcutta, 1786, BL, IOR H/Misc/799, 59–169; *Transactions of the Agricultural and Horticultural Society of India*, I: 8, 31, 130, 220, 225.

35. Davis, *Late Victorian Holocausts* 介绍了殖民统治对印度粮食安全的影响，发人深省。

36. Joseph Banks to the Court of Directors, 1789, and George Sinclair, 1798, both in Banks, *The Indian and Pacific Correspondence of Sir Joseph Banks*, II: 396, V: 24; Bazar prices in Calcutta, June 1791, Robert Kyd papers, BL, MSS EUR/F95/2, 193b; Tennant, *Indian Recreations*, II: 153–154 (first quote); Minute of William Bentinck, 12 Nov. 1803, BL, IOR F/4/179, 5r–6r; Benjamin Heyne to William Bentinck, Bangalore 21 Jan. 1805, BL, IOR P/242/73, 684–688; Whitelaw Ainslie, 'Treatise on the edible vegetables of India', 12 Sept. 1810, BL, IOR/F/4/379/9495; Arnold, 'Agriculture and "Improvement" in Early Colonial India', 519; Mazumdar, 'The Impact of New World Food Crops'; Lang, *Notes of a Potato Watcher*, 31–33; Narayanan, 'Cultures of Food and Gastronomy in Mughal and Post-Mughal India', 91–92 (second quote), 119–131.

37. Roy, 'Meat-Eating, Masculinity, and Renunciation in India', 66 (second quote); Gandhi, *Affective Communities*; Roy, 'A Dietetics of Virile Emergency', 258–259 (first quote); Ray, *Culinary Culture in Colonial India*.

38. Shrikant Botre, personal communication, Oct. 2018.

39. Crosby, *The Columbian Exchange*; Crosby, *Ecological Imperialism*; Foster and Cordell, eds., *Chillies to Chocolate*; Mazumdar, 'The Impact of New World Food Crops'; Carney, *Black Rice*; McCann, *Maize and Grace*; Earle, *The Body of the Conquistador*; Carney and Rosomoff, *In the Shadow of Slavery*.

40. Galloway, 'Agricultural Reform and the Enlightenment in Late Colonial Brazil'; Mackay, *In the Wake of Cook*; Osborne, *Nature, the Exotic, and the Science of French Colonialism*; Miller and Reill, eds., *Visions of Empire*; McClellan III and Regourd, 'The Colonial Machine'; Drayton, *Nature's Government*; Spary, *Utopia's Garden*; Touchet, *Botanique & Colonisation*; Castro-Gómez, *La hybris del punto cero*; Schiebinger and Swan, eds., *Colonial Botany*; Vos, 'Natural History and the Pursuit of Empire'; Schiebinger, *Plants and Empire*; McClellan III, *Colonialism and Science*; Bleichmar, *Visible Empire*; Jonsson, *Enlightenment's Frontier*.

41. Watt, *Dictionary of the Economic Products of India*, VI: 3, 266–272.

42. Frost, 'The Antipodean Exchange'; Bligh, *A Voyage to the South Sea*, 49; Malte-

Brun, *Universal Geography*, 3: 551.

43. Cunningham, 'Part of Two Letters to the Publisher from Mr James Cunningham', 1203; Saint Pierre, *A Voyage to the Isle of France*, 131; *Asiatic Journal and Monthly Register for British India and its Dependencies* 18 (1824), 113, 388, 608; 22 (1826), 575; *The South African Magazine* 3 (1869), 263; Dyer, *The West Coast of Africa as Seen from the Deck of a Man-of-War*, 61, 122; Richthofen, *Ferdinand von Richthofen's Tagebücher aus China*, II: 123–139; Beinart and Middleton, 'Plant Transfers in Historical Perspective', 14 (quote).

44. McCook, *States of Nature*, 59.

45. [Butler], *The History of the Bermudaes*, 30; Hughes, *The American Physitian*, 14; Browne, *Civil and Natural History of Jamaica*, 175; Lunan, *Hortus Jamaicensis*, II: 92–93; Parry, 'Plantations and Provision Grounds', 14.

46. Penn, 'A Further Account', 74; *New London County, Selectmen, Agreeable to an Act of Assembly of the State of Connecticut, for Regulating the Prices of Labour; London Gazette*, 2 June 1778, 3; 5 July 1783, 1; Wilson, 'Americans Learn to Grow the Irish Potato' (quote 345–346); Coates, *The Metamorphosis of Landscape and Community in Early Quebec*.

47. Glaiser, 'French and Indian War Diary of Benjamin Glaiser', 84 (second quote); Washington, *The Writings of George Washington*, IV: 180; Sarah Fayerweather, Manuscript cookbook, 26 June 1764, SL; Anonymous Cookbook, c.1780?, SL; William Bache, 'Oration on the History, Culture and Qualities of the Potatoe. Delivered at the Publick Commencement in the University of Pennsylvania, on the 8th of July, 1790', *Massachusetts Spy*, 23 Dec. 1790; Mrs Mathew [Belinda] Clarkson, manuscript receipt book, 1793, Marion King Schlefer Recipe Collection, SL (first quote); Simmons, *American Cookery*, 10 (quote); Wilson, 'Americans Learn to Grow the Irish Potato'; Mt. Pleasant, 'The Paradoxes of Plows and Productivity', 473.

48. Ray, *Culinary Culture in Colonial India*, 45; Mackenzie, 'Contested Ground', 701.

49. Josephine, interviewed by her granddaughter Iranga Tcheko, 2017, personal communication.

50. Worboys, 'The Discovery of Colonial Malnutrition between the Wars', 222–223 (quote); Davis, *Late Victorian Holocausts*.

51. *Colonial Times and Tasmanian Advertiser*, 1 Dec. 1826.

52. Smith and Choules, *Origin and History of Missions*, II: 301; Laufer, *American Plant Migration, part 1*, 89.

53. Eastwick, *Journal of a Diplomate's (sic) Three Years' Residence in Persia*, I: 258.

54. Petrie, *Chefs of Industry*.

55. Smith and Choules, *Origin and History of Missions*, I: 438 (quote); Beinart and Middleton, 'Plant Transfers in Historical Perspective'. 关于土豆在非洲的不均衡分布，可以参考 McCann, *Stirring the Pot*; FAO, 'Potato World: Africa'。

56. Viqueira Albán, *Propriety and Permissiveness in Bourbon Mexico*; Adelman, *Republic of Capital*; Lafuente, 'Enlightenment in an Imperial Context'; Voekel, *Alone Before God*; Silva, *La ilustración en el Virreinato de Nueva Granada*; Castro-Gómez, *La hybris del punto cero*; Nieto Olarte, *Orden Natural y orden social*; Safier, *Measuring the World*; Bleichmar et al., eds., *Science in the Spanish and Portuguese Empires*; Meléndez, *Deviant and Useful Citizens*, 45–46; Soule, *The Bishop's Utopia*; Paquette, ed., *Enlightened Reform*.

57. Zeta Quinde, *El pensamiento ilustrado en el Mercurio peruano*.

58. Gerbi, *The Dispute of the New World*; Pagden, *Spanish Imperialism and the Political Imagination*; Brading, *The First America*; Cañizares-Esguerra, *How to Write the History of the New World*. 关于获奖文章，参见 Caradonna, *The Enlightenment in Practice*, 156–159。

59. Bowles, *Introducción a la historia natural y de la geografía física de España*, 230–231; 'Explicación de la voz batata para incluir en un diccionario de la lengua', *Memorial literario, instructivo y curioso de la corte de Madrid* 21 (1790), 362; Francisco González Laguna, 'Memoria de las plants extrañas que se cultivan en Lima introducidasen los últimos 30 años hasta el de 1794', MP 11 (1794), 10 June 1794 and 13 June 1794, 163, 165–177; Doyle, *Tratado sobre el cultivo, uso y utilidades de las patatas o papas, corregido y considerablemente aumentado* (1804), 105; *SAA*, 30 Mar. 1797, 28 Mar. 1805, 4 and 11 July 1805, 20 Feb. and 6 Mar., 1806; Gómez de Ortega, *Elementos teóricos-prácticos de agricultura*, II: 137; Estéban Boutalou, 'Memoria sobre las patatas', *SAA* 19 (1806); Alonso de Herrera, *Agricultura general*, II: 248.

60. *Coup d'oeil sur les quatre concours*, 14. 胭脂虫（Cochineal）是一种源自中美洲甲虫的红色染料。

61. Santo Thomas, *Grammatica*, 159v.

62. Andagoya, 'Relación que da el Adelantado de Andagoya de las tierras y provincias que abajo se hará mención', 1545, *Pascual de Andagoya*, 138–139; Cieza de León, *Parte primera de la chrónica del Perú*, book 1, chap. 40; 'Descripción y relación de la Provincia de los Yauyos', 1586, 'Descripción de la tierra del repartimiento de San Francisco de Atunrucanay Laramanti', 1586, and 'Relación de la Provincia de los Collaguas', all in *Relaciones geográficas de las Indias*, ed. Jiménez de la Espada, I: 156, 234, 586; Acosta, *Natural and Moral History of the Indies*, 148, 201–202; Molina, *Compendio de la historia*

civil del reyno de Chile, 120–121, 213; Naranjo Vargas, 'La comida andina antes del encuentro'.

63. Molina, *Relación de las fábulas y ritos de los incas*, 62–63; Acosta, *Natural and Moral History*, 262; García, *Orígen de los indios del Nuevo Mundo*, 59–169; Guaman Poma de Ayala, 'El primer nueva corónica y buen gobierno', esp. chaps. 11, 12, 37; Spalding, *Huarochirí*, 63, 262; Silverblatt, *Moon, Sun, and Witches*; Weismantel, *Food, Gender and Poverty in the Ecuadorian Andes*; Salomon and Urioste, *The Huarochirí Manuscript*, 54–73, 120, 131; Harrison, *Signs, Songs and Memory in the Andes*.

64. Apuntes de los gastos del Monasterio de Santa Tereza de Jesús ... desde 19 de abril de 1794, año 2; and Sobre seglares de Santa Catalina, y alimentos de religiosas, refectorio, etc., 1 Mar. 1796; both in AAA, Santa Teresa legajo 2, and Santa Catalina legajo 8 respectively; Leon Pinelo, *Question moral si el chocolate quebranta el ayuno elesiástico*, 57, 63; Pilcher, *¡Que Vivan los Tamales!*; Earle, *The Body of the Conquistador*.

65. 'Idea general del Perú', MP 1 (1791); 'La Province of Caxatambo' and 'Descripción de la Provincia de Chachapoyas'; both in MP 5 (1792), 6, 190, 194, 225; Carrió de la Vandera, *El lazarillo de ciegos caminantes desde Buenos Aires*, 301, 335; Joseph Ignacio Lequanda, 'Descripción de Caxamarca', MP 10 (1794), 202; Francisco López, 'Descripción de Porco', MP 11 (1794), 19, 29–30, 37–38, 63, 68–72, 77, 87–89, 92.

66. Davalos, *De morbis nonnullis Limae, grassantibus ipsorumque therapeia*, 11–12. 达瓦洛斯被迫在法国入学，因为秘鲁的大学不会接受像他这样的被归类为混血儿的学生。丹尼斯·兰迪斯（Dennis Landis）翻译了拉丁文。

67. Panacio Montano, 'Medicina práctica', MP 1 (1791), 45–47.

68. Unanue, *Observaciones sobre el clima del Lima*, 153. 关于胀气（wind），也可参见 Valle y Caviedes, 'Defensas que hace un ventoso al pedo', *Obra completa*, 277。

69. Chambers, 'Little Middle Ground', 45.

70. Francisco López, 'Descripción de Porco', *MP* 11 (1794), 101–102; *MP* 11 (1794), 106; Manuel Espinavete López, 'Descripción de la Provincia de Abancay', *MP* 12 (1795), 131, 137, 145–146, 156–157; Spalding, *Huarochirí*, 134, 163, 196–197; Haitin, 'Prices, the Lima Market, and the Cultural Crisis of the Late Eighteenth Century in Peru'.

71. Withers, *Placing the Enlightenment*, 14 (quote); Conrad, 'The Enlightenment in Global History'; Conrad, *What is Global History?*, 76; Premo, *The Enlightenment on Trial*.

72. Laufer, *American Plant Migration, part 1*, 74; Jia, 'Weather Shocks, Sweet Potatoes and Peasant Revolts'.

73. Struys, *Drie aanmerkelyke en zeer rampspoedige Reizen*, 58; Richthofen, *Ferdinand von Richthofen's Tagebücher aus China*, II: 117, 156, 165, 174, 230, 244; Davies, *Yün-nan*, 233; Laufer, *American Plant Migration, part 1*, 69–79; Ho, 'The

Introduction of American Food Plants into China', 191–201; Murray, 'New World Food Crops in China', 320–368 (346 quote); Gitomer, *Potato and Sweetpotato in China*, 7–12; Lo and Barrett, 'Cooking up Fine Remedies', 416.

74. Will and Wong, *Nourish the People*; Bray, *Technology and Gender*, 36–37; Bray, 'Chinese Literati and the Transmission of Technological Knowledge'.

75. Swislocki, 'Nutritional Governmentality'.

76. Lee, 'Taste in Numbers'; Swislocki, 'Nutritional Governmentality'.

77. Schmalzer, *Red Revolution, Green Revolution*.

78. Jansky et al., 'Potato Production and Breeding in China'. 蒋介石国民政府也资助了土豆研究，参见 Van de Ven, *War and Nationalism in China*, 260–262。

79. Gitomer, *Potato and Sweetpotato in China*, 21; Schmalzer, *Red Revolution, Green Revolution*, 88, 114.

80. Gitomer, *Potato and Sweetpotato in China*; Xiaoping Sun, personal communication, 29 Sept. 2017.

81. 例如，可参见 Klein, 'Connecting with the Countryside?', 121; Ingebretson, 'The *Tuhao* and the Bureaucrat', 246。

82. Gitomer, *Potato and Sweetpotato in China*; Reader, *Potato*, 267–278; Jansky et al., 'Potato Production and Breeding in China'; China News Service, 'China to Boost Potato Cultivation'. Jing, 'Introduction' discusses 'population quality'.

83. Garnett and Wilkes, *Appetite for Change*.

84. Jansky et al., 'Potato Production and Breeding in China'; Ministry of Agriculture and Rural Affairs, 'China to Position Potato as Staple Food'; China News Service, 'China to Boost Potato Cultivation'; Ying, 'The Great Potato Debate'; 'Vice Minister of the Party at Ministry of Agriculture Emphasizes: Strengthen Potato Industry'.

85. Sala, 'Tudou for the Tuhao'. 安妮·格里森（Anne Gerritsen）、黄璐（Huang Lu）和克莱尔·唐（Claire Tang）帮助翻译了这句口号。

86. Chen, 'Pushing the Potato'.

87. Qu and Xie, eds., *How the Chinese Eat Potatoes*, 23.

88. 例如，可参见 Wang, 'The Post-Communist Personality'; Yan, 'The Chinese Path to Individualization'。

89. *Asiatic Journal and Monthly Register for British India and its Dependencies* 18 (1824), 113, 389.

90. Beinart and Middleton, 'Plant Transfers in Historical Perspective'.

91. Trentmann, *Empire of Things*.

第五章　资本主义的土豆

1800年12月,在法国革命引发的动荡中,《伦敦公报》(*London Gazette*)刊登了一封据报道是寄给约克郡韦克菲尔德(Wakefield)镇治安法官的信。信中警告说,如果穷人的需要没得到解决,"这座城镇很快就会成为废墟"。信中还写道:"如果你真的只给我们四分之一品脱的一条面包的定量,那小心你的小命儿,该死的乔治三世和比利·皮特(Billy Pitt)也会他娘的永远变成定量的一份,你的红鲱鱼土豆、你自己、与它有关的,都去死吧。"[1] 针对君主和首相的威胁受到高度重视,忧心忡忡的政府悬赏重金并最终将写信者抓获。

18世纪晚期对土豆的大力推广,赋予了土豆重要的意义。土豆一直被视为幸福、个人福祉和健康的源泉而被推荐给穷人劳动者,它与市场经济的发展、社会组织旧模式的解体密切相关。韦克菲尔德的写信人选择土豆为攻击的对象,他(她)借助土豆认识到并否定他(她)和土豆都身处其中的政治、经济模式。

在整个19世纪,土豆已经深深扎根于对资本主义优点的争论中。对于对新的治理模式持怀疑态度的人而言,土豆只不过是一种剥削工具。1830年在肯特郡,有一条写着"我们不能靠土豆生活"的横幅在风中飘扬。² 靠土豆生活，是以简单的形式表达他们在新制造业经济中生活更加艰难，处在只能吃土豆和无休止劳动的危险中的感受。

到了 19 世纪 20 年代，认为大量充满活力的人口能确保商业繁荣的 18 世纪的信念逐渐消失。托马斯·罗伯特·马尔萨斯对灾难性人口增长的悲观看法，把这种转变囊括在政治经济学的认识中。此后，关于土豆对国家福祉贡献能力的悲观情绪开始蔓延。土豆非但不会促进贸易和经济交流，且反而会成为现代化的障碍，因为它恰恰养活了资本主义旨在消灭的那部分人口。吃土豆的爱尔兰农民代表着那个被淘汰的世界。在 19 世纪，爱尔兰的悲惨例子，被许多欧洲国家的作家们用作土豆扼杀企业、工业和经济实力的证明。吃土豆的农民，此时从经济成功国家的榜样变成了一种障碍。

欧洲各国转而关注城市无产阶级的粮食需求。产业工人的数量不断增多，如今他们的饮食状况至关重要。城市居民营养不良、社会动荡和国家效率之间的关系，引起了关心民族工业实力的科学家、政治家的担忧。包括热量、蛋白质和脂肪等的营养科学的新语言，将普通人的饮食习惯与国家的经济福祉紧密联系起来，而劳动人民的饮食习惯，再一次被认定为经济成功的障碍。然而，此刻的问题是他们吃土豆太多，而不是太少。专家认为，大量食用土豆是低效率员工的饮食习惯。从这些不经济的、营养低劣的土豆，到韦克菲尔德写信人诅咒的土豆和红鲱鱼，谈论土豆，是将个人的生活与市场经济的发展紧密相连并具体化的一种途径，也是对它的影响表达意见的一种途径。简而言之，谈论土豆，为谈论资本主义提供了一种途径。

该死的土豆

正如韦克菲尔德的写信人所说的那样,到19世纪初,英国穷人把土豆与市场向资本主义和非个人的转变联系在一起,这是一种不受欢迎的转变。土豆汤等"湿软食物"曾被坚持不懈地推荐给穷人,它们成为政治经济学新模式的一部分,这种模式正在取代义务与家长作风相结合的旧纽带。一些学者认为,正是出于这个原因,英国的劳动人民断然拒绝吃土豆。[3] 实际上,19世纪早期的英国劳动人民很乐意吃土豆。他们在菜园里种植土豆、在市场上购买土豆,并以各种各样的方式烹饪土豆。在粮食骚乱时期,劳动人民将土豆与肉类、黄油等合意的食物一起进行储备,由此可见他们对土豆的喜爱。在1800年5月的一次起义中,伯明翰的群众没收了土豆仓库中库存的土豆,然后以他们认为合理的价格卖掉。同年,曼彻斯特也发生了同样的暴动,一些人在土豆市场的冲突中被踩伤。饥饿的人们如果在富有邻居的菜园里找到了土豆,也会把它们吃掉。1801年,德文郡的土地所有者决定,由于"这个教区内的不少菜园、耕地和果园里的卷心菜、胡萝卜、土豆、萝卜经常遭到抢劫",他们将建立一个基金来起诉违法者。[4] 简而言之,土豆是劳动阶层饮食的重要组成部分,它和其他食物一起被食用。如1800年在英格兰东南部城市刘易斯(Lewes)贴出的一封匿名信中所写,土豆与面包、奶酪、黄油和肉是"穷人想买的五种基本食物"。[5]

许多劳动者反对的,是以吃土豆代替他们的主食小麦面包的建议。韦克菲尔德的写信人对土豆和红鲱鱼的苛责,是对精英政治家和自封的慈善家的热情尝试的回应,他们不仅仅把这类食物作为健康饮食的补充,而且还作为面包的替代品来进行推广。

18世纪许多土豆推广者的目标,正是说服穷人劳动者以食用土豆来代替面包。18世纪人口的增长,与治国新理念一起,推动着人们寻找更高产的作物来替代作为欧洲普通饮食基础的低产量谷物。土豆成为优势明显的领头羊。倡导者们声称,土豆是做面包的谷物的完美替代品,它们甚至能以面包的形式被食用。将低麸质的土豆制成酵母面包的挑战,吸引了许多国家好奇的家庭实验者和训练有素的科学家。自18世纪50年代以来,声称能产生良好效果的食谱就已经流传开来,而且倡导者们一直持续地推荐土豆面包,认为它既美味又有营养。同样,有科学头脑的人也在尝试制造土豆淀粉,它可以用来代替小麦淀粉来给假发扑粉,他们还尝试把土豆进行蒸馏,做成白兰地或烧酒,同样也是为了减少谷物供应的压力。[6]

到18世纪90年代,断断续续的推荐被有组织的推广所取代。随着军事冲突席卷欧洲大陆,加之数年的歉收使得粮食供应减少,寻找替代方案成为了政治要务。在英国,几个世纪以来对商业生产面包成分进行规定的面包法于1795年被修改,允许将土豆、黑麦、大麦、燕麦和玉米添加进面包。[7] 由政府资助的农业委员会成立了"土豆委员会",致力于持续关注含土豆及大麦等廉价谷物的混合面包的生产。委员会聘请了几位伦敦的面包师,对以商业规模生产这类面包进行可行性实验。"这种探索可能带来的公共利益,"他们表示,"将不需任何建议而立即被呈给委员会。"[8] 出于同样的担忧,其他欧洲国家也启动了类似的项目。[9]

贵族家庭被劝诫要以身作则,禁止白面包出现在餐桌上。吃全麦面包或由谷物与土豆混合制成的面包,成为了一种爱国行为。1795年,农学家、政治家谢菲尔德勋爵(Lord Sheffield)在苏塞克斯(Sussex)的庄园里自豪地宣告:"我禁止我们家目前使用任何(其他)面粉。"[10]

这些行为的背后是这样一种假设：既然下层阶级自然会模仿富人的行为，那么看到贵族吃混合面包，就足以鼓励每个人都效仿。"所有下层阶级都会明白。他们会钦佩、崇敬……他们中的许多人会逐渐模仿。"保守党政治家奥克兰勋爵（Lord Auckland）如此认为。历史学家比阿特丽斯·韦伯（Beatrice Webb）和西德尼·韦伯（Sidney Webb）夫妇在1795年指出："我们发现几乎所有的公共机构，无论是教区委员会、法律委员会、镇议会还是季审法庭，直到枢密院和议会本身，都通过正式决议自愿同意限制个人小麦消费。"政治宣传家汉娜·莫尔（Hannah More）的说教小说中的虚构人物们也是这样做的。[11]1795年发表在《泰晤士报》上的一封信很好地总结了给"富人"的建议：树立好榜样，减少奢侈食物的消费，多吃全谷物、蔬菜汤、燕麦和土豆。[12]

然而，这些劝诫的真正目标，是劳动人民和穷人。面包是普通民众饮食的构成部分，富人们担心，面包的缺乏会引发威胁到韦克菲尔德的那种动乱。海峡对岸的事件让他们倍觉担忧。这些担忧并非不合理，歉收导致小麦的价格在1794年到1795年间翻了一番，劳动人民发现最普通的大麦面包和奶酪也愈发买不起了。土豆往往不在考虑之列。在英国的许多地方，有组织的团体拦截粮食运输，从商人那里征用食品，因为这样可以使它们以更低的价格出售。自1795年4月起，纵火等骚乱事件也持续增加。[13]1795年致信《泰晤士报》的作者特别提到了对这些问题的担心。他力劝穷人当心法国革命的可怕榜样，不要暴动，厉行"节约"，接受以燕麦、大麦和土豆为基础的饮食。首相威廉·皮特也提出了类似的建议，他向议会保证土豆和玉米面包"既美味又营养"，并把它们推荐给全国人民。[14]（参见图14）

很多证据表明，穷人并不愿吃这样的面包。失望的上流社会人士报告了全国各地穷人的不服从行为，认为他们"口味太精细"而不愿吃

图 14 讽刺漫画表现的受推荐的"面包替代品"。威廉·皮特和他的大臣们在一间房里举行宴会,墙上贴着警告即将发生饥荒的通知。一张表上列出了一些他们可能考虑的面包替代品:鹿肉、烤牛肉、海龟汤、香槟。这些贵客面前放着一小篮土豆,上面写着"慈善施舍用的土豆面包"。漫画讽刺富人鼓励穷人在小麦短缺时出于爱国而吃土豆。

混合面包。[15] 牧师威廉·布茨(William Butts)哀叹说,格莱姆斯福德(Glemsford)郡萨福克(Suffolk)村劳动者的"挑剔",令他们拒绝白面包之外的一切。[16] 埃塞克斯(Essex)郡的农民谴责了坚持吃小麦面粉的农户"铺张浪费"的行为。[17] 斯托克波特(Stockport)的律师霍兰德·沃森(Holland Watson)在文章中抱怨道:

> 为了低价向穷人分发燕麦片和其他食品,一份非常慷慨的认购协议已得到签订。我很遗憾地说,这一措施并没有给老百姓带来多少满足,他们仍然吵吵嚷嚷要继续吃小麦面包,尽管他们得到的所有保证是他们再也吃不到了。[18]

第五章 资本主义的土豆

斯托克波特人因不满这样的保证而发动了暴乱，这种行为激怒了地主。牛津郡季审法庭主任查尔斯·威洛比爵士（Sir Charles Willoughby）抱怨说，他只能通过提供补贴的小麦来引导他的劳动者吃全麦面包，"条件是他们要做出优质的黑面包，并且只去除最粗糙的麸皮"。[19]

劳动人民会因为想效仿上流社会的消费习惯而不喜欢白面包。那些抱怨表明，鼓励效仿之努力的失败，令历届英国政府，以及地主和季审法庭的主任都相当失望。劳动人民反而认为，白面包是最有营养的。1767年，一位伦敦面包师介绍道，"用到的每一磅精面粉，营养几乎占一半。因此，我周围的穷人更喜欢精面粉，并说最好的是最便宜的"。在面包中加入土豆和其他膨胀的食材，只会稀释面包所含的营养。灵福德（Lingford）济贫院的居民们说，这种面包还会引发"肚子里的抱怨"。18世纪上半叶小麦价格的下跌有助于城市穷人获得这种更有营养的粮食，并且他们也不愿意放弃它。[20]

正是这些混合面包运动，以及第三章讨论的廉价汤推广活动，令韦克菲尔德的写信人这样的劳动人民，对土豆、红鲱鱼等富人推荐给他们的新食物感到厌恶。从这个角度而言，土豆本身并不令人反感。它和肉一起食用，可改变以面包为主的单调饮食。但在它代替面包被食用，而且除了一小片鲱鱼之外别无他物时，就会被认为是不够的。[21] 在这种情况下，拒不支持以土豆代替面包的做法，就变成了一种政治行为。鼓励吃土豆，是更宏大雄心的一部分，这种雄心要把经济格局转变为无产阶级工人充满其中，而他们当中的每个人要自己追求自己的福祉。"你的红鲱鱼土豆、你自己、与它有关的，都去死吧。"他们这样作答。[22]

土豆与新资本主义分配制度之间的这种关系，在威廉·科贝特（William Cobbett）的著作中尤为清晰。科贝特是一位政治活动家、农民、记者，晚年还是奥尔德姆的国会议员。虽然科贝特最初支持皮特政府的整体计划和对雅各宾派的反对，但从 19 世纪早期开始，他却对政府的财政政策日益不满，也更加不满于政府缺乏对农村劳工悲惨处境的关注。实际收入在降低、可用的公共土地在流失、拾荒等传统的维持生计的做法被禁止，这一切都削弱了家庭自给自足的能力，而农业就业政策更是风雨飘摇。1802 年，科贝特开始编辑周报《政治记录》(Political Register)。该报在 1816 年前直言不讳地提出了议会改革、男性普选权、提高农村劳动者工资、降低税收，以及在政治上更多地关注农村劳工需求的必要性。科贝特得到英格兰南部的农村劳工和工匠的深深认同，他遍游当地，不仅倾听"绅士、农民、商人、雇工、劳工、妇女、孩子们"的意见，还面向大批群众演讲。他尖锐地批评他亲眼目睹的商业化农业、机械化，以及对利润的过分追求所造成的农村生活的变化。他撰写了大量文章，赞美传统农村文化，并一直抱有一种希望，认为它能够将农户和他们的劳工的需要进行调和，从而恢复他所认为的曾经存在于农村的和谐。[23]

科贝特看不起土豆。它们是"懒散、肮脏、悲惨和奴性"的根源。[24]吃土豆的爱尔兰农民昭示着英国劳工未来的可怕景象。科贝特认为，土豆让爱尔兰人活下去，是为了忍受地主的剥削。英国地主对土豆的热情，是他们试图"把英国劳工带到爱尔兰人的状态中去，而爱尔兰人的生活方式，在食物方面，和猪，而且是和喂养不佳的猪差不多"。[25] 在科贝特看来，爱尔兰人和（勉强）养活他们的土豆之间少有差别。两者都肮脏、不文明、野蛮。他描绘了一幅爱尔兰农舍生活的凄凉画面，那里的居民"用爪子从土里刨出（土豆），不洗就扔进锅里，然后和院子里的牲畜们一起在肮脏的木板上吃土豆。[26] 总之，他确信"爱尔兰人的苦难和堕落，主要是因为把土豆当作几乎唯一的食物"。[27]

科贝特感到惋惜的是，土豆和茶在英国已经取代了劳动者饮食中有益健康的啤酒、培根和面包。[28] 此外，他还把土豆与卫理公会等宗教的胡言乱语联系起来，在他看来，这些宗教鼓励穷人改变自己，而不是改变社会。他警告说，要当心"那个滑头的卫理公会教派小偷，他会劝你靠土豆为生"。[29] 被新自由主义意识形态当作核心的对于幸福的承诺，在他看来尤其不诚实。他抱怨说，农民们把土豆田看成"社会底层人民的福气"，但实际上这让他们的收入保持在低水平。[30] 科贝特谴责削弱济贫法的提案，济贫法几个世纪以来一直为体弱多病、失业和贫苦的人支起了一张安全网，他特别指出认为这些改变会带来更大的幸福是一种幻想。在1834年议会针对英国社会救济体系的大规模修订进行辩论时，科贝特反对该修订案促进幸福的主张，谴责它"厚颜无耻至极"。他写道：

> 以它会使劳动人民的生活更幸福的借口来支持这个法案，真是厚颜无耻至极，特别是这当中还有对起草法案的律师的指示，说"**想要劳动人民以比如今他们赖以生存的食物更差的为生**"。在世界上所见过的所有无耻行为中，以它将使劳动人民生活得更好为理由而支持该法案，是最无耻的行为。

他认为，该法案的真正目的是迫使英国的劳动人民靠"土豆和海草"生活，却谎称这种差劲饮食会让他们更幸福。[31]

科贝特认为，土豆不过是一种劣等作物，它消耗地力，也无法为家畜提供草料。它没有营养，在喝茶时吃更是如此。最重要的是，它和劳动人民的无产阶级化和贫困化脱不了干系。"这是一个不可否认的事实，"他说，"以这种块茎替代面包来吃的那部分人，是不幸而可怜的。"科贝特坚持认为，任何把土豆带回自己家的雇工，都应该被当场解雇。[32]

1830年的斯温上尉起义（the Captain Swing uprising）让科贝特正好有了机会来强调土豆、土地资本主义和农村贫困之间的关系。牵动了成千上万农村劳动者、惊恐万状的地主、大量城市居民乃至政府的这场动乱，本质上是对英国农业转型为全面资本主义制度的抗议。到19世纪20年代，正如历史学家艾瑞克·霍布斯鲍姆（Eric Hobsbawm）①和乔治·鲁德（George Rudé）②所言，许多农村劳动者"不仅是完全的，而且是未充分就业、贫穷化的无产阶级"。[33]抗议者抱怨薪酬下降、失业、吝啬的贫困救济款。他们尤其针对的是减少劳动力需求的脱粒机等设备。在20多个郡里，大批（主要是）男性在村里游行，要求增加薪酬、销毁脱粒机，还有人放火焚烧谷仓和农舍。这些要求有时候以虚构的反叛领袖"斯温上尉"签署的恐吓信提出，这些事件因此以他命名。起义在南部和东部尤为激烈，这些地区深受农业繁荣、农户与劳工之间旧社会关系瓦解的影响。不过，英格兰的大多数郡都经历了一些动荡。

科贝特不仅预测到了斯温暴乱，并且还在起义期间走遍了南部各郡，在挤满农业劳工的村公所发表演讲。他对他们要求的支持，导致他以煽动暴力的罪名被逮捕，不过对他的审判后来遭到失败。于科贝特而言，土豆象征着骚乱者揭竿而起所反抗的悲惨处境。农村劳动者，值得赞扬的是，"决意不能沦为土豆"。他们游行、示威，是"为了保护自己不要沦落到如此可怕的境地"。"沦为土豆"是科贝特对他所谴责的整个体系的一种概括。[34]

同样，对于暴乱者而言，土豆也代表着他们及家人所忍受的贫困

---

① 艾瑞克·霍布斯鲍姆（1917—2012），英国著名历史学家、英国皇家科学院院士。代表作有"年代四部曲"等。——译注
② 乔治·鲁德（1910—1993），英国著名马克思主义史学家和社会史家。主要著作有《法国大革命中的群众》《革命欧洲（1783—1815）》等。——译注

状况。在肯特郡的萨尼特（Thanet）岛上，游行者高举的横幅上写道："我们不会靠土豆生活。"在苏塞克斯郡的普尔伯勒（Pulborough）村，三四十个劳工聚集在教堂的休息室里，向前来倾听他们诉说的农户解释说，他们已经"靠吃土豆饿得够久了，必须要改变了"。他们要求增加薪酬，并如愿以偿。在东苏塞克斯郡的灵默（Ringmer），一群劳工去向地主表达不满时，也同样抱怨他们单调的土豆饮食。"我们日夜辛劳，背包里只有土豆，解渴的唯一饮料是冰冷的泉水，我们被迫如此已经太久，我们没有理由抱怨吗？"他们如此问道，随后要求增加工资、立即解雇当地济贫院的院长。[35] 靠土豆为生，是对新资本主义经济下农村穷人苦难的高度概括。

这并不意味着抗议者拒绝吃土豆。尽管科贝特把土豆看成像茶壶那样恶毒入侵劳动者农舍的东西，但大多数骚乱者并不持同样观点。在肯特郡奥尔平顿（Orpington）村附近放火烧毁谷仓的劳工高兴地发现"我们只想要些土豆，这把火煮土豆再好不过了"。威尔特郡的村民表示，希望未来他们的孩子能"吃一肚子土豆睡觉，而不是吃半肚子土豆哭"。[36] 约瑟夫·阿奇（Joseph Arch）是英国第一家农业劳工联盟的创始人，在起义期间，他还是沃里克郡（Warwickshire）的一个孩子。他回忆说，土豆是一种令人向往的商品。土豆供应不足，种土豆的人都把它们"囤积起来"自己吃。[37] 确实，威廉·科贝特愤怒地威胁说要解雇所有有土豆的劳工，这表明土豆很受乡下劳工的欢迎，他希望保护他们以免因为土豆饮食而贬低了身份。[38] 劳工们却希望继续吃土豆，但更多是在他们自己的选择下。他们拒绝的不是土豆，而是沦为土豆。

沦为土豆，依然是对资本主义经济中工人阶级贫困状况强有力的概括。1838 年的请愿者们坚持认为，真正为英国种粮食的劳工，应该得到比"**土豆和破布**"更好的东西。[39] 在上个世纪，精英阶层常建议穷人靠土豆快乐地生活，但这种建议越来越多地被贴上嘲讽的标签。科贝特并不是唯一谴责这种说法的人。在1842—1843年那个严酷的冬天里，牛津大学默顿学院（Merton College）院长罗伯特·布洛克·马舍姆（Robert Bullock Marsham）提出，如果劳工们买不起面包，他们至少可以"享受土豆"，他的这番话被大肆嘲笑。[40]对于抽象市场的批评者而言，土豆是其冷酷无情的缩影。

吃土豆的民族能有什么进步呢？

虽然科贝特谴责土豆是资本主义的剥削工具，但资本家们却有不同的看法。于他们而言，土豆是市场规律的障碍，而不是促进因素。自由资本主义的倡导者期待着未来的理性经济实践，也期待着将劳动作为一种商品进行合理交换。从这个角度而言，多产的土豆给市场带来了太多的自主权。17世纪政治哲学家威廉·配第认为，土豆让爱尔兰人过上了懒散的生活，19世纪的自由主义者也抱怨说，土豆恰恰促成了他们希望消灭的不经济行为。因此，在土豆是促进还是阻碍资本主义市场关系传播的这个问题上，科贝特和自由主义者产生了严重分歧。但他们却一致认为，爱尔兰为他们意见相左的分析都提供了最清晰的例证。

在18世纪，爱尔兰常被誉为因土豆的好处而使健康、勤劳的劳动人口增加的典范。18世纪70年代，英国农业作家亚瑟·杨在游历爱尔兰期间，称赞土豆是一种促进人口增长、提高健康水平的营养食物。它更多的诱人之处，是将爱尔兰劳工与市场隔绝开来。杨发现，英国劳工受到谷物价格的支配，因为大多数人的食物都是购买来的，而不是自己种植的。而爱尔兰却不同，"穷人与价格无关，他们依赖的不是价格，而是蔬菜作物（即土豆）有稳定的收成"。"那么哪种是最好的呢？"杨问道:"是英国人,还是有土豆园和奶牛的爱尔兰人?"[41] 对杨来说,答案是明确的。

到了19世纪20年代,土豆让家庭能够独立于市场条件之外维持生计的能力,对一些观察家已经没有那么大的吸引力了。正如历史学家大卫·劳埃德(David Lloyd)所言,这种转变反映了英国经济和治国之道的变化。随着英国经济从根本上的耕地模式向进一步工业化和更广泛资本主义框架的转变,18世纪对建立自给自足劳动力的强调被另一种信念所取代,即财富和权力来源于经济体系的平稳运行,而不仅仅是庞大的勤奋工作的人口数量。市场经济不仅需要健康的人口,而且需要围绕有偿劳动的具体形式、雇主与劳动力之间关系的特定社会组织,以及市场在确定价格和工资方面的主导地位。劳埃德指出,土豆越来越多地与"道德、政治和资本主义发展相抵触的经济特征"相关联。受到杨称赞的自力更生、爱吃土豆的爱尔兰劳动者,已经变成"可耻的、潜在破坏资本再生所需要的经济和政治形式的另类"。[42]

托马斯·罗伯特·马尔萨斯的著作就是这种变化的晴雨表。马尔萨斯于1766年出生在萨里郡(Surry),成年后的大部分时间都在东印度学院(East India College)担任历史和政治经济学教授,该学院是一家培养未来殖民官员的机构。作为英国圣公会的牧师,他也在学院里布道。他在一生中发表了一系列具有煽动性、广受争论的关于经济和政治问题的著作,它们反映了他的道德信念和他对经济组织性质的想法。在19世纪30年代,他的思想产生了重大的影响,导致1834年的新济贫法(New Poor Law)大幅度削减贫困救助津贴,将教区的救济局限在定义很窄的"值得"救济的穷人身上,该法案被科贝特谴责为"厚颜无耻至极"。[43]

马尔萨斯最著名的著作是《人口论》(*Essay on the Principles of*

Population），它于1798年首次出版，在随后的几十年里被反复重印、大量修订。《人口论》很好地抓住了人们日益增长的疑虑，即仅凭健康的人口是否就能促进经济增长。马尔萨斯之所以想写这篇文章，显然是因为他与父亲就人类社会的幸福能力进行的交谈，而年轻的马尔萨斯对此持悲观态度。他的总体目标，用历史学家阿列桑德洛·荣卡格利亚（Alessandro Roncaglia）的话来说，就是"坚称改善广大劳工处境的任何尝试都是徒劳的"。[44] 马尔萨斯坚持认为，由于人口规模受到粮食供应的限制，确保劳动人民有充足粮食的任何努力都会导致不可持续的人口增长，并不可避免地导致某种形式的崩溃。他解释说，济贫法"因此可谓在某种程度上创造了它们所供养的穷人"。废除这些法律，将使穷人不愿生孩子，从而"促进广大人类的幸福"。[45] 这篇论著受欢迎的程度，从后来的版本数量，以及马尔萨斯后来在制定公共政策方面的成功可以看得出来，这反映出它能使关于政治稳定、经济成功和人口规模之间关系的新观点变得明朗而具体。

马尔萨斯对土豆不乐观是毫不奇怪的。由于土豆的产量比小麦高许多，它导致人口以前所未有的速度增长，而与18世纪的作家不同的是，马尔萨斯并未将其视为一种优势。他相信土豆体制导致的爱尔兰人口增长将达到其自然极限。充足的粮食供应所保证的，并非人口的增长，而可能是一场即将发生的灾难。他在1803年写道，他认为很可能出现的土豆歉收，会导致"无可比拟的可怕"状况。[46]

像科贝特一样，马尔萨斯把土豆与退化联系在一起。土豆与"人的无知和野蛮"，再加上肮脏的生活条件、衣物的不足和糟糕的卫生条件，使爱尔兰人口一直处于萧条而悲惨的状态。如果同样的体制扩展到英格兰，"爱尔兰的烂衣破屋将会随之而来"。土豆消费的增加、土豆田的传播将"对这个国家底层人民的幸福造成他们从未受到过的最残酷而致命的打击"。⁴⁷他在各种论著中剖析"以土豆为主食的极端罪恶",并将土豆的广泛食用与爱尔兰劳动者落后而堕落的状态联系在一起(尽管他绝不认为土豆有唯一的责任)。信天主教的劳动者的土豆种植造成了不可持续的人口增长,但并没有降低其他商品的价格,因此导致乡下劳工无法购买他们过上富裕生活所需的任何其他商品。在马尔萨斯看来,土豆对爱尔兰人是无益的,并且给其他国家的人开了一个不祥的先例。增加土豆的消费不可能带来任何好处。⁴⁸

苏格兰经济学家约翰·雷姆赛·麦克库洛赫(John Ramsay McCulloch)①在1824年《大英百科全书》(*Encyclopaedia Britannica*)的一个条目中,同样清楚地表达了爱尔兰对土豆的依赖及其现有的土地所有制的悲观影响。麦克库洛赫是马尔萨斯的朋友、经济学家大卫·李嘉图(David Ricardo)的门徒。他的关于"农舍制度"条目的文章,目的是驳倒英国劳工得到土地种植土豆或养牛能获益的意见。在麦克库洛赫看来,这样的提议完全是误导性的,因为它会导致英格兰的经济降至爱尔兰或法国农村的水平。他观察到,在那些地方,大量小块土地当中不合理的劳动和资本分配扼杀了一切工业精神,压制了对经济增长至关重要的自我完善的渴望。麦克库洛赫特别引用了爱尔兰的例子,他指出,由于乡下劳工不得不通过自己的劳动来满足所有需求,结果"人们接近野蛮状态"。由于缺乏与外部世界推动力量的接触,停滞不前的乡下劳工"陷入了冷漠、迟钝和愚呆的状态"。孤立和无知阻碍了人文科学的形成,使乡下人甚至无法认识到他们的悲惨处境。因此,他指出,"他

---

① 约翰·雷姆赛·麦克库洛赫(1789—1864),英国经济学家、统计学家。伦敦大学学院教授。麦克库洛赫最著名的著作是教科书《政治经济学原理:这门科学产生和发展的概述》(1825),另还有著作《论赋税和公债制度的原理及实际影响》(1845)、《政治经济学文献》(1845)等。——译注

们不能集体行动，于是，因反抗不够他们不得不屈服于压迫者的枷锁"。⁴⁹ 土豆所带来的自主权，根本就不是自主权。

由于土豆连续歉收，爱尔兰在1845年至1849年发生了可怕的饥荒。让英国政府满意的是，这证明了它自身政策所大力鼓励的农业形式实际上是不可持续的。自1845年起，土豆晚疫病的侵袭破坏了欧洲许多地区的土豆收成。在欧洲大陆，土豆作物的损失造成成千上万人死亡，特别是在荷兰和普鲁士等黑麦和小麦收成也受影响的地区。在比利时，1845年的土豆收成比往年低了87%，而1846年的黑麦收成则减少了50%。结果导致大约4万人死亡。⁵⁰ 然而，爱尔兰的灾难规模，却完全是另一回事。

在爱尔兰，如杨、马尔萨斯等人所见，土豆将土地和光能转换成热量的超级能力，使得农村家庭能够依靠小块土地维持生活。在英国殖民统治下，随着商业化小麦、乳制品、肉类生产的扩大，他们被迫蜷缩在这样的小块土地上。到19世纪40年代，大约40%的人口几乎完全以土豆为生，或如果有足够的土地饲养一头奶牛，他们就以土豆加一点酪乳为生。爱尔兰农村的穷人每天只吃3到5公斤土豆，除此之外几乎没有其他食物。烹饪用的泥炭唾手可得，加之土豆在潮湿的爱尔兰土壤中产量极高，因此整个家庭都能靠科贝特所称的可怜的土豆饮食维持生活。⁵¹ 在安第斯山脉的一座山谷里，可能种植有一百多种不同类型的土豆，而19世纪爱尔兰种植的土豆，大部分却都是单一的黄肉品种，被称为"爱尔兰大土豆"。依赖于单一的栽培品种极大地增加了疾病的易感性。当1845年庄稼歉收，接着1846年、1848年和1849年相继歉收之后，超过100万人死亡。由于饥荒和因饥荒而另有100万人移民到美洲，爱尔兰的800万人口在双重打击下遭到极大破坏。⁵²

饥荒是由晚疫病的到来引发的，但英国政府的反应却极大地加重了灾难的规模，他们把这场危机视为重塑爱尔兰社会的良机。伦敦财政部行政总监查尔斯·特里维廉（Charles Trevelyan）等官员认为，爱尔兰的整体经济结构是对现代资本主义实践的公然侮辱。在英属西印度群岛废除奴隶制后，特里维廉继承了大量的政府补贴金，他于19世纪20年代在东印度学院师从马尔萨斯。在进入财政部之前，他在印度做了很多年的殖民官员。在政府处理饥荒的方法上，他的影响极大。他希望，土豆经济的崩溃会促使爱尔兰的小农场主离开他们的小块土地，成为无产阶级的一员。他认为这会是一个巨大的进步，完全值得通过饥荒这种"短暂的邪恶"来实现。它还会清除那些低效、萎靡的爱尔兰地主阶级，英国人认为他们应该对灾难性的人道主义危机负责。在他看来，英国政府最不应该做的事，就是通过援助受灾的爱尔兰人来支撑这个陈旧的体系。对于特里维廉这样的自由主义者而言，土豆是迈向现代化的障碍，是通往经济理性道路上的绊脚石，是国家的敌人。"靠土豆为生的国家希望何在？"他反感地疾呼。[53]

特里维廉明确认为饥荒是有好处的。他问道，爱尔兰劳动者将来如何养活自己？他只看到了一种可能性："这些阶级所占据的地方再也不能维持，他们必须靠自己的劳动所获得的薪酬来生活。"爱尔兰人将不再依靠自己种植的土豆，而是靠"用收入来购买的"谷物为生。换句话说，他们将成为农村无产阶级。他们将继续靠农业为生，但"农业根据大幅改善的新条件进行"。[54]特里维廉和许多其他英国自由派人士一致认为，饥荒由此提供了"极佳机会"，将爱尔兰从自给自足的土豆食用者的社会，变成爱尔兰人消费"更高级食物"、养成使他们具备购买这类食物所必需的"勤奋和进步习惯"的社会。一位国会议员在1846年说："即使在最痛苦的天命中，也会有慰藉之处，甚至常有庆贺之机。"[55]以议会的眼光看来，土豆带来的饥荒，行之有效地迫使身体健全的爱尔兰人成为农村无产阶级。而土豆却反而导致，并且也代表着，他们没能更早地进入现代世界。

19世纪50年代早期，最伟大的资本主义理论家卡尔·马克思在伦敦撰文时运用了这些概念，把政治上迟钝的法国农民比作装在袋子里的土豆。他在《路易·波拿巴的雾月十八日》(*Eighteenth Brumaire of Louis Napoleon*)中写道，法兰西民族不过是由一群互不相干的农民家庭组成的，如同一堆拼成一袋的单个土豆。两者都不能被认为拥有更大的集体特征。麦克库洛赫认为，爱尔兰农业体系所造成的隔离，意味着佃农们"不能集体行动，于是，因反抗不够他们不得不屈服于压迫者的枷锁"。马克思持有和他相同的观点，认为贫困小农自给自足的生存状态，把他们彼此隔绝开来，也把他们同现代世界隔绝开来。由于每座农场基本上都是自给自足的，小农没有机会发展任何一种阶级意识。结果，他们无法认识到，也更不能发展自己的利益，因而一直停留在政治生活的边缘，被困在一种陈旧而有害的生活模式中，被暴君和煽动者的花言巧语所俘虏。他发现，"他们不能代表自己，他们必须被代表"。马克思用土豆来比喻这种与现代世界隔绝的状态，也就毫不令人奇怪了。[56]

马克思几乎无需去伦敦，就能体会到土豆对进步的阻碍。至19世纪中叶，在欧洲的许多地方，土豆已经成为缺乏社交和消费欲望的停滞世界的缩影，而且它以前资本主义的生产方式的持续为特征。"满足于吃土豆的人如何指望有进步？"法国医师兼记者阿梅代·德尚布尔（Amédée Dechambre）问道。他发现，"极度清醒"和过度消费对国家进步的损害是相同的。[57]在17世纪，配第等作家不喜欢土豆是因为它使爱尔兰农民得以逃避政府的管辖。而到了19世纪，土豆的罪行却是助长了对经济现实的逃避。（参见图15）

158

图 15 土豆与资本主义理性。自由贸易杂志《斗争》(*The Struggle*) 在想要解释关税的破坏性影响时，指向了土豆。这幅 1843 年的漫画展示的是七个被关在笼子里的男人，代表着失去了进口廉价食品的英国。在收成不好的情况下，较富裕的人不会受到影响，但较穷的人被迫多吃土豆，少吃牛肉和面包。标题解释说，5 号男人"是个劳工，他不再吃面包，主要吃土豆"。该杂志暗示，如果劳动人民抵制自由市场，沦落到只能吃土豆的地步，那他们无法责怪任何人，只能怪自己。

土豆的懒惰之血

在欧洲各地，认为土豆阻碍了资本主义行为准则的观点，贯穿在工业化和城市化的发展中。在整个 19 世纪，英国、法国的工业化和城市化的增长超过了一倍，德国也紧随其后。[58] 工作性质、社会组织的这些变化，引发了人们对它们影响道德、社会凝聚力和人民健康的担忧。人们担忧的焦点在于产业劳动人口的体力和活力，尤其担忧城市工人的血肉之躯无法支撑产业所需的不间断劳动力。"疲劳"成为国家效率和经济成功的大敌。食物，作为驱动人类发动机的燃料，对于维持工人身体的正常运转，从而确保民族工业的竞争力至关重要。

159

从 19 世纪下半叶开始，与不断发展的工业化相呼应的对热力学的科学研究，让人们认识到，人体就像是一台发动机，而劳动力，正如意大利政治家、经济学家弗朗西斯科·尼蒂（Francesco Nitti）解释说，"只不过是转化为机械能的热能而已"。而热能，只能通过工人体内"燃烧食物而产生"。因此，"饮食不良的国家，其工作活力也差"。[59] 历史学家安森·拉宾巴赫（Anson Rabinbach）指出，至19世纪中叶，科学家和政治家已达成一致，认为了解这台发动机的燃料需求，对于一个国家的经济、政治成功至关重要。[60] 卡尔·冯·沃特（Carl von Voit）、威尔伯·阿特沃特（Wilbur Atwater）等许多研究者设计了一系列巧妙的食物化学实验，能够把食物输入与能量输出关联起来。19世纪70年代热量单位卡路里的发明，就是对食物与工作之间这种关系的简明表达：卡路里衡量的是一种特定食物能让其食用者消耗的能量的总数。这一点，加之对人类消化生理学的科学共识的出现，使食物化学成为一种表达劳动者饮食与他们生产力之间关系的有效语言。与通过卡路里等数值测量来量化食物的营养价值相呼应的，是19世纪正在进行的建立数学方法来衡量工作的努力，它进一步细化了食物投入和劳动产出之间的关系。因此，膳食体系的整体营养，以热量和化学成分为形式，可从数学上与食用者的健康、活力关联在一起，而食用者的健康、活力又对经济效率产生影响。由来已久的劳动者的饮食习惯在很大程度上影响国家的经济、军事成功的观念，通过表达这种关系的数学、科学语言的出现而得到极大增强。[61]

食品化学实验通常明确聚焦于产业工人、士兵等被视为对国家成功特别重要的人群的能量需求。譬如，巴黎大学（University of Paris）生理学教授阿曼德·戈蒂埃（Armand Gautier）曾致力于计算以上两类人在完成一系列任务时所需要的食物能量。普鲁士医生威廉·希尔德斯海姆（Wilhelm Hildesheim）在1856年的研究中，也同样将重点放在军队饮食的营养充分性上。⁶²18世纪的作家们满足于主张将丰富、健康的食物供应与国家的财富、权力概括地联系起来，而19世纪的科学家们，已能够证明不同食物的化学成分，与它们能够提供的热力学工作量之间用数值精确表示的相关性。

颇有影响力的德国生理学家马克斯·鲁布纳曾是普鲁士政府在军事供应和监狱饮食方面的顾问，还担任普鲁士帝国健康委员会（Imperial Health Board）的主管。他撰写了大量文章，论述营养对国家福祉的重要性。他起先在马尔堡（Marburg）和柏林担任教授，后来在柏林的恺撒·威廉学会（Kaiser Wilhelm Society）担任科学发展创会理事。他强调确保德国城市劳动者膳食营养的必要性。他认为，营养是"个人生理和心理表现的基础，从而也是国家生产力和公共卫生的基础"。⁶³城市劳动者的健康和饮食直接关系到国家的经济和政治利益。鲁布纳同戈蒂埃等许多人一样，对城市无产阶级和士兵的营养需求特别感兴趣，他持续将注意力放在分析这些关键群体的饮食习惯上，不断强调现代饮食对劳动者身体的破坏性影响，并出版了旨在解释合理营养原则的通俗科学著作。通过这些活动，鲁布纳和他的同事们帮助树立了这种新营养范例的政治、经济和社会重要性。

就土豆对工业效率的贡献而言，研究结果并不令人鼓舞。19世纪，随着土豆更多地进入许多欧洲人的饮食，它的商业种植规模稳步扩大。到1850年，普鲁士的土豆产量增加了13倍，以吨计的总产量大大超过了谷物的联合总产量。上西里西亚（Upper Silesia）等地的劳动者据说是靠土豆、黑面包和杜松子酒维持生活。1854年，医师弗里德里希·蒂德曼（Friedrich Tiedemann）针对德意志中南部下层阶级的饮食状况的观察也得到相同结果。⁶⁴到19世纪中叶，佛兰德斯几乎所有的小农户都种植了土豆和黑麦，荷兰各省的土豆种植面积占据了高达20%的农业用地。根据内政部长在 1845 年的说法，土豆是"荷兰下层甚至部分中产阶级最普遍、最重要的食物"。乌得勒支（Utrecht）的劳动者一天吃三顿土豆。[65] 据报道，在丹麦日德兰半岛中部地区（Mid Jutland），土豆是贫农的主食（或许能配上一两勺煮卷心菜）。在瑞典，到 19 世纪 40 年代，土豆的产量占庄稼总产量的近 40%。[66] 据报道，在 19 世纪 80 年代，俄罗斯的农民每年消耗的土豆为 185 公斤。"农民的食物实在是太千篇一律了，"一位观察家注意到，"天天吃黑麦面包、土豆、卷心菜，剩下的食物只是这三种食物的调料。"[67] 在英国，土豆长期以来一直是普通人饮食的重要组成部分，由于商业种植的显著增长，土豆覆盖了 25% 的可耕地。英国科学促进会在 1881 年的报告中称，土豆"无疑是（居民）最重要的蔬菜食物"。和在别处一样，它们在劳动人民的饮食中占据突出地位。[68]（参见烤土豆食谱）

营养学这门新科学的代言人，大部分对土豆的日益普及持悲观态度。蒂德曼在 1854 年写道，真正可悲的，是下层阶级的土豆消费如此明显的增长，因为这种缺乏营养的块茎，对佝偻病和结核病病例每年的增加，以及"残废者和痴呆者"数量的明显增加负有唯一的责任。按照另一位德国医师的说法，德国劳动者多土豆的饮食，将不可避免地导致他们退化，以及年复一年身体和智力能力的衰退。他预言他的祖国将面临灾难性的后果。[69] 土豆已远不是 18 世纪的推广者所认为的营养丰富的动力源泉，而是被视为营养低劣的食物，越来越多地受到谴责。实验室分析显示，它的蛋白质含量比肉类低，而营养学家认为，它的化学成分与人体的化学成分极不相配。多土豆的饮食使得食用者行动迟缓，虚弱无力，不适合现代劳动的要求。按照荷兰医师、政治家雅各布·摩莱萧特（Jacob Moleschott）的观点，一位劳动者如果光吃土豆，只需 14 天，就会完全丧失工作能力。[70] 摩莱萧特于 1845 年在海德堡大学获得了医学学位,随后在瑞士、德国和意大利的几所大学任教,后来他入籍意大利,并从 1876 年起在罗马担任参议员。摩莱萧特在科学和政策问题上的不同文章,都强调消除贫困、确保劳动人民能够负担得起营养丰富的食物、使他们有尊严地生活的重要性。他还强调食物在工业生产和国家进步中的中心地位。如许多其他的食品化学家一样,摩莱萧特认为,大量的蛋白质和磷是营养食物的要素。正因此,他认为土豆是劣质的肉类替代品。在他看来很不幸的是,土豆在普通德国人的饮食中,却恰恰经常扮演着这个角色。[71]

烤土豆

埃琳娜·莫洛科韦兹(Elena Molokhovets)极受欢迎的《给年轻家庭主妇的礼物》(*A Gift to Young Housewives*)一书于 1861 年在圣彼得堡首次出版。这本已经出版了 20 多个版本的烹饪书,旨在帮助富人家的女性打理家务、管理仆人队伍,并设法解决中产阶级社交的其他期望。许多菜肴需要的食材远远超出了大多数俄罗斯人的能力。莫洛科韦兹的杏仁复活节面包需要 70 个鸡蛋。不过,也有一些反映大多数人日常饮食的普通食物。这个简单的炭灰烤土豆食谱选自 1897 年版。

烤土豆

"将土豆清理干净,勿用水洗。将它们放入俄罗斯火炉的木炭中,或在灶火中烘烤,频繁翻转以免烤煳。早餐食用,分别加盐和黄油。"

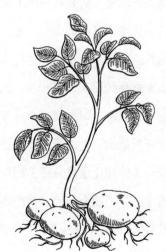

摩莱萧特坚持认为，以土豆为主的饮食难免会导致疾病和虚弱。对于经常依赖土豆的劳动者，它提供的营养并不够。在1858年的《食物讲解》（Lectures on Food）中，他问道："土豆的懒惰之血，怎能给予肌肉工作的力量，或者给予大脑希望的冲动？"而且，对土豆的依赖是导致爱尔兰贫穷并永远屈从于英国的原因。土豆的化学性能，使爱尔兰人的身体产生"无力的绝望"。而相反的是，英国殖民者的血管里，流淌的是肉食者充满活力的血液。"可怜的爱尔兰，贫穷产生新的贫穷，"他叹息道，"你们无法取胜！"国家的实力和经济的成功，受到土豆饮食致命的阻碍，而不是推动。[72]

另一些有影响力的营养学家，和摩莱萧特一样，也把以土豆为主的饮食的前景看得很暗淡。德国化学家卡尔·冯·沃特对于氮和蛋白质的开创性营养实验，令他本人确信，面包应该成为大部分劳动者的主食，而土豆和其他蔬菜不应超过30%。[73] 沃特的计算表明，人类的发动机要有效运转，每天至少需要118克蛋白质。其他许多营养学家也赞成以这个"沃特标准"为最低值。[74] 在劳动者的饮食中，用低蛋白质的土豆取代更有营养的肉类，这样的吃法肯定是危险的。弗朗西斯科·尼蒂确信，比利时工业的快速发展，主要源于产业工人的饮食从以土豆为主转向以肉类为主。他指出，在1853年至1891年间，他们的人均肉类消费从15.920公斤攀升至20.770公斤，而土豆消费则从22.573公斤下降到17.866公斤。从土豆到肉类的饮食转变，同样也是爱尔兰劳动者在家乡"懒惰、虚弱、古怪"，移民到美国后却变成精力充沛而多产的工人的原因。[75]

因此，新的营养理论，为比较不同国家的工业生产率提供了坚实的框架。尼蒂相信，美国和英国工业生产率更高的原因是肉类的高消费，而德国和法国工人表现较差的原因是土豆等"重碳酸盐"食品的高消费。他制作了一系列对比图表，显示饮食与士兵及产业工人身体健壮之间的密切关系。[76] 德国学者、社会评论家格哈特·冯·舒尔茨-格弗尼茨（Gerhart von Schulze-Gävernitz）在对德国和英国的棉花工业进行详细比较研究时得出了类似的结论。舒尔茨-格弗尼茨的家族本身就从事棉布印花生意，他的祖父曾在曼彻斯特当过学徒。19世纪80年代，年轻的舒尔茨-格弗尼茨游历了英格兰北部，参观了棉纺厂，考察了制造业的细节。饮食也属于他分析的一个重要部分。他询问了奥尔德姆和兰开夏郡其他地区的操作工人每周的预算，并听取了他们对过去半个世纪工作和饮食习惯变化的看法。他的研究使他确信，英国工人的生活水平要高得多，而他们的高级饮食既是这种优势的原因，也是这种优势的反映。英国棉纺织业的工人每天吃肉和小麦面包。相比之下，德国工人则以黑面包和土豆为生。其结果是，英国操作工人的工作效率是德国操作工人的两到三倍，至少他算出来如此。[77] 一个世纪前腓特烈大帝（以及英国农业委员会）如此热情拥护的土豆，现在却被认为是德国在经济产出和国家实力上落后于英国的原因。

结论

在布莱克浦（Blackpool）停留期间，年轻的格哈特·冯·舒尔茨-格弗尼茨在一家客栈吃饭时，与住在此处的一位老人攀谈起来。一边喝着茶，吃着火腿、羊肉、土豆和豌豆，这位曾经做过棉纺工人的老人一边讲起了他的生平故事。他的父母都是兰开夏郡的手工织布工，19世纪20年代他的童年是极其贫困的。他早年在纺织厂工作，后来在动力织布机厂工作。他的父母和兄弟姐妹靠土豆和燕麦饼维持生活。他回忆说，直到很久以后，他搬到曼彻斯特时，才第一次见到小麦面包。年轻时，他曾是宪章运动的热情支持者，宪章运动是工人阶级运动，要求实现男性普选、废除担任政治职务对财产的要求，以及对选举制度进行其他改革。这位老人对产业工人所面临的条件的尖锐批评、对宪章主义信念的长久忠诚，令舒尔茨-格弗尼茨印象深刻。

与舒尔茨-格弗尼茨对话的这位老人认为，自那些艰难的日子以来，工人的境况总体而言有了很大改善。他滔滔不绝地讲述工会带来的好处，当他说到普通人连在沙滩上看业余足球比赛的闲暇时间都没有，更没有一天假日的时代，围坐在桌边的年轻工人们都大为震惊。"但是，如果您希望看到这种变化的真正迹象，"这位头发花白的老者眉开眼笑地继续说道，"兰开夏郡的优势就摆在您面前的桌子上。"接着老人"拿着一片小麦面包，得意洋洋地站起来"。[78]

要面包，而不要作为穷人饮食的土豆——对于纺织厂老工人而言，一片小麦面包象征工人阶级为生活水平作出的奋斗，因为对于他，对于其他很多工人，面包代表着与他那在"几乎不能说话和走路"时就被束缚在纺织厂、每天工作13个小时并被禁止加入工会、以土豆和燕麦饼维持的贫困生活所截然相反的生活。在19世纪，小麦面包和土豆餐形成的鲜明对比，成为工业化以及它与经济增长、政治改革的关系所引发的紧张局势的核心。如布莱克浦这位纺织工人这样的产业体制的批评者，以及如特里维廉这样的热情的资本主义者，即使在解释土豆使资本主义前进的作用上存在着严重的分歧，但他们都一致认为，土豆饮食概括的是劳动人民与资本主义制度之间的关系。日常饮食，以及土豆在日常饮食中的地位，为谈论工人、国家和新工业秩序的关系提供了一种话语。

18世纪形成的民众日常饮食与国家富强的关系，为19世纪的讨论奠定了基础，这些讨论越来越多地把重点放在培养强健产业工人的目标上。供养城市无产阶级的挑战在于，如何将这些人类机器所需要的营养食物与他们的工资所能掌控的营养资源相匹配。平衡劳动力市场价值与工人所能消耗的能量之间的关系，对整个国家具有直接的重要性。正如弗朗西斯科·尼蒂在1896年所坚称的，为劳动者提供更好的饮食直接导致"整体财富更多更有益的增长"。[79]肉类等富含蛋白质的食物会对它有所促进；富含淀粉而低蛋白质的土豆饮食，却会阻碍工业化并削弱国家的军事力量。（参见图16）

而同时，舒尔茨-格弗尼茨遇到的那位纺织厂工人，并没有拒绝他晚餐的羊肉、火腿和豌豆所配上的热乎乎的煮土豆。这些土豆的象征意义，与兰开夏郡和小麦的优势形成鲜明对比的土豆是不同的。土豆成为无产阶级贫穷象征的同时，也成为工人阶级餐桌上更正常、更受欢迎的

省份	Provinces.	生产性人口占总人口的比例 Percentage of the Productive Population to the entire Population.	蛋白质在劳动者饮食中的绝对含量 Absolute Amount of Albuminoids in the Workman's food.	从土豆中获得的比例 Proportion obtained from potatoes.	1851—1855年每1000例死亡中死于肺结核的数量 Deaths from Phthisis out of 1,000 Deaths 1851-55.	每1000人中免于服兵役者的数量 Exemptions from Military Service per Thousand.
		%	克 grammes	%		
西佛兰德斯	West Flanders	58	75	11·58	194	147
东佛兰德斯	East Flanders	52	62·8	18·18	226	128
埃诺	Hainault	42	59·9	3·88	171	75
林堡	Limburg	40	75	25·59	240	128
安特卫普	Antwerp	37	74	31·43	200	156
布拉班特	Brabant	37	75	17·30	194	131
列日	Liège	36	72	13·75	141	77
那慕尔	Namur	31	49	1·96	149	64
卢森堡	Luxemburg	30	79	17·26	122	82

图16 弗朗西斯科·尼蒂的图表：高土豆摄入量与疾病的关系。19世纪许多分析人士认为，吃土豆过多会降低工人的精力和活力。这张图表来自意大利经济学家弗朗西斯科·尼蒂关于"食物与国家劳动力"的研究，它将食用土豆的普遍程度与肺结核的高发病率联系起来。吃土豆的地区免于兵役的人数也更多，这是国家直接关心的问题。

食物。正如一百年前那位匿名写信人所言，它们是"穷人想买的五种基本食物"之一。[80] 土豆既能反映新工业无产阶级的赤贫生活，也能反映工联主义（trade unionism）和工人阶级行动所赢得的舒适生活，土豆的这种能力，说明了在思考人与现代国家关系时食物的重要性。

注释

1. *London Gazette*, 27 Dec. 1800. 四分之一品脱的一条面包重约四磅。

2. Cobbett, *Cobbett's Weekly Political Register*, 24 Mar. 1832, 786–787.

3. Salaman, *History and Social Influence of the Potato*, 601–602; Thompson, 'The Moral Economy of an English Crowd'; Gallagher and Greenblatt, 'The Potato in the Materialist Imagination'.

4. Salaman, *History and Social Influence of the Potato*, 521–522; Booth, 'Food Riots in the North-West of England', 90; Wells, *Wretched Faces*, 95, 98, 164, 167 (quote), 312; Bohstedt, 'The Myth of the Feminine Food Riot', 24, 41, 49; Rule and Wells, *Crime, Protest and Popular Politics in Southern England*, 32; Bohstedt, *The Politics of Provisions*, 184, 203–204, 241.

5. Styles, 'Custom or Consumption?', 103.

6. Skytte, 'Ron at utaf potatoes brånna brännavin', 231–232; James Stonhouse, 'Expedients for alleviating the Distress occasioned by the present Dearness of Corn', Northampton, *Universal Magazine* 21 (1757), 270; Young, *Farmer's Letters to the People of England*, 193–194; Engel, *Traité*; Henry, *The Complete English Farmer*, 114; Corrado, *Il credenziere di bueongusto*, 39–40; Béguillet, *Traité des subsistances*, VI: 507–512; Parmentier, *Les pommes de terre, considérées relativement à la santé & à l'économie*, 32, 176–180; James Anderson, 'Of Ardent Spirits Afforded by Potatoes', *Letters and Papers on Agriculture* 4 (1788), 43–52; *Verhandlungen und Schriften*, 72; Lindroth, *Kungl. Svenska vetenskapsakademiens historia*, 261–263; Kisbán, 'The Beginnings of Potato Cultivation in Transylvania and Hungary', 180–181; Spary, *Feeding France*.

7. Webb and Webb, 'The Assize of Bread'; Stern, 'The Bread Crisis in Britain'; Wells, *Wretched Faces*, 210; Petersen, *Bread and the British Economy*.

8. Committee on Doctor Gordon's Discoveries in the Art of Dying, & Making Potatoe Bread, 23 Aug. 1794, SR RASE B/X, fols. 2–3, MRELSC. 也可参见 *European Magazine and London Review*, Aug. 1790; Board of Agriculture Minutes of Committee on Potatoes,

Committee on the Scarcity of Grain, and the High price of Provisions, and Committee of Correspondence and Expenditure, 23 Aug. 1794–27 Nov. 1795, SR RASE B/X, fols.1–45, MRELSC; *Annals of Agriculture* 24 (1795); *Account of the Experiments tried by the Board of Agriculture in the Composition of Various sorts of Bread*, Board of Agriculture; William Augustus Howard, Grays Inn Great Square, 14 July 1795, TNA, HO 42/35, fols. 157–158; Board of Agriculture, *Hints Respecting the Culture and the Use of Potatoes*。

9. Parmentier, *Traité sur la culture et les usages des pommes de terre, de la patate, et du topinambour*; Delle Piane, *De' pomi di terra*; 'Report from Saxon Electoral Society of Agriculture on the Cultivation of Potatoes', *Communications to the Board of Agriculture*, Board of Agriculture, 295; Doyle, *Instrucción formada de orden del Consejo por D. Enrique Doyle; Doyle, Tratado sobre el cultivo, uso y utilidades de las patatas* (1797); Dionisio Escudo, Ampudia, 20 Feb. 1798, *SAA* 3, 19 Apr. 1798 (1798), 253; *Instrucção sobre a cultura das batatas*; Amoretti, *Della coltivazione delle patate*; Birembaut, 'L'École gratuite de boulangerie'; Kaplan, *Bread, Politics and Political Economy*; Spary, *Feeding France*.

10. Lord Sheffield, Sheffield, 28 June 1795, TNA, HO 42/35, fols. 51–52. 也可参见 Henry Shelley to Duke of Richmond, Lewes, 28 June 1795, TNA, HO 42/35, fols. 57–58。

11. More, *The Way to Plenty*; Webb and Webb, 'The Assize of Bread', 208–209 (second quote); Webb, 'Not So Pleasant to the Taste', 7 (first quote). For examples see the printed declarations made at the Staffordshire quarter sessions, 16 July 1795, the circular published in Newmarket-upon-Trent, 22 July 1795, the printed 'engagement entered into by the lords of his majesty's privy council, and others, in order to diminish the consumption of wheat in their respective families', Manchester, 6 July 1795, and the Minute from Hamilton, 4 Aug. 1795; all in TNA, HO 42/35, fols. 255, 264, 312, 401.

12. 'Plain Man', *Times*, 11 July 1795, 3.

13. Thompson, *The Making of the English Working Class*; Wells, *Wretched Faces*, 406–457.

14. *Times*, 4 Nov. 1795, 2.

15. *Annals of Agriculture* 24 (1795), 204.

16. *Annals of Agriculture* 24 (1795), 137.

17. *Annals of Agriculture* 24 (1795), 284–286. 也可参见 Henry Curzon, Waterperry House, Wheatley, Oxon, 5 July 1795, TNA, HO 42/35, fol. 96; Webb and Webb, 'The Assize of Bread', 210; Salaman, *History and Social Influence of the Potato*, 504–506; Rule and Wells, *Crime, Protest and Popular Politics*, 22–24; Earle, 'The Political Economy of Nutrition'。

18. Holland Watson to War Office, Stockport, 5 Aug. 1795, TNA, WO 1/1094.

19. Sir Charles Willoughby to my lord Duke, Baldon House, 5 July 1795, TNA, HO 42/35, fols. 91–93. 也可参见 Tapwell, *A Friendly Address to the Poor of Great Britain*。

20. Mr Peacock's Evidence as to the bread, and Mr Smith's Evidence, both in GL, Ms 7801, box 2; *Times*, 4 Nov. 1795; Eden, *The State of the Poor*, I: 526; Thomas Turton to Duke of Portland, Starborough Castle, 7 Feb. 1801, TNA, HO 42/61, fol. 118 (quote); Fourth Report, 17 Dec. 1800, *House of Commons Sessional Papers the Eighteenth Century*, 402–403; Hanway, *The Great Advantage of Eating Pure and Genuine Bread*, 10–11; *Parliamentary History of England, from the Earliest Period to the Year 1803*, 238; Petersen, *Bread and the British Economy*, 15–35 (33, quote); Fox, 'Food, Drink and Social Distinction in Early Modern England', 180; Earle, 'The Political Economy of Nutrition'.

21. 关于加入土豆 / 肉类和单吃面包的对比，参见 Koenker, 'Was Bread Giffen?'

22. *London Gazette*, 27 Dec. 1800.

23. Cobbett, *Cobbett's Weekly Register*, 14 Dec. 1822, 686 (quote); Dyck, *William Cobbett and Rural Popular Culture*; and Dyck, 'William Cobbett and the Rural Radical Platform', 191.

24. Cobbett, *Cottage Economy*, para. 80.

25. Cobbett, *Cottage Economy*, para. 77; Cobbett, *Cobbett's Two-Penny Trash* 1 (1831), 200; Gallagher and Greenblatt, 'The Potato in the Materialist Imagination'.

26. Cobbett, *Cottage Economy*, para. 79.

27. Cobbett, *Cottage Economy*, para. 99; Cobbett, *Cobbett in Ireland*, esp. 82, 92–95, 125.

28. Cobbett, *Cottage Economy*; Dyck, *William Cobbett and Rural Popular Culture*.

29. Dyck, *William Cobbett and Rural Popular Culture*, 98–106. 或可参见 Cobbett, *Cobbett in Ireland*, 110。

30. *Annals of Agriculture* 32 (1799), 602 (quote); Cobbett, *Cottage Economy*, para. 77; Dyck, *William Cobbett and Rural Popular Culture*, 116–117. 关于把土豆菜园视作"从根本上使农业工人进一步服从雇主的剥削性农业制度的一部分"，参见 Burchardt, 'Land and the Laborer', 684。

31. Cobbett, *Cobbett in Ireland*, 108–109.

32. Cobbett, *Cottage Economy*, para. 99; Cobbett, *Cobbett in Ireland*, 95, 125（这项禁令不包括用于填充鹅的土豆）; Dyck, *William Cobbett and Rural Popular Culture*, 210–211。

33. Hobsbawm and Rudé, *Captain Swing*, 15 (quote), 47; Holland, 'Swing Revisited'.

34. Cobbett, *Cobbett's Weekly Political Register*, 16 June 1832, 652; Rule and Wells, *Crime, Protest and Popular Politics*, 11.

35. Cobbett, *Cobbett's Weekly Political Register*, 20 Nov. 1830, 787, 11 Dec. 1830,

956, 24 Mar. 1832, 786–787, 14 Apr. 1832, 93; Hammond and Hammond, *The Village Labourer*, 227–229; Dyck, *William Cobbett and Rural Popular Culture*, 166.

36. *Times*, 17 Sept. 1830, 3, 22 Nov. 1830, 3, 23 Nov. 1830, 3, 30 Dec. 1830, 3; Hobsbawm and Rudé, *Captain Swing*, 99, 123.

37. Arch, *From Ploughtail to Parliament*, 12–13. 也可参见 Zylberberg, 'Fuel Prices, Regional Diets and Cooking Habits'。

38. "如果你依然渴望这该死的块茎，没有它，爱尔兰不会落到如今的状况，因为它，面包被逐出了这里的工人家庭。如果你依然渴望这'粗鄙的'食物，你只能去其他地方找它。"他写信给他的农场雇工说。Cobbett, *Cobbett in Ireland*, 95.

39. Rule and Wells, *Crime, Protest and Popular Politics*, 117.

40. Gurney, '"Rejoicing in Potatoes"'.

41. Young, *Farmer's Letters to the People of England*, 202–203; Young, *A Tour in Ireland*, II: 112–114 (quote); Young, *The Question of Scarcity Plainly Stated*, 77; Lloyd, 'The Political Economy of the Potato'.

42. Tribe, *Land, Labour and Economic Discourse*; Lloyd, 'The Political Economy of the Potato', 313 (quote).

43. Dean, *The Constitution of Poverty*; Vernon, *Hunger*.

44. Roncaglia, *The Wealth of Ideas*, 163; 一种不太一样的解读，参见 Bashford and Chaplin, *The New Worlds of Thomas Robert Malthus*。

45. Malthus, *An Essay on the Principle of Population* (1798), 38–39.

46. Malthus, *An Essay on the Principle of Population. The 1803 Edition*, 453.

47. Malthus, *An Essay on the Principle of Population. The 1803 Edition*, 265, 454, 450, respectively.

48. Malthus, 'Newenham on the State of Ireland', 164; Lloyd, 'The Political Economy of the Potato'.

49. McCulloch, 'Cottage System', 382–383, 386.

50. Vanhaute et al., 'The European Subsistence Crisis of 1848–1850', 26.

51. Clarkson and Crawford, *Feast and Famine*, 59, 93; Vanhaute et al., 'The European Subsistence Crisis of 1848–1850', 22–23.

52. 对此的概览请参见 Ó Gráda, 'Ireland's Great Famine: An Overview', and Daly, 'Something Old and Something New'。

53. Trevelyan, *The Irish Crisis*, 2; Gray, *Famine, Land and Politics*; Gray, 'The European Food Crisis and the Relief of Irish Famine, 1845–1850'; Lloyd, 'The Political Economy of the Potato'.

54. Trevelyan, *The Irish Crisis*, 229–320. "佛兰芒人像爱尔兰的佃农一样，绝望而固执地墨守着他的那块土地，同样无法或不愿意识到，只有当他成为一个有报酬的劳动者，而不再是一个贫穷的所有者，他的状况才能得到改善。"《泰晤士报》坚称。'Flanders and Ireland', *Times*, 29 Nov. 1847, 5; Gray, 'The European Food Crisis and the Relief of Irish Famine, 1845–1850', 104.

55. Gray, *Famine, Land and Politics*, 103, 119.

56. Marx, 'The Eighteenth Brumaire of Louis Napoleon', 101. 马克思认为自己吃土豆是极其贫困的证明，Stedman Jones, *Karl Marx*。

57. Dechambre, ed., *Dictionnaire encyclopédique des sciences medicales*, 229–231.

58. Broadberry et al., 'Industry', 169–170; Malanima, 'Urbanization', 244.

59. Nitti, 'The Food and Labour-Power of Nations', 35, 38.

60. Rabinbach, *The Human Motor*.

61. Rabinbach, *The Human Motor*; Carpenter, *Protein and Energy*; Kamminga and Cunningham, eds., *The Science and Culture of Nutrition*; Cullather, 'The Foreign Policy of the Calorie'; Treitel, 'Max Rubner and the Biopolitics of Rational Nutrition'; Biltekoff, *Eating Right in America*; Simmons, *Vital Minimum*.

62. Hildesheim, *Die Normal-Diät*; Meinert, *Armee- und Volks-Ernährung*; Rabinbach, *The Human Motor*, 128–130.

63. Rubner, *Volksernährungsfragen*, 48–52 (48, quote); Rabinbach, *The Human Motor*, 262–264; Treitel, 'Max Rubner and the Biopolitics of Rational Nutrition', 14 (trans.); Treitel, 'Food Science/Food Politics'.

64. Meinert, *Armee- und Volks-Ernährung*; Wiegelmann, *Alltags- und Festspeisen in Mitteleuropa*, 87–90; Dickler, 'Organization and Change in Productivity in Eastern Prussia'; Teuteberg, 'Der Verzehr von Nahrungsmitteln in Deutschland pro Kopf und Jahr'; Komlos, 'The New World's Contribution to Food Consumption', 68–74; Bass, 'The Crisis in Prussia', 187–189; Mahlerwein, 'The Consequences of the Potato Blight in Southern Germany', 214–216.

65. Blum, *The End of the Old Order*, 271–276; Bekaert, 'Caloric Consumption in Industrializing Belgium'; Vanhaute, '"So Worthy an Example to Ireland"', 123; Paping and Tassenaar, 'The Consequences of the Potato Disease in the Netherlands 1845–1860', 157 (quote).

66. Morell, 'Diet in Sweden during Industrialization, 1870–1939'; Henriksen, 'A Disaster Seen from the Periphery', 297–298; Gadd, 'On the Edge of a Crisis', 321–322.

67. Smith and Christian, *Bread and Salt*, 254–255 (quote), 278–287.

68. Caird, *Our Daily Food*, 35; Burnett, *Plenty and Want*, 113 (quote), 151; Overton, *Agricultural Revolution in England*, 102–103.

69. Tiedemann, *Geschichte des Tabaks und anderer ähnlicher Genußmittel*, 388–389; Meinert, *Armee- und Volks-Ernährung*, I: 139, II: 186–188, 358.

70. Moleschott, *Der Kreislauf des Lebens*, II: 570.

71. Moleschott, *Lehre der Nahrungsmittel*, 115–120, 184; Moleschott, *Der Kreislauf des Lebens*, II:568–570.

72. Moleschott, *Der Kreislauf des Lebens*, II: 268–270; Moleschott, *Lehre der Nahrungsmittel*, 80, 115–120 (119 quote), 184; Meinert, *Armee- und Volks-Ernährung*, II: 189; Kamminga, 'Nutrition for the People, or the Fate of Jacob Moleschott's Contest for a Humanist Science', 26; Meneghello, *Jacob Moleschott*, 63, 67, 83, 117, 133–137, 309.

73. Rabinbach, *The Human Motor*, 129. 或参见 Pissling, *Gesundheitslehre für das Volk*, 68; Hildesheim, *Die Normal-Diät*, 37; Meinert, *Armee- und Volks-Ernährung*, II: 188。

74. 目前的建议量大约是这个数值的一半。

75. Nitti, 'The Food and Labour-Power of Nations', 41, 57.

76. Nitti, 'The Food and Labour-Power of Nations'.

77. Schulze-Gävernitz, *The Cotton Trade in England and on the Continent*, 135–138 (quote), 176–189; Rabinbach, *The Human Motor*, 129.

78. Schulze-Gävernitz, *The Cotton Trade in England and on the Continent*, 200–201.

79. Nitti, 'The Food and Labour-Power of Nations', 60–61.

80. Styles, 'Custom or Consumption?', 103.

第六章　安全的土豆

土豆皮特

"跟我来夫人，我给你看一两样东西。"一只叫土豆皮特（Potato Pete）的卡通土豆如此许诺道。土豆皮特是1940年英国出版的一份小册子中的主角，是食品部（the Ministry of Food）的发明创造。作为战时工作的一部分，食品部大力鼓励土豆消费。《土豆皮特食谱》（*Potato Pete's Recipe Book*）旨在帮助家庭主妇将土豆融入日常饮食的每个阶段，包括早餐和制作布丁。食谱上说，医学建议：每人每天至少吃12盎司（理想情况下是16盎司）土豆。（相当于两个大烤土豆。）它提供了咸土豆华夫饼、土豆鱼蛋糕，当然还有土豆汤的食谱，并且还包括土豆营养价值的信息。这本小册子提供了一些实用的提示，告诉人们如何通过定量配给来让人更长久地享用数量有限的脂肪和牛奶。土豆皮特调皮的注解和双关语，使这本小册子变得生动起来。它将带皮煮土豆的建议粉饰为："好品味需要我穿上外衣。"它在辣芥末土豆的食谱旁写着"我是个心狠手辣的家伙"。在书的封底，土豆皮特穿着一双高筒靴、衣冠楚楚地和两位时髦女性手挽手走在队伍前面，一长串开心的家庭主妇（还有几个男人）跟在后面。[1]（参见图17）

图 17　土豆皮特。如历史学家莉齐·科林厄姆（Lizzie Collingham）所说，土豆是二战的味道。在英国，全国性提高生产的运动非常成功，但促进消费却并不简单。《土豆皮特食谱》试图把一本以土豆为特色的食谱集转变为一部浪漫小戏剧，让土豆皮特扮演一位不太让人信服的主角。它在提供布丁食谱的那一页上向主妇们恳求说："让我做你的甜蜜情人吧。"其他食谱里也有更不雅的言语。

通过土豆皮特的恶搞，我们得知，至20世纪中叶，土豆再次成为改变日常饮食习惯的国家计划颇受欢迎的一部分。英国食品部对土豆的兴趣反映了二战所造成的对国家粮食供应的压力，但这种压力并不仅仅源于这些紧迫的意外事件。自20世纪10年代开始，饥饿、贫困和治国之间关系的概念正在发生改变，而对国家安全的担忧也以新的方式聚焦在农业、粮食供应、日常饮食习惯与国家的关系上。本章描述了土豆与国家福祉在20世纪被重新联系起来的这种发展。

在19世纪末，饥饿对政治和社会的影响，是让世界各地政治家感到忧心忡忡的问题。历史学家詹姆斯·弗农指出，贫困日益被看成一种公共威胁，而不再是个人的灾难，需要"共同社会责任和行动的新规范"。[2] 当时对营养科学和疾病传播机制正在进行的研究，与对损害工人阶级健康的酗酒和疾病的担心相互交织，在科学家和政治家当中形成了一个共识，那就是饥饿和饮食不当可能带来的政治挑战，不再是仅由不需协同的个人便能解决的问题。这些对国家实力和安全的威胁，必须通过国家层面的计划予以化解。

为此，官员们设计了一系列富有想象力的计划和政策，旨在改变国民的饮食习惯，从而确保整个国家的安全。他们这样做的根基奠定于18世纪，当时的政治思想家已开始认为，普通人的饮食与国家的富强休戚相关。与18世纪的前人相比，20世纪的国家能更好地将这些概念转化为有效的政策。尽管18世纪的治国模式坚持认为，国家会通过劳动人民多吃营养丰富的食物而受益，但政客们却鲜有手段来真正改变日常饮食习惯。正如政治人类学家詹姆斯·C.斯科特所言，18世纪的国家缺乏"始终如一的强权，精细的管理网，或能在社会工程中进行更具侵入性实验的详细知识"。[3] 因此，20世纪对饮食的政治重要性的关注，与其说是治国目标的根本改变，不如说是国家为实现其目标所能部署的技术的转变。20世纪和近代早期的国家有许多共同的抱负，但现代国家拥有治国的手段，包括统计局、公立教育等，用斯科特的话说，这些"将他们在这条路上带至所有17世纪的君主梦寐以求却难以到达的远方"。[4] 改变饮食习惯的计划，是这个手段的重要组成部分。

自20世纪初开始，世界各国多地都在试验新的方法，通过改变饮食方式和饮食内容来改善人们的健康。在一些国家，这些方案反映了一个极大扩展的概念，即国家有责任确保人民的福祉，由此走向现代福利国家。但在那些并没采取明确福利主义做法的国家，也同样有改变流行饮食习惯的计划出现。"家庭经济"课程向世界各地的女学生教授人体的营养需求、食材中的维生素含量，以及不同的烹饪技术对做出的菜品营养性的影响等内容，希望她们会使用这些知识来改善自己和家人的饮食。在欧洲、亚洲和美洲的许多国家，也开始有学校食堂和政府资助的低成本餐厅为儿童和工人提供营养均衡的膳食。[5]

这些方案的社会政治尺度是显而易见的。它们的目的是改善参与者的健康状况，使他们成为更好的公民。在英国，从1904年开始的学校晚餐计划，其各个方面都经过精心设计，以指导孩子们建立正确的饮食和文化习惯，如克己、讲卫生等。这些措施，如一位自由民主党议员所说，是"一流的帝国主义"，因为健康的人口确保了英国军事和经济的成功。[6]在墨西哥，20世纪20年代上台的左翼政府制定了一项雄心勃勃的公共餐厅计划，旨在反复灌输摄入富含营养食物的饮食习惯和现代化的行为。新成立的公共援助及公共卫生部（Ministries of Public Assistance and Public Health）下属的医师们认为，工人阶级饮食中的低蛋白质含量和其他营养缺陷导致生产力低下，从而削弱了墨西哥。为解决该问题，政府建立了遍布各处的"国家餐厅"，为工人阶级家庭提供有补贴的、营养丰富的膳食。这些机构还开办烹饪和营养课程，目的都是为了将顾客转变为有用的现代公民。同样的雄心，也导致了英属印度殖民政府和20世纪初日本政客的态度，他们也同样希望通过改变大众饮食来提高生产力、增进繁荣。[7]

这类计划把人口的膳食营养与国家的强大联系在一起，它们反映出18世纪政治思想家的倡议在许多方面都在延续。士兵、可能成为士兵的人、劳动者的健康、活力，仍然是被特别关注的目标。这些计划在各处都受到劳动人民不同程度的欢迎，他们感激其所提供的饮食和社会机会，但反对它们的评判语气和规训意图。

粮食、公共健康和第一次世界大战

第一次世界大战的爆发，成为创造这种政治饮食工具的转折点。数百万士兵的调动，需要精准的补给体系来给予这些新兵必需的食物，行政专业知识和制度性基础设施因此得以创建，它们使得日常饮食习惯可得到更持久的干预。这场战争还把人们的注意力集中在营养不良带来的政治和军事后果上。

许多官员认为，新兵缺乏营养的健康状况削弱了军事力量。在英国，这种担忧是之前布尔战争（Boer War）的启示带来的。当时的征兵官员判定，有 40% 至 60% 的新兵不适合服役。1916 年后，为给可招募的新兵分级而成立的医学委员会，也得出了同样令人不安的结论。英国首相大卫·劳合·乔治（David Lloyd George）认为，健康状况的不佳，使英国失去了至少 100 万名可招募的新兵。他坚持认为，按照委员会使用的分类系统，"你不能用 C-3 级人口来维持 A-1 级帝国"。[8] 培养 A-1 级人口，需要政府采取行动来确保人们饮食得当。招募新兵的军官们认为，对健康饮食基本原则的无知，加之糟糕的预算技能，是新兵中普遍存在令人沮丧的营养不良现象的原因。1918 年，曼彻斯特的一位军事专员表示，孩子"体质差"的主要原因，是他们的母亲"对最简单的家庭经济和烹饪方式的完全无知"。他相信，为工人阶级家庭的女孩们引入强制性家政科学课程，将会消除或至少大大减少不适合服兵役男性的数量。[9] 正如印度的殖民者把饥荒归咎于当地人的落后一样，英国的劳动人民，也被认为要为他们常生活在其中的恶劣环境承担责任。

另有一些人则认为，造成营养不良的主要原因是贫困而不是无知，但这并没有减少人们的担忧。巧克力制造商朗特里的继承人西博姆·朗特里（Seebohm Rowntree），针对贫困和家庭支出，在世纪之交的约克郡进行了一项广泛的调查。他仔细汇编了家庭饮食的信息，使他能够证明，在约克郡，贫困导致劳动阶级消耗的食物"比科学家证明的保持体能所需的食物少25%左右"。[10] 这些营养不足的人，使得征兵军官感到担忧，并且削弱了英国的战斗力。大规模的征兵将折磨劳动人民的健康欠佳问题暴露在聚光灯下，也凸显了穷人营养不良对整个国家的影响。

还有一则广为流传的共识是，政府需采取行动协调农业生产与战时粮食需求。交战各国纷纷设立新部委，显然是为了应对这些挑战。英国于1916年成立了食品控制部（the Ministry of Food Control），奥匈帝国成立了联合粮食委员会（Joint Food Committee），美国在参战时成立了燃料与食品管理局（Fuel and Food Administration），德意志帝国建立了战时食品办公室（War Food Office）、战时小麦局（War Wheat Corporation）和帝国土豆办公室（Imperial Potato Office）。这些新机构与国有化铁路、码头和军需品工业一起，极大地增强了国家对经济生产和日常生活的控制。它们尤其为指导食物生产和消费提供了手段。[11]

如德国土豆办公室所指出的那样，土豆在这些大计划中占据了重要的一部分。[12] 国家层面对土豆的重新关注，既反映了它在欧洲许多地区饮食中的重要性，也反映了对营养的科学认识的变化。20世纪初，营养学家对蛋白质的目光短浅的关注，开始被更复杂的、对有关好营养的化学成分的看法所取代。土豆因此恢复了作为健康食物的些许光彩。"一战"爆发之前不久，食品科学研究的领军人物威尔伯·阿特沃特（Wilbur Atwater）断言："简单、量大、标准的食材，如较便宜的肉、鱼、牛奶、面粉、玉米粉、燕麦片、黄豆、土豆，与任何昂贵的食材一样容易消化、富含营养，并同样满足人们保持健康所需。"[13] 劳动人民得到保证，土豆同牛排等更贵的食物一样营养丰富。因此，营养学家们一口咬定，工人阶级不需要由于错误地认为更贵的食物更健康、更能维持生命而为此耗尽他们有限的资源，也没有任何理由去为提高收入抗争而使饮食更奢侈。

战时的英国向农民们颁布了种植令，规定他们须种植的小麦和土豆的面积。从1917年起，英国政府还购买全年收获的土豆，并以固定价格出售。这些提高产量的工作是成功的，土豆的收成比战前水平高出近50%，人均每周消费量从3.67磅跃升至5.26磅。[14] 其他规定还限制了面粉在面包和糕点中的使用，将餐馆和酒店允许使用的分量固定下来，并把浪费粮食定为犯罪。到1918年，政府的食品规定文件自身的重量都达到半磅。违规者会被告上法庭。比如在1917年，艾米丽·埃文斯（Emily Evans）因被指控以超过法定最高价的价格出售土豆而出现在彭布罗克郡（Pembrokeshire）的被告席上。[15]

各国支持战争的宣传都试图鼓励更多的土豆消费。1918年，艾奥瓦州小镇马科基塔（Maquoketa）的一个橱窗内的口号是"参军入伍，吃土豆锄德皇"。美国政府希望增加国内土豆消费，以抵消出口给欧洲盟国的2000万蒲式耳小麦。因此，土豆经常出现在新成立的食品管理局发布的不含面粉的战时"胜利食谱"中。一本战时食谱上这样鼓动着："吃的土豆有淀粉，战士才能往前行；吃下一块烤土豆，小麦才能装满舱。吃下土豆省小麦，打败德国坏皇帝。"[16] 马科基塔的斯塔克与卢克西（Staack & Luckiesh）药店的老板们，利用他们的商店橱窗为这个建议展现了富有想象力的视觉舞台。在窗口的前面，是一支舞动着军刀、挥动着星条旗的土豆小队，领导他们的，是代表美国驻欧洲远征军首领潘兴将军（General Pershing）的一颗大土豆。说明标签上写着赞扬这支"最新战斗部队：爱国土豆（Potatriots）"编队的话语。"吃土豆，省小麦"是海报上明确传达的信息。在店内，人们能找到合适的食谱。（参

图18 "吃土豆锄德皇"的橱窗展示。"土豆是个好士兵,带皮一起吃掉它",爱荷华州一家药店的橱窗上贴着这样的告示。所有参战国的政府都宣传土豆,尽管还出现了供不应求、实行定量配给的情况。美国不仅建议烹调者连皮吃土豆,还教导如何制作土豆面包和其他爱国的小麦替代品。

见图18）

在德国,土豆是第一种受到价格上限管制的食品,而且官员们把维持土豆供应看成最重要的事务。政治家们有理由担心,土豆供养减少会降低人口的工作能力,并可能引发政治动荡。正如1915年年初柏林的一位官员所言,"土豆问题是最重要、最迫切的,因为土豆对贫困人口起着如此重要的作用"。除了控制土豆价格外,政府还要求面包师在制作面包时尽量多用土豆。(为了维持小麦供应,面包从1915年1月起实行定量配给。)这样制作出来的面包被称为 K-brot（K面包）,K 同时代表"Krieg"（战争）和"Kartoffel"（土豆）。然而,到了1915年3月,土豆本身也供不应求了，也开始与面包和面粉一起实行定量供应。老百姓发现越来越难以获得足够的食物。若干城市建立了公共食堂，结果是成败参半。向学童提供营养早餐的特别方案也得到了尝试。生产军需品和其他原料的工厂建起了食堂，它们的厨房得到定量的肉类和土豆，以确保关键行业的工人有充足的食物。[17]

德国政府确保土豆等基本粮食稳定供应的工作，因政策设计不善和连续的粮食歉收而遭到了破坏。英国1914年起对德国实施的封锁使情况进一步恶化，因为德国之前三分之一的粮食依赖进口。到1917年，德国人面临着真正的饥饿。为帝国政府当顾问的营养学家马克斯·鲁布纳认为，民众正在经历"缓慢的饥饿"。[18]试图通过从国外进口土豆来解决短缺问题，不过是把问题转移到了出口国身上。在中立国家瑞典，土豆出口到德国引发了严重抗议，首先是针对配给，然后是针对国内粮食供应减少。不满日益升级。工人委员会开始呼吁分配土地以实现小规模土豆种植。随着土地改革运动的进行，要求立法限制工作时间以及其他实行社会政策的呼吁也开始出现。1917年瑞典"土豆革命"期间，大约有25万人参加了抗议活动。[19]

在整个欧洲，这些历程有助于重新定义粮食、人口与国家之间的关系。老百姓能更清晰地表达国家应该确保他们获得足够食物的期望。历史学家西蒙·汉考克（Simon Hancock）指出："日常生活的界限已经改变。"[20]在德国，因土豆、面包、黄油和肉类的短缺，公众对战争的支持率降低，最终导致政府崩溃，并于1918年11月垮台。[21]与此同时，战时定量配给、政府经营的食堂的建立、农业和消费者委员会的成立，以及其他改变大众饮食的尝试发展出来的手段，对日常饮食习惯的干预远比18世纪更大。确保土豆的充足供应，并鼓励人们多吃土豆，就包含在这些新手段当中。

培育更好的土豆

至20世纪早期,植物育种技术已经改变了欧美的商业化农业。大学和研究机构的科学农学项目创造了新的杂交品种,它们高产、耐虫害、能在具有挑战性的环境条件下茁壮成长,还具备其他好品质。专门从事土豆育种研究的实验室也开始出现。剑桥大学于1912年建立了土豆育种研究所,该所后来推出了许多获得商业成功的土豆品种,如"马里斯派柏"(Maris Piper)土豆。康奈尔大学是美国主要的植物育种中心,它从20世纪早期就开始开设土豆育种课程。[22] 玉米、小麦等其他作物的新品种试验的类似项目同时也在进行。因此,成功的杂交品种在生产力和适应性方面彻底改变了欧美的商业农业。

对土豆的育种工作主要关注的是产量和抗病害,如80年前导致欧洲满目疮痍的晚疫病。培育这种作物的方法之一,是将已定植的商业土豆品种与不太知名的、具有相关品质的非商业土豆品种杂交。出于这个原因,非商业的和野生的品种非常走俏,尽管训练有素的农学家们普遍不看好与它们相关的"传统"农业。育种者的猜测很正确,搜寻这些被忽视的地方品种或地区栽培品种的最佳地点,是安第斯山脉。科学家们希望能对存在于智利、玻利维亚、秘鲁、厄瓜多尔和哥伦比亚的许多野生土豆品种和地方品种尚未开发的育种潜力加以利用。南美洲的传统土豆和野土豆,由此为可转化为现代土豆种的品种提供了原料。[23]

利用这一资源培育新的商业品种的工作开始于19世纪,在很大程度上受到个体种植户和育种者的推动。美国目前通常用于烤土豆和工业薯条的"赤褐布尔班克"(Russet Burbank)品种,就是这样一次南美之旅的最终结果。[24] 至20世纪30年代,收集有前景的地方品种的探险,得到了农业专门研究机构在基础设施上的正式支持,并享受政府的直接资助。埃尔文·鲍尔(Erwin Baur)是位于慕赫堡(Müncheburg)的威廉大帝植物育种研究所(Kaiser Wilhelm Institute for Plant Breeding)的所长,他于1931年前往南美,带回了数千个土豆标本。[25] 同年,俄罗斯遗传学家尼古拉·伊万诺维奇·瓦维洛夫(Nikolai Ivanovich Vavilov)两次到拉丁美洲进行探险,收集了大量的土豆和许多其他作物。[26] 1938年,在植物学家E. K. 鲍尔斯(E. K. Balls)和W. 巴尔福·古尔利(W. Balfour Gourlay)的带领下,大英帝国土豆采集探险队也启程前往拉丁美洲。采集者们在安第斯参观了农村市场,与当地人交谈,并咨询了科学家和育种者。[27]

并不是只有欧美的科学家们才重新对土豆产生兴趣。被欧洲探险队寄予希望,以利用其专业知识的安第斯生物学家们,与20世纪20年代末建立的实验育种站和国家研究项目也是不无关系的。20世纪40年代,秘鲁新成立的农业部开始了一系列以传统品种为重点的育种项目,随后启动了一项国家土豆改良计划和其他几种作物的改良计划。曾在玻利维亚和美国学习农学的生物学家卡洛斯·奥乔亚(Carlos Ochoa)担任了秘鲁国家土豆计划的主管。奥乔亚培育了许多杂交土豆新品种,包括现在秘鲁种植最普遍的商业品种"复兴"(Renacimiento)。玻利维亚和哥伦比亚的农业部,也从20世纪30年代末开始资助土豆采集和实验农场。[28] 中国科学家管家骥在国民政府的支持下,于20世纪30年代设立了土豆研究项目,后来在成都中央农业实验所开设了土豆育种课程。[29] 印度的中央土豆研究所于1949年在巴特那(Patna)成立,合并了几所成立于20世纪30年代的小型育种站。印度后来在开发适应亚热带条件的新品种方面发挥了主导作用,现在是仅次于中国的世界第二大土豆出产国。

这些远足所收集的品种和通过实验开发的品种,保存在20世纪40年代始建的国家级马铃薯资源库中。奥乔亚得到的品种储存在农业部设在曼塔罗山谷(Mantaro Valley)的试验站中,后来的秘鲁国家种质资源库便是以此为基础建立起来的。大英帝国土豆采集探险队的材料,被保存在大英帝国马铃薯资源库中,它后来更名为英国马铃薯种质资源库(Commonwealth Potato Collection)。其他品种被保存在美国马铃薯基因库(US Potato Genebank)、德国马铃薯种质资源库(German Groß Lüsewitz Potato Collection),以及一些合作资源库,如位于瓦赫宁根(Wageningen)的荷兰-德国马铃薯资源库(Dutch-German Potato Collection)等。[30]

这些国家级中心的建立,就像资助收集土豆的国家探险队一样,表明各国认识到了土豆作为管理国家粮食供应之资源的重要性。第一次世界大战,尤其让欧洲国家对需要建立健全的机制来确保这一点毫不迟疑,而到了20世纪30年代,植物育种科学已经能够随时提供帮助了。[31]当1939年战争重新降临欧洲时,各国政府利用了这些机制来应对它们和其公民所面临的前所未有的挑战。和从前一样,土豆加入了为战争作贡献的队伍。

战争的味道

在第二次世界大战的全球灾难中,粮食是核心的问题。德国和日本发动侵略战争跟粮食供应不足有一定关系。战争爆发后,要养活调动起来的近20亿人,还包括从事农业、驾驶运输船、在关键行业作业,和以其他方式为战争作贡献的数百万人,各国更加面临着巨大的挑战。正如历史学家莉齐·科林厄姆所证明的那样,无论在国家粮食政策上,

还是在个人生存技能上，土豆都是举足轻重的。³²

战时的英国政府积极推广土豆，以减少对粮食进口的依赖，同时提高国民的整体健康水平。在20世纪20年代到30年代间，土豆消费的停滞甚至下降，引起了已把土豆视为健康饮食基石的政府营养学家的担忧。³³ 由于"战争比和平时期需要更好的体质和健康状态"，官员们认为有必要从根本上改变国民的饮食习惯。《公共健康》(Public Health)杂志在1941年曾抱怨说，"英国人的饮食是随意的，它现在的构成，很大程度上取决于机会，以及随着工业化和繁荣而发展起来的口味。众所周知，国内的某些地区，无论贫富，都存在着食物摄取不足的问题"。³⁴ 合理饮食，包括大量提高土豆消费，是个人的义务，也是国家的需要。

在1939年，英国严重依赖进口粮食，如从加拿大和阿根廷进口用于生产面包和饲养牲畜的数百万吨小麦。官员们认为，为这些目的进口粮食，是对英国脆弱的航运能力的高风险而低效的利用。政府鼓励以土豆和用国产小麦制成的全谷物面包为基础的饮食，以代替白面包和进口谷物喂养的新鲜肉畜。（肉类罐头等能量密集型加工食品从属国进口，作为这一推荐饮食的补充。）相应地，在全国范围的"翻耕"运动中，大约240万公顷的牧场被转变为可耕种的农场，农户在郡战争执行委员会的指导下增加了用于种植土豆的农田面积。所有家庭都被要求养兔子、养鸡，尤其是种植土豆。土豆被宣传为能量和维生素C的极佳来源，适合英国的农业条件，适合家庭种植户的能力。食品部的马铃薯司在增加产量方面做得非常成功。到战争结束时，土豆种植面积比1939年翻了一番。成功如此巨大，以至于个人消费都跟不上了。³⁵ 土豆皮特调皮的倡议，就是这项将1940年积累的50万吨剩余土豆进行消耗的大运动的一部分。营养学家通过广播和海报宣传来强化皮特传递的信息。英国政府如它在前几个世纪所做的那样，也鼓励殖民地和盟国地区种植土豆。伊拉克、叙利亚、埃及、巴勒斯坦和塞浦路斯都出台了土豆种植计划。以英国在肯尼亚的基地为例，这样的工作使英国得到了基库尤（Kikuyu）农民种植的当地土豆的供应。[36]

当苏联在1941年加入战争时，它养活自己的工作面临着巨大的困难。斯大林的集体化计划导致了乌克兰1933年严重的饥荒，造成大量死亡，农业生产陷入混乱。在粮食需求更大的时刻，战时动员却使农村失去了1900万农业劳动者。因此，到1942年，谷物和土豆的收成较战前都下降了三分之一。尽管实行了定量配给制度，士兵和平民却一直都在挨饿。德国军队占领乌克兰的重要农业地带后，局势更加恶化。至少有300万苏联人——可能更多——在战争中饿死。[37]分配和供给的问题迫使士兵们自己寻找食物；在装配线上，骨瘦如柴的工人因饥饿而倒下。苏联官员的应对，是敦促每个人都种植土豆。从19世纪开始，面包和土豆就已成为大众饮食的支柱，而且与谷物不同的是，土豆相对容易在小块土地上种植。报纸提供农艺建议，工厂管理者也努力为工人们提供配额。结果到了1943年，莫斯科变得"漂浮在土豆作物的绿色海洋中"了。包括苏联军队在内的许多组织，在当地的荒地上建立农场，以补充微薄的国家口粮。到1944年，这些附属农场出产了260多万吨土豆和其他蔬菜，足以保证每天额外提供250卡路里的热量给那些有幸得到它们的劳动者。这些工作，为大约2500万人提供了基本的食物来源。[38]

无论有无自上而下的鼓励，全球各地的人们，都求助于土豆来继续活命。苏联农民在战争期间几乎完全依靠自己种植的粮食，他们的土豆消费量增加了一倍多。一位俄罗斯人回忆，"他们早餐、午餐和下午茶都吃土豆；用所有方法吃——烤、油炸、做土豆饼、做汤，但最常见的就是简单煮一下"。[39]住在罗马城外乡下的可怜劳工乔瓦尼·塔索尼

（Giovanni Tassoni）一家，在占领的德国军队于1943年开始征用粮食后，将整个菜园都用于种植土豆。在华沙犹太人区忍饥挨饿的犹太人，悄悄潜入城市的腹地，希望能挖到一些土豆，偷带回去给家人吃。在1.2万公里之外的新几内亚，被最高指挥官抛弃的孤立的日本士兵们，试图偷村民的土豆果腹。正如科林厄姆所说，家庭自产土豆、冻土豆、脱水土豆、腐烂的土豆、碎土豆、熟土豆、偷来的土豆、定量配给的土豆，就是二战的食物，也是二战的味道。[40]

在南太平洋的日本士兵，也对土豆无不熟悉，因为自20世纪20年代以来，作为重塑国民饮食习惯的更广泛运动的一部分，日本政府就已鼓励食用土豆。按照最新的营养研究，日本军方制定了一项全面的大众饮食计划，大量使用猪油、肉类、土豆和洋葱，他们认为这是供养士兵的最佳（也是最经济的）方式。新颖的菜肴，如油炸土豆饼或肉饼（*korokke*），在军人中很受欢迎，并使得日本人熟悉了西式菜系。在战争期间，广播节目和杂志通过提供食谱和烹饪建议而强化了这一过程，一切都是为了改变国民的饮食习惯。街坊邻里的组织协会，也重申政府的讯息，同时推广新的烹饪方式，20世纪40年代建立的公共餐厅也是如此。[41]

在德国，纳粹同样也试图改变饮食习惯，以创建他们想要的那种政体。纳粹的政治哲学将德国国家健康与德国人的个人健康等同。因而公民的义务，就是要照顾好自己。"健康等于责任"于1939年成为党的官方口号。马克思主义者认为应该允许人们对自己的身体做任何他们想做的事。希特勒青年团的成员在训练手册中得到的提醒是，对于好的纳粹党员，营养"不是个人问题！"[42]战时的定量配给，是不给多余的那部分人口提供食物，而给合适的德国人提供营养。犹太人、精神病人、政治对手，以及其他许多被纳粹国家否认的人，完全得不到充足的配给。

被占领地区的犹太人每天仅得到420卡路里的食物，而从事重体力劳动的德国工人分配到的，是这个数量的十倍。[43]

定量配给还有另一个目的，那就是通过减少德国对进口的依赖来实现其"营养自由"。一战的饥饿和内乱，依然笼罩着纳粹的记忆，他们决心不再重蹈前几届政府失败的粮食政策的覆辙——它们被认为是德国战败的原因之一。从1933年起，国家社会主义运动取得了一些成功，使德国在主要粮食上实现了自给自足。到1939年，几乎所有的糖、肉、谷物和土豆都由国内出产。最后这项成就尤为重要，因为正如纳粹宣传者所坚持的那样，德国人是"土豆的人民"，他们在日常饮食中把土豆作为基础。因此，土豆的充足供应，被视为对德国的独立和军事野心的实现极其重要。[44] 德国在1941年入侵苏联时，苏联著名的植物育种机构是它最觊觎的资产之一。德国科学家接管了苏联的大部分研究所，而且在1943年之后摧毁或带走了成千上万的植物样本，包括土豆的育种材料，他们以"国家利益"的名义为这些行为进行辩解。[45]

纳粹把土豆作为极力宣传的对象。无数的广播、杂志和培训课程都在宣传如何烹制这种"营养丰富、能填饱肚子而又便宜"的蔬菜。在战争期间，土豆每年的消费量增加了一倍多，从1200万吨增加到3200万吨。吃土豆被视为爱国姿态而得到鼓励，家庭主妇得到建议如何以最健康、最经济的方式烹制土豆，以及如何用经过特别处理的土豆皮喂猪。[46] 土豆消费能够增加，得益于1939年以后德国和其占领地上土豆耕种的增加。战俘和其他被强迫的劳动力被组织起来收获这些战利品，它们被运回德国，导致被占领地的粮食供应进一步减少。譬如，波兰在1942年就被指定向德国提供15万吨土豆、60万吨谷物和3万吨肉类。[47]

纳粹认为人与土豆关系密切，因此他们的目标不仅仅是确保土豆的持续供应，还希望放大人们食用的土豆的"德国性"。国家在植物育种项目上投入了大量资金，以培育出抗马铃薯癌肿病、晚疫病等病害的健壮、适合本地的品种。《帝国批准品种名录》(*Imperial List of Approved Varieties*)由此而产生，它规定了可以进行商业销售的土豆种类。1941年，销售甚至种植没有列入名录的土豆都属于违法行为。在20世纪10年代，德国农民种植了大约1500种不同的土豆，而到1941年，他们被允许种植的品种只有74种了，其他品种都遭到禁止。正如纳粹规定哪些民族和人种可以居住在德国的土地上一样，他们也规定了哪些土豆允许被种植。这样做的目的是培育出独特的国家粮食，用来给国民身体提供营养。通过种植和食用如此极具德国性的土豆，个人不仅提高了国家的经济自主性，也使自己的身体更具德国性。最佳的办法，是把这些土豆和新的"土生土长"型猪肉一起食用，这种猪本身就是以本地土豆为食的。[48]德国的粮食政策及其对土豆的态度，反映了纳粹意识形态更大的目标。

在美国，战时补给的手段同样也受到更大的意识形态背景的影响。在德国，政府希望把德国人吃的土豆和猪肉变得更加本土化。而在美国，政府的工作反映了中央主导的国家营养计划的益处，与保证个人主义之间的紧张关系。保护民众的健康和效率是战时的核心问题，在确保这一点上，土豆依然发挥了作用。美国政府的指导方针强调土豆的多种好处，它们在如1943年的健康饮食可视化"基本七类"等营养图表中得到突出体现。这张饼状图将食物分为构成饮食的七大类，"土豆和其他蔬菜水果"构成了一个大类。

与此同时，罗斯福政府强调，它的目的是让公民在知情的情况下作出饮食选择，而不是规定人们吃什么。它的建议通过广播、报纸杂志以及宣传小册子传播，绝不是为了破坏美国"主动和民主的习惯"。[49]美国饮食习惯委员会（Committee on Food Habits）肯定，"在美国，选择食物是成年的标志之一"，这与 200 年前伊曼努尔·康德的观点遥相呼应。缺乏饮食选择，会使成年人退化到"孩子的状态，导致出现依赖态度和士气低落"。换句话说，过于指令性的饮食指南，可能会使美国民众易于接受法西斯主义。[50] 因此，任何土豆消费量的增加，都应通过教育和说服运动，而不是通过更直接的干预来实现。

1943 年 3 月，定量配给从糖和咖啡扩展到更大范围的其他食品，媒体随之发起了一场宣传攻势，旨在让家庭主妇们相信，这项措施提供了展示爱国主义的机会，并不构成政府对私人生活的侵犯。物价管理办公室聘请了一位顾问（南达科他州[South Dakota]的菲利普·L. 克罗利[Philip L. Crowlie]夫人），同时战争信息办公室也成功地游说广告商对配给计划进行宣传。该办公室还向全国性和区域性报纸的妇女版面编辑提供故事，强调这些措施所允许的自由，而不是强调它们施加的限制。例如，《纽约时报》解释说，该计划的目的是给家庭主妇们"尽可能多的选择"。《每日邮报》发布了包括肝泥香肠、奶油土豆和菠菜在内的食谱，希望读者能借此机会展示个人烹饪创造力和爱国主义情怀。外观漂亮的迷人菜品，再加上烹饪技巧，把战时的烹饪转变为女性在家庭建设艺术上才华的展现。[51]（参见肉冻土豆沙拉食谱）

*

二战期间，对于陷入这场全球战争的许多人而言，土豆是活下去的关键。它的重要性反映了它为独立于更大的统治框架的普通民众提供服务的非凡能力。莫斯科和贝德福德（Bedford）的土豆种植者，因对土豆丰富淀粉的共同依赖而被联系在一起。而同时，土豆在国家供应计划中的不同地位，又反映出敌对国家在现代化和治国方略上采取的不同模式。在英国，土豆巩固了战时社会主义，这种制度允许直接干预公众的饮食习惯，结果是公共健康得到显著改善。在美国，吃土豆被认为是展示个人创造力的机会。在这些情况当中，土豆的突出地位，反映了战时动员对粮食供应的超常需求。

在战争末期，在防止进一步冲突的国际讨论中，粮食成为中心问题。1945年后成立的世界银行、联合国粮农组织（FAO）等国际机构，反映了这样一种信念：政治不稳定与贫困，尤其与饥饿和营养不良密切相关。然而，在雅尔塔和布雷顿森林的酒店中被设想出来的世界新秩序，没有给予不起眼的土豆在根除这些罪恶上的多少功劳。

肉冻土豆沙拉

整个二战期间，《纽约时报》一直提供每周菜单计划，并定期对食物进行报道。该报的第一位美食编辑玛格特·墨菲（Margot Murphy）在她的专栏中把政府的建议转化为具体的食谱。1943年传来了土豆大丰收的消息，随之而来的是各种土豆菜肴的食谱，包括这道分层肉和土豆沙拉。它用照片对每一步进行说明，最后一张照片是布置优雅的餐桌上摆放着一盘脱模的菜，周围点缀着鸡蛋和蔬菜沙拉。墨菲将其描述为在炎热的天气里食用的"凉爽而诱人的主菜"。

肉冻土豆沙拉

第一步　在一只碗中制作混合物A：将2杯煮熟的小肉块——牛肉、羊肉、红肠、香肠、午餐肉等均可——与1.5杯芹菜粒或碎卷心菜、1.5杯切成薄片的萝卜、1/4杯切碎的腌黄瓜混在一起。

第二步　在第二只碗中制作混合物B：将2杯煮熟后放凉并切成小块的土豆，与半杯黄瓜粒、半茶匙盐、半茶匙美式芥末、半茶匙白糖、1/3杯蛋黄酱和碎大蒜、醋各一汤勺混在一起。

第三步 在第三只碗中,放入1.5汤勺明胶,用1/3杯凉水使其软化,加入2杯开水,加入3块肉汤冻,搅拌至完全溶解。在吐司烤盘底部饰以5条切片黄瓜,薄薄敷上一层明胶混合物,放凉至凝固。将剩余肉冻放凉至变成浆体,将其中2/3加入混合物A,1/3加入混合物B。

第四步 吐司烤盘底部明胶凝固之后,依次将混合物A的一半、混合物B、剩余混合物A倒入,放凉至完全凝固。

第五步 明胶凝固之后,将其倒入大浅盘中,用生菜、土豆片和熟鸡蛋做点缀,配上蛋黄酱食用。本食谱适合于8人食用,如4人食用,将所有食材减半烹制。

188 粮食安全与土豆

联合国粮农组织于1945年的成立,标志着改善世界人口营养健康是一项紧迫的政治和经济目标这一共识的国际制度化。成员们称他们的目标是:

> 提高各自管辖范围内人民的营养水平和生活水平，确保提高所有粮食和农产品的生产和分配效率，改善农村人口的状况，从而促进世界经济的发展，确保人类免于饥饿。[52]

在随后的几十年里，更多的机构和计划得以建立以支持这些目标，其中包括联合国儿童基金会（UNICEF）（1946年）、世界粮食计划署（World Food Programme）（1961年）、世界粮食理事会（World Food Council）（1974年）、粮食安全援助计划（Food Security Assistance Programme）（1975年）、国际农业发展基金（International Fund for Agricultural Development）（1976年）和世界粮食安全协约（World Food Security Compact）（1985年）等。这些机构一致坚信，营养状况良好的全球人口将会增强政治的稳定，并推动全球经济不断扩张。如历史学家尼克·库拉瑟（Nick Cullather）所解释的那样："战后秩序的构建从粮食开始。"[53]

这些信念建立在之前的基础上，同时也反映了战后的发展。欧洲和美国的专家预测，世界人口将很快超过粮食供应所能支持的水平，引发一场全球性的马尔萨斯灾难。他们对贫困和人口过剩威胁世界稳定的担忧，似乎在印度、中国、拉丁美洲等地的民族主义运动崛起中得到了证实。惊恐的西方政府，希望减少贫困和积极的现代化计划，能够抑制社会主义的吸引力。[54] 而在得到更好供养的人口当中，又会产生效率更高的劳动者，推动经济飞速发展。因此，能够养活全国人口的现代化农业，是经济发展和稳定的关键。健康、充满活力的人口会影响国家财富和稳定的观点，可以追溯到启蒙运动时期。亚当·斯密在1776年曾发现："人在吃得差的时候工作强过吃得好的时候，心灰意冷的时候工作强过精神抖擞的时候，经常生病的时候工作强过一直健康的时候，这些似乎是不太可能的。"[55]

营养与经济、政治稳定之间的这些关系是"粮食安全"概念的基础。这个词1974年开始出现在联合国文件中，指的是当时不久前的粮食歉收而造成的粮食短缺威胁。讨论的重点是粮食储备的需要，以及通过发展计划增加产量的重要性。几年后的1979年，联合国粮农组织发表了《为粮食安全奋斗》(The Struggle for Food Security)，强调粮食生产的增加应伴随（或先于）社会变革。这个概念继续演变，从最初关注宏观经济和农业政策，到更多地关注如何使用粮食，而不是粮食总量问题。需要解决的粮食安全问题的范围，在家庭、地区、国家和全球经济体系之间波动。然而，在这些不断变化的概念背后，是一个共识：各国人口的饮食健康，是一个具有全球政治和经济重要性的问题。1996年联合国国际粮食峰会上发表的《世界粮食安全罗马宣言》(Rome Declaration on World Food Security)很好地抓住了这一政策共识：粮食无保障会威胁"国际社会自身的稳定"，并会侵犯基本人权。[56]

粮食安全的首要原则源于"发展"的概念。发展过去是（现在也是）一项旨在促进贫穷国家工业化和经济增长的国际战略，否则这些国家就会缺乏发展所需的经济资源。它的根本目标是通过直接的投资和干预计划，将所谓的落后、过度庞大的农业人口转变为工业劳动力。从20世纪50年代开始，由西方政府、世界银行等国际机构以及洛克菲勒基金会（Rockefeller Foundation）等私人组织支持的雄心勃勃的发展计划，在世界许多地方开始推行。这个概念的前提，是认为经过适当培训的顾问带来的技术和科学干预，将成功地推动这样的转变。仔细建模、建立从个案研究中获得的知识库、培训当地专家等都是重要的因素。发展的实践和技术，就像发展（development）这个词本身一样，暗含着成熟的过程，或从传统经济实践到现代经济实践的过渡。不过，如库拉瑟所指出的那样，"传统"实践实际上往往是近代殖民干预的结果。[57]

农业变革是发展的根本。发展计划除了刺激工业的方案外，还把目标放在使农村农业从低效率的自给自足转向盈利的商业化生产上。重点主要集中在农村信贷系统、机械化和种子新品种传播，以及它们发挥潜力所必需的化肥和杀虫剂上。培育这些新品种的工作，在"发达"和"发展中"国家的合作者之间于多种层面上进行。印度、中国等地的科学家到欧美学习，而西方科学家则在发展中国家的研究所长期或短期任职。这种松散的研究项目网络，取得了一些显著的成功，例如康奈尔大学培养的一位中国台湾植物遗传学家，培育出了 IR8 型半矮秆杂交水稻。IR8 的产量显著高于现有的商业品种，且不会因谷粒的重量而倒塌。根据一些估计，部分出于这个原因，发展中国家的粮食供应在 1960 年至 1990 年间大约增长了 12%。所谓的"绿色革命"为减少饥饿作出了实实在在的贡献，它的倡导者希望它会提高全球粮食的安全，并减少左翼政治运动在世界上最贫穷的人中的吸引力。[58] 因此可以说，粮食供应是整个发展计划的重中之重。

在发展和粮食安全的最初概念中，人们到底吃什么食物、形成饮食习惯的文化习俗等，都算不上是重要的因素。印度农学家有时也提出，育种者需要考虑到小麦新品种"做面饼（chapati）的品质"，但是国际专家认为这种担忧是"细微的"。[59] 非商业性、自给自足的作物引起的关注更少。联合国粮农组织最初的关注点主要是谷物。一些成员建议将它的职权范围扩大到包括"在消费中起着重要作用的其他基本粮食"，这时他们头脑中想的是能进行商业交易的商品如食糖、奶粉和肉类，而不是块根和块茎作物。联合国粮农组织的数据显示，在发展中国家，块根和块茎作物在人均卡路里摄入量中占比多达 50%。[60] 第四章曾经提到，在毛泽东时代的中国，农民们向农学家们恳求的是如何改良甘薯的品种，而没有把重点放在杂交水稻上。1951 年，哥伦比亚农业部部长也曾游说洛克菲勒基金会将土豆纳入其区域研究项目。在20世纪60年代的肯尼亚，尽管山药是农村妇女的基本食物来源，但农业官员并不认为它是一种"重点作物"。一位老农妇试图告诉农业部的推广人员她在山药种植方面的专业知识，她却被告知，他们的"推广计划中没有山药的位置"。这类粮食在粮食安全的早期讨论中极少受到重视。[61]

与国际发展计划有关的农学家普遍承认将注意力放在参与商业贸易的谷物上有其弊端。20世纪70年代，有报告开始强调支持热带块根作物和大蕉研究的必要性，并承认，这些被广泛食用的食物，还没有享受到与它们在全球饮食中的重要性相称的研究水平。1971年，联合国粮农组织承认，总体而言，"这些作物几乎都是在发展中国家种植和消费的，几乎没有进入世界贸易。因此，发展中国家的研究项目很少关注它们"。虽然已经进行了一些工作，但联合国粮农组织承认"还有很多工作要做"。[62] 现代杂交作物的成功也引起了恐慌。它们在商业化农业中无处不在，增加了单一病原体或害虫摧毁大部分作物的可能性。因此，发展机构开始讨论，有必要通过保护可能有益于生成新商业品种的地方品种，来阻止世界"遗传资源"的"退化"。这些植物——或者更准确地说，它们包含的遗传物质——因此被框定为未来发展项目的全球资源。[63]

出于这些原因，发展机构为土豆研究提供了一些资金。例如，从20世纪50年代开始，秘鲁农学家卡洛斯·奥乔亚就利用洛克菲勒基金会的资助，在安第斯高地收集传统土豆品种。[64] 该基金会还一直考虑建立一家与已有的研究水稻和玉米的国际机构等同的土豆研究中心。秘鲁的国家马铃薯研究项目，如奥乔亚领头的那些项目，也致力于把土豆纳入发展项目中。然而，国际机构坚持认为，在全球对抗饥饿的斗争中，他们所称的"爱尔兰土豆"基本上无关痛痒。例如，洛克菲勒基金会1970年的一份报告就将土豆划分为与解决世界大多数地区的粮食安全问题"无关"的品种。[65] 土豆与热带的块茎、块根作物，继续成为国际发展计划雷达中的盲点。

至20世纪90年代，由于非政府组织和活动人士的鼓动，对自给农业的这种漠视难以继续下去。1992年，联合国自身也承认，农村的"知识、创新和实践"在保护生物多样性、确保粮食安全方面发挥了重要作用。[66] 因此，关于粮食安全、国家马铃薯育种计划以及土豆作为数百万人日常主食的持续重要性，在国际范围内展开了讨论，最后达成了共识：土豆是确保国家和全球粮食安全的重要资源。联合国决定将2008年定为"国际土豆年"（IYP），它充分体现了土豆作为确保粮食安全的武器的这一新地位。将一整年用一种粮食来命名，即使不是完全史无前例，至少也是不同寻常的。第一个这样的"国际年"是1959—1960年的"世界难民年"。后来还有一些年份被分配给全球妇女地位（1975年）、世界扫盲（1990年）、土著人（1993年）等主题，但在2008年之前，唯一专门用粮食命名的年份是2004年"国际水稻年"。联合国把2008年定为国际土豆年，主要是因为土豆对粮食安全的贡献。这种块茎在国际土豆年的文献中被牢牢地定位为"高度推荐的粮食安全作物"。联合国粮农组织承认，在过去，国际发展计划忽视了土豆对全球粮食安全的贡献，国际土豆年旨在纠正这种疏忽。土豆有助于粮食安全，是因为它被广泛食用，因而是"一种真正的全球粮食"，还因为它有营养，容易种植，而且——关键的是——因为它不是一种国际贸易商品。由于土豆市场一般都是地方性的，购买者不会受到全球价格波动的影响。与之形成对比的是，谷物价格自1990年以来已经显著上涨，给发展中国家带来了严重的饥饿问题。而且，许多人种植土豆是为了自己食用，因此能够完全避开市场。联合国粮农组织预测，正是由于这些原因，在解决全球不断增长的人口对粮食需求的工作中，土豆将成为重要的组成部

图19　2008年国际土豆年标志。2008年,联合国宣布土豆为"隐匿的宝藏",是解决粮食短缺和饥饿问题的良方。它曾经藏在眼皮底下。自20世纪70年代以来,联合国的分析人士已经认识到土豆等块茎类作物对世界许多地区日常饮食的重要性,但几十年来,他们的注意力一直集中在小麦和大米等全球贸易大宗商品上。国际土豆年试图扭转这种不平衡。

分。[67]导致土豆20世纪70年代被联合国粮农组织忽视的那些特征,现在却被认为是土豆在确保粮食安全的斗争中最重要的条件。(参见图19)

　　国际土豆年活动与联合国粮农组织以前做法的不同之处,并不仅仅在于它把重点放在土豆上。国际土豆年反映了关于自给农业对粮食安全贡献的截然不同的理解。早期的发展模式是将重点放在商业性农业上,而自给农业不适合这种模式——这种模式在农田播种单一作物时效率最高,因单一作物可以统一使用杀虫剂和化肥,并在最佳时机收获。小规模农业的典型特点是在多块土地上种植多种不同的作物,这种耕作方式不太适合现代商业性农业的技术。不仅如此,农民自身往往被定位为要被发展所取代的过时农业类型的一部分。战后的许多发展顾问慨叹农民未能实施推荐的政策，这与近代早期和殖民时期轻视小农农业技能的倾向如出一辙。1970年的一份研究报告指出："如何克服人们具有破坏性的保守主义，并进行国家要生存下去所必须实施的激烈的土地改革，是坦桑尼亚政治领导人不得不面对的最困难的问题之一。"[68]小农户经常因保守或反对变革而受到惩罚。而且，他们的农业知识很少得到认可。在秘鲁，洛克菲勒基金会努力将其马铃薯育种计划与高原地区使用的"落后"和"不科学"的做法隔离开来。支持安第斯山区农业技术的秘鲁农学家受到批评，未能得到资金的帮助。虽然不是所有的发展倡导者都持这种态度，但小农往往是"需要农业专家加以解决的问题"。[69]

与早期把小农看成问题的观点形成鲜明对比的是，联合国粮农组织如今对小农农业的灵活和适应性赞赏有加。联合国粮农组织的出版物强调，小农耕种体系在维持土豆遗传多样性方面发挥着重要作用，因此"安第斯山脉的小农正在为保障世界粮食安全提供帮助"。联合国粮农组织同时还强调了妇女对这个进程的重要贡献。国际土豆年网站上有许多农民人工种植和收获土豆的照片，仅有一张照片显示机械化操作，那就是一辆正在装载人工收获的土豆的卡车。[70]使用简单工具进行小规模土豆种植，已成为粮农组织解决粮食安全问题的手段之一。

总部设在利马的国际马铃薯中心（International Potato Center，简称CIP，源于其西班牙语名字的首字母缩写）也传达了类似的信息。作为通过种子库、研究和推广项目来保存主要粮食作物遗传多样性的国际协作工作的一部分，它于1971年在洛克菲勒基金会的资助下成立。CIP最初的计划是培育改良的土豆品种，同时也培训科学家和发布最新的研究信息。如今，它已成为现代土豆育种最重要的知识宝库，在论及土豆对全球粮食安全的贡献时，它具有举足轻重的发言权。

与联合国粮农组织一样，CIP 的早期出版物重点关注的是农艺方面的挑战，如生产抗真菌和抗线虫品种、制定有效的收获后贮藏战略以及建立国际研究协作的重要性等。中心并未考虑小农技术和女性在全球粮食安全中的贡献，也未过多关注土豆在该地区历史上的长期重要性，除了偶尔在年度报告中使用哥伦布时期前的装饰图案。[71] 如今这些方面都是 CIP 工作的显著特色。它的网站把土豆描述为"面对人口增长和饥饿比例上升的粮食安全方面的关键作物"，把安第斯农民视为过去和现在在保护土豆生物多样性方面的核心角色。[72] 它的许多出版物都强调土豆与可追溯千年的生动鲜活的土著文化之间的联系。从这个角度来看，传统农业实践是土豆的现代性不可少的一部分。（参见图 20）

不仅如此，如今国际马铃薯中心还将安第斯土豆品种列为秘鲁"极珍贵遗产"的一部分。[73] 也就是说，土豆，是秘鲁作为一个国家的文化独特性的一部分。事实上，创立联合国国际土豆年的想法，正是源于秘鲁政府多年来的积极游说。在推动创立国际土豆年的同时，它还将 5 月 30 日定为全国土豆日，并设立了全国马铃薯大会（National Potato Congress）和各种区域性土豆节。这些后来的工作，强调了土豆深厚的秘鲁根基。根据 2005 年设立国家土豆日的立法，秘鲁是土豆最主要的发源地，并一直拥有最广泛的遗传多样性，秘鲁媒体很快将这些因素转化为土豆"100% 秘鲁"的主张。[74] 正如社会学家米卡拉·德索西（Michaela DeSoucey）所注意到的那样，这种饮食民族主义"战略性地把对国家身份的考虑与国家作为文化遗产保护者的观念结合起来"。[75] 不出所料的是，当邻近的智利在第二年试图把近 300 种土豆登记为智利土豆时，秘鲁政府作出了强烈反应。秘鲁外交部部长坚称："土豆起源于秘鲁南部已得到广泛认可，因此它是我们国家的文化遗产。"[76]

图 20　国际马铃薯中心出版物中的安第斯土豆农户。来自秘鲁高地安培（Ampay）的村民们聚集在一起，向大地母亲献上祭品，标志着播种季节的开始。右侧是几个与图 2 所示之物类似的脚犁。与之前关注"科学土豆"不同，CIP 如今颂扬的是土豆在秘鲁文化和社会中的深厚根基。这张照片是一本 2001 年刊物的封面。

把土豆作为秘鲁国家遗产一部分的这类官方庆祝活动，与日益增长的把安第斯土豆作为手工食品的美食兴趣产生了共鸣。在过去，土豆等与土著居民有关的食物，与种植和食用它们的安第斯人一样地位低下。然而，自 21 世纪初以来，土豆的（然而不一定是安第斯农民的）地位显著提高。"新安第斯菜系"的出现，为独特的地方食物创建了一个市场，让它们能够被出售给城市消费者和渴望品尝正宗秘鲁美食的游客。新安第斯菜系重新诠释了老秘鲁菜肴，它们使用源自分子烹饪和类似的国际餐厅潮流的新技术来制作新颖的菜肴，如"quinotto"（用藜麦做的意大利调味饭）或人们熟悉的秘鲁菜"*causa*"（塞满肉或其他馅料的土豆泥馅饼)的"分解"版。加斯顿·阿克瑞奥(Gastón Acurio)等秘鲁厨师,因对当地食材和香料的开创性运用而享誉全球。出于这些原因,利马开始被誉为美食之都。秘鲁种类繁多的土豆,很容易被囊括在关于真实性、本土性的故事当中,这些故事都强调了土豆与该地区及其历史的深厚文化联系。[77]

作为回应,国际马铃薯中心启动了一项研究项目,以研究在秘鲁美食复兴中起到显著作用的颜色鲜艳的土豆,并与秘鲁烹饪学院合作开发新菜系。[78] 并非只有秘鲁对本地土豆品种进行国家美食主义颂扬。从丹麦到厄瓜多尔,许多国家也同样设立了国家土豆日,或寻求国际立法对特定品种进行保护。北爱尔兰的库默(Comber)新土豆、爱达荷土豆等许多品种,现在都在商标条例和欧盟条例(这些条例把特定品种的生产限制在特定地区)下进行了注册。国际监管机构通过这样的方式,给予土豆作为一种国家遗产的正式地位,从而帮助土豆实现国家化。[79] 土豆作为一种被忽视的地方性粮食资源的悠久历史,令它如今能够以糖或玉米等其他主要商品在很大程度上没能做到的方式,在全球粮食体系与烹饪传承的概念之间游走。

在秘鲁,土著社区试图利用这种将土豆视为一种国家文化遗产的概念,以巩固让他们自己获得国家承认的主张。国际马铃薯中心的土著农民是安第斯土豆遗传多样性的重要保管人的坚持主张,对于这些族群是极其有帮助的。举例来说,2004年,在国家立法认可"土著居民集体知识"的基础上,皮萨克谷(Pisac Valley)的一个族群与国际马铃薯中心签署了一项"本土马铃薯归国协议"。该协议的内容是,国际马铃薯中心向皮萨克族群转让若干土豆标本,目的是让最早种植土豆的该历史地区参与协作研究,同时帮助当地人种植土豆供自己食用。该协议还为该地区提供了获得少量收入的途径。族群的"土豆公园"为游客提供为期5天的"土豆之旅",游客还可以在一家展示当地土豆的餐厅里享受美食。美食民族主义、对粮食安全的担忧,以及对世界生物多样性所有权的争论,使安第斯土豆成为今天秘鲁文化和经济资本的源泉。[80]

结论

20世纪以来,土豆作为一种政治工具的重要性在稳步上升。至"一战"期间,再没人指责土豆产生懒惰的土豆血,而是把它誉为健康的主食。各国政府为满足战时人民的需要,积极鼓励把食用土豆作为爱国行为。国家马铃薯育种计划和国家资助的采集探险,反映出人们确信:对于20世纪70年代所称的粮食安全问题,土豆是能够提供保障的。今天许多通过商业交易能够买到的土豆品种,都是这类国家支持计划的结果。

这些20世纪的膳食干预计划,在许多方面都和18世纪政治家认为确保民众的好营养很重要的模糊主张是不同的。19世纪和20世纪初发生的事件,改变了民众和政府之间的关系,而政府在战后的岁月里拥有的技术能力远超18世纪的君主。战时欧洲出台的国家配给计划,是18世纪90年代提出的、劳动人民应考虑多吃土豆的热心建议无法比拟的。18世纪的土豆倡导者,也不一定会赞同20世纪的福利主义者尝试通过减少社会和经济不平等来结束饥饿的做法。但与此同时,个人的饮食习惯是国力和生产力的基础这一主张,在坚持认为工人的好营养对国家繁荣至关重要的18世纪作家听来,却也并不陌生,这些想法正是诞生于18世纪。正如社会学家米切尔·迪恩(Mitchell Dean)对我们的提醒,如果"不否认过去两个世纪国家转型的明显现实",18世纪和20世纪的治理理念之间可能存在着连续性。[81]

在20世纪，尽管商业育种和促进消费计划成为国家粮食安全计划的一部分，但在战后年月出现的经济发展模式，却很少关注发展中国家的小规模农户的农业实践，包括那些拥有种植土豆专业技术的农户，而他

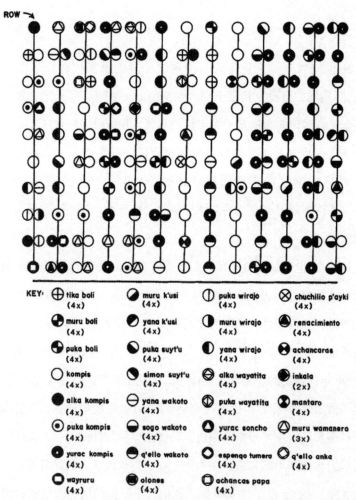

图21 安第斯土豆田的示意图。安第斯农民通常种植数十种不同的土豆。一块农田里可多达近50种。这幅图描绘了这样一块农田。虽然这种布局可能显得杂乱无章，但事实并非如此。种植反映了土壤和海拔的特殊要求，并仔细将土豆种与微观环境相匹配。这块田中包括的"复兴"（Renacimiento）品种，是秘鲁国家马铃薯计划的主任卡洛斯·奥乔亚开发的一个极其成功的商业品种，田里还有更传统的地方品种。

们的技术后来被认定为保护粮食安全的重要资源。直到最近，小农户和小农农业才被纳入国际粮食安全模式当中。联合国粮农组织将2008年定为国际土豆年，标志着把"传统"农业耕作视为创新和相关知识基地的新认识。在近代早期欧洲传播了土豆种植的小农专业知识，在20世纪中叶的许多历史学家眼中是被无视的。同样地，保存马铃薯基因多样性的专业知识，也直到近年来，才受到关心粮食安全的国际组织的重视。

联合国粮农组织和国际马铃薯中心重视小规模土豆种植农户的专业知识，这样的做法是正确的。像大学培养的科学家一样，这些耕种者往往是实验主义者，为了确定和开发新品种和新的耕种方法，他们观察、解释、做田间实验、评估和操作。他们与当地的其他人讨论自己的经验，并按照交谈的内容调整自己的做法。大多数在安第斯山区耕种土豆的农民，会保持12到15块不同的土地进行连续耕种，但更多土地没有被用于连续耕种。他们也会与邻居交换土豆种和土地。通过这样的交换，一个农民有可能获得多达100个不同的土豆品种。将特定的土豆种，与特定土地的土壤和环境要求相匹配，需要大量实用的农学知识，并会导致不断的实验。正是这种持续的评估和创新，导致了安第斯山脉土豆品种显著的多样性。在20世纪60年代，农学家认为秘鲁有大约1400种本地土豆品种；目前估计这个数字在2700到4500之间。[82]（参见图21）

反映于国家土豆日的庆祝、国际地理标志公约和国家美食主义的对传统土豆品种的继承，将国家层面的对人民日常饮食习惯的关心，与传统耕作方式是全球粮食安全问题的部分解决之道的认识，组合成一个关于文化遗产的故事。食物传达民族主义叙事的强大能力，使土豆成为这些关系的恰当象征。[83]土豆仿佛在告诉我们，作为个体我们是谁，作为集体我们又走向何方。

注释

1. *Potato Pete's Recipe Book*.
2. Vernon, 'The Ethics of Hunger and the Assembly of Society', 723; Vernon, *Hunger*.
3. Scott, *Seeing Like a State*, 88.
4. Scott, *Seeing Like a State*, 343; Helstocky, 'The State, Health, and Nutrition'.
5. Thane, 'The Working Class and State "Welfare" in Britain'; Shapiro, *Perfection Salad*, 137–138; Pedersen, *Family Dependence, and the Origins of the Welfare State*; Daunton, *Wealth and Welfare*; Kaufmann, *Variations of the Welfare State*; Aguilar, 'Cooking Modernity'; Pohl-Valero, 'Food Science, Race, and the Nation in Colombia'.
6. Pedersen, *Family Dependence, and the Origins of the Welfare State*, 53 (quote); Vernon, 'The Ethics of Hunger and the Assembly of Society'.
7. Worboys, 'The Discovery of Colonial Malnutrition between the Wars'; Arnold, 'The "Discovery" of Malnutrition and Diet in Colonial India'; Kamminga and Cunningham, eds., *The Science and Culture of Nutrition*; Davis, *Home Fires Burning*; Helstocky, 'The State, Health, and Nutrition'; Drinot, 'Food, Race and Working-Class Identity'; Cwiertka, 'Militarization of Nutrition in Wartime Japan'; Andresenand Elvbakken, 'From Poor Law Society to the Welfare State'; Aguilar, 'Cooking Modernity'; Cullather, 'The Foreign Policy of the Calorie', 359; Nally, 'The Biopolitics of Food Provisioning'; Swislocki, 'Nutritional Governmentality'; Pohl-Valero, '"La raza entra por la boca"'.
8. Winter, 'Military Fitness and Civilian Health in Britain', 212.
9. Ministry of National Service, 1917–1919, *Report upon the Physical Examination of Men of Military Age by National Service Medical Boards*, 43–44 (quote). 1905年全国女工工会（National Union of Women Workers）会议上，一位代表坚持说："孩子们遭受的不是食物不足，而是食物不当、住房和通风条件差、个人卫生状况差、习惯不规律、睡眠不足、自制力缺乏。"核心问题则是"母亲的无知"。Pedersen, *Family Dependence, and the Origins of the Welfare State*, 52. 也可参见 Zweiniger-Bargielowska, *Managing the Body*。
10. Rowntree, *Poverty*, 303–304.
11. Beckett, 'Total War'.
12. Burnett, *Plenty and Want*, 255; Petránová, 'The Rationing System in Bohemia during the First World War'; Essemyr, 'Food Policies in Sweden during the World Wars'; Scholliers, 'The Policy of Survival'; Davis, *Home Fires Burning*; Bonzon, 'Consumption and Total Warfare in Paris (1914–1918)'; De Nijs, 'Food Provision and Food Retailing in

The Hague, 1914–1930'.

13. Langworthy, *Potatoes and Other Root Crops as Food*; Atwater, *Principles of Nutrition and Nutritive Value of Food*, 44 (quote).

14. Salaman, *History and Social Influence of the Potato*, 576–577; Dewey, *British Agriculture in the First World War*, 206, 244–248; Crowe, 'Profitable Ploughing of the Uplands?', 209.

15. Hancock, 'The Social Impact of the First World War in Pembrokeshire', 61.

16. Cullather, *The Hungry World*, 21; Farnham et al., eds., *Choice War Time Recipes*, 56.

17. Davis, *Home Fires Burning*, 59.

18. Davis, *Home Fires Burning*, 169.

19. Blomqvist, 'Sweden's Potato Revolution'; Håkan Blomqvist, personal communication, May 2018.

20. Hancock, 'The Social Impact of the First World War in Pembrokeshire', 61.

21. Davis, *Home Fires Burning*.

22. Murphy and Kass, *Evolution of Plant Breeding at Cornell University*; Brown, 'Notes on the History of the Potato Breeding Institute, Trumpington'.

23. Ochoa, 'Potato Collecting Expeditions in Chile, Bolivia and Peru'.

24. Bethke et al., 'History and Origin of Russet Burbank (Netted Gem)'. "赤褐布尔班克"（Russet Burbank）起源于 C. E. 古德里奇（C. E. Goodrich）1851 年收集的一种被他称为粗紫辣椒（Rough Purple Chili）的土豆。

25. Elina et al., 'Plant Breeding on the Front', 165.

26. Hawkes, 'N.I. Vavilov—The Man and his Work', 3–6.

27. Hawkes, *Potato Collecting Expeditions in Mexico and South America*, 8; Christiansen González, *El cultivo de la papa en el Perú*; Shepherd, 'Imperial Science'; Reader, *Potato*, 231–239; Scott, 'Plants, People, and the Conservation of Biodiversity of Potatoes in Peru'.

28. Hawkes, *Potato Collecting Expeditions in Mexico and South America*, 8; Christiansen González, *El cultivo de la papa en el Perú*; Shepherd, 'Imperial Science'; Scott, 'Plants, People, and the Conservation of Biodiversity of Potatoes in Peru'.

29. Qu and Xie, eds., *How the Chinese Eat Potatoes*, 374–378.

30. Scott, 'Plants, People, and the Conservation of Biodiversity of Potatoes in Peru', 25; Sood et al., 'History of Potato Breeding'.

31. Helstocky, 'The State, Health, and Nutrition', 1582–1583; Davis, *Home Fires Burning*; Daunton, *Wealth and Welfare*; Leisering, 'Nation State and Social Policy'.

32. Collingham, *The Taste of War*, 8.

33. Boyd-Orr, *Food, Health and Income*.

34. 'Review of *The Nation's Larder*', 14–15.

35. Salaman, *History and Social Influence of the Potato*, 578–581; Hammond, *Food and Agriculture in Britain*, 33–37, 79–86; Burnett, *Plenty and Want*, 307; Martin, *The Development of Modern Agriculture*, 40; Collingham, *The Taste of War*, 29, 90, 101, 361–363, 384–414.

36. Collingham, *The Taste of War*, 137, 404.

37. Moskoff, *The Bread of Affliction*; Collingham, *The Taste of War*, 219–227, 317.

38. Moskoff, *The Bread of Affliction*; Barber and Harrison, *The Soviet Home Front*, 77–93; Ries, 'Potato Ontology'; Collingham, *The Taste of War*, 220–225, 323, 334, 337 (quote), 341.

39. Barber and Harrison, *The Soviet Home Front*, 85–86.

40. Piktin, *The House that Giacomo Built*, 57; Collingham, *The Taste of War*, 14, 70, 173–174, 207, 296–297, 501.

41. Pauer, 'Neighbourhood Associations and Food Distribution in Japanese Cities'; Cwiertka, 'Popularizing a Military Diet in Wartime and Postwar Japan'; Collingham, *The Taste of War*, 277–281.

42. Procter, *The Nazi War on Cancer*, 124.

43. Collingham, *The Taste of War*, 193, 357–359.

44. Procter, *The Nazi War on Cancer*; Cole, 'Feeding the Volk', 160.

45. Elina et al., 'Plant Breeding on the Front'. 土豆是德国农林业生物研究所（Biologisches Reichanstalt für Land- und Forstwirtschaft）的第一批实验对象，该机构于1905年并入德国内政部，Saraiva, *Fascist Pigs*, 74–75。

46. Reagin, *Sweeping the German Nation*, 158–159; Cole, 'Feeding the Volk', 160–183 (160 quote); Saraiva, *Fascist Pigs*, 128–129.

47. 例如，1941年9月至1942年8月期间驻扎在白俄罗斯的陆军集团军中心食用的90%的土豆都是通过这种方式获得的，Collingham, *The Taste of War*, 29, 156–64, 186, 211–213。

48. Saraiva, *Fascist Pigs*, 7.

49. Bentley, *Eating for Victory*, 68 (quote); Collingham, *The Taste of War*, 416–429; Biltekoff, *Eating Right in America*.

50. Bentley, *Eating for Victory*, 63–64.

51. Yang, 'Creating the Kitchen Patriot', 62.

52. FAO, 'Constitution'.

53. Whitaker, 'Food Entitlements', 1589–1590; Staples, *The Birth of Development*; Shaw, *World Food Security*; Schanbacer, *The Politics of Food*; Cullather, *The Hungry*

World, 34 (quote); Gibson, *The Feeding of Nations*.

54. 关于在发展中国家饥饿使人民支持共产主义的经典陈述，请参见 Truman, 'Inaugural Address'。

55. Smith, *An Inquiry into the Nature and Causes of the Wealth of Nations*, I.viii.45.

56. United Nations, *Report of the Forty-Ninth Session of the Committee on Commodity Problems*, paras. 41–67; FAO, *Report of the Council of FAO*; FAO, *Approaches to World Food Security*; Pottier, *The Anthropology of Food*; Carr, 'Postmodern Conceptualizations, Modernist Applications'; Jarosz, 'Defining World Hunger'; FAO, Rome Declaration on World Food Security (quote).

57. Cullather, 'Development? It's History'; Cullather, *The Hungry World*.

58. George, *How the Other Half Dies*; Escobar, 'Power and Visibility'; Cullather, *The Hungry World*; Pingali, 'Green Revolution', 12303.

59. Cullather, *The Hungry World*, 201.

60. United Nations, *Report of the Forty-Ninth Session of the Committee on Commodity Problems*, paras. 41–67; FAO, *Roots, Tubers, Plantains and Bananas in Human Nutrition*, table 3.6.

61. Jacob George Harrar, 'Notes on meeting with Colombian Minister of Agriculture ［Antonio Angel Escobar］', Bogotá, 9 Aug. 1951, RF, Rockefeller Foundation Records, Officers Diaries, PG 12, 1951 February 11–1952 February 6, 90; Moris, *Extension Alternatives in Tropical Africa*, 55 (quote).

62. FAO, 'FAO Overview'.

63. Ochoa, 'Potato Collecting Expeditions in Chile, Bolivia and Peru'; Fenzi and Bonneuil, 'From "Genetic Resources" to "Ecosystems Services"'.

64. Shepherd, 'Imperial Science'; Scott, 'Plants, People, and the Conservation of Biodiversity of Potatoes in Peru', 25.

65. Rockefeller Foundation, 'Accelerating Agricultural Modernization in Developing Nations'; FAO, 'FAO Overview' (quote).

66. Fenzi and Bonneuil, 'From "Genetic Resources" to "Ecosystems Services"', 78.

67. FAO, 'Why Potato?'; FAO, International Year of the Potato 2008.

68. Escobar, 'Power and Visibility'; Scott, *Seeing Like a State*, 241; Mackenzie, 'Contested Ground'; Beinart, 'African History and Environmental History'; Watts, 'Development and Governmentality'; Nally, 'The Biopolitics of Food Provisioning'; Coulson, *Tanzania*, 124–125 (quote).

69. Netting, *Smallholders, Householders*; Scott, *Seeing Like a State*, 241 (quote); Pottier, *The Anthropology of Food*; Shepherd, 'Imperial Science'; Cullather, *The Hungry World*.

70. FAO, 'Potato and Biodiversity' and 'Potato and Gender'.

71. 例如，可参见 Centro Internacional de la Papa, *Informe anual 1975*; Centro Internacional de la Papa, *Informe anual 1979*。

72. Centro Internacional de la Papa, 'Potato' (quote); Scott, 'Plants, People, and the Conservation of Biodiversity of Potatoes in Peru'.

73. Centro Internacional de la Papa, 'Native Potato Varieties'.

74. Congreso de la República, Resolución Suprema, 287, 797; 'Perú: Hoy se celebra el Día Nacional de la Papa', SERVINDI; FAO, 'Background'; Scott, 'Plants, People, and the Conservation of Biodiversity of Potatoes in Peru'.

75. DeSoucey, 'Gastronationalism', 447.

76. BBC News, 'Potatoes Spark Chile–Peru Dispute'.

77. Matta, 'Valuing Native Eating'.

78. Scott, 'Plants, People, and the Conservation of Biodiversity of Potatoes in Peru'; Schlüter, 'Promoting Regional Cuisine as Intangible Cultural Heritage in Latin America; Palomino Gonzales, 'Gourmetización del alimento andino'.

79. Faulhaber, 'Cured Meat and Idaho Potatoes'; DeSoucey, 'Gastronationalism'; Cassidy, 'Comber Spuds get European Protected Status'.

80. Pottier, *Anthropology of Food*, 176–179; Carneiro Dias and da Costa, 'Indigenous Claims to Native Crops and Plant Genebanks'; Scott, 'Plants, People, and the Conservation of Biodiversity of Potatoes in Peru', 30; Parque de la Papa.

81. Dean, *The Constitution of Poverty*, 6.

82. Ploeg, 'Potatoes and Knowledge'; Mayer, *The Articulated Peasant*; De Jong et al., *The Complete Book of Potatoes*; Scott, 'Plants, People, and the Conservation of Biodiversity of Potatoes in Peru', 22.

83. 例如，请参见 Wilk, 'Beauty and the Feast', 296; Palmer, 'From Theory to Practice'; Helstocky, 'Recipe for the Nation'; O'Connor, 'The King's Christmas Pudding'; DeSoucey, 'Gastronationalism'。

总结　帕蒙蒂埃、农民和个人责任

土豆英雄

巴黎的帕蒙蒂埃大道（Poulevard Parmentier）从协和广场向东延伸至拉雪兹神父公墓（Père Lachaise Cemetery），公墓里安葬着马塞尔·普鲁斯特（Marcel Proust）、弗雷德里克·肖邦（Frédéric Chopin）和吉姆·莫里森（Jim Morrison），这里还有安托万·奥古斯丁·帕蒙蒂埃的墓穴。这位18世纪最著名的土豆推广者，被他所推广的块茎作物环抱着。到墓地的游客通常会在墓上留下祭品。在帕蒙蒂埃的墓上，他们会留下土豆。（参见图22）

长久以来，帕蒙蒂埃一直被誉为把土豆带到法国的人。帕蒙蒂埃大道的铭牌上清楚地说明，是他"把土豆种植引进了法国"。帕蒙蒂埃于1737年出生在一座乡下小镇，后来在巴黎接受了药剂师培训。据他后来说，他对土豆的兴趣可以追溯到18世纪50年代，那时正值七年战争，他在法国军队服役。最初是在当战俘时，后来是在法兰克福生活时，他开始思考土豆作为面包的补充甚至替代它的潜力。回到法国后，他开始研究土豆的化学性质。1771年，他在一项竞赛（由贝桑松科学、文学和艺术学院［Besançon Academy of Science, Literature and Arts］赞

图22 拉雪兹神父公墓内装饰着土豆的帕蒙蒂埃墓。安托万·奥古斯丁·帕蒙蒂埃和土豆之间的纽带在他死后延续了200年。他的坟墓位于巴黎的拉雪兹神父公墓，墓上装饰着作为还愿的土豆。法国各地的家庭都吃"帕蒙蒂埃烤土豆泥"（Hachis Parmentier），即顶上撒有土豆的烤牛肉馅土豆泥馅饼。把土豆带到法国的名声无疑是他的，虽然也许功劳不完全是他的。

助）中，以关于粮食短缺时期使用谷物替代品的论文获得最佳奖，当时的法国正经历着这种短缺。他提出用土豆作为替代品。这是第一篇推荐土豆的公开论文，在这篇1771年获奖的论文之后，一系列关于土豆作为人类食物、动物饲料和工业淀粉的五花八门用途的论文也紧跟着发表了。[1]

从18世纪70年代到他1813年去世，帕蒙蒂埃孜孜不倦地试验土豆等食物的化学成分，试图了解它们的营养特性，目的，正如他反复强调的那样，是为他的国家提供更安全的基本食物。帕蒙蒂埃花了几十年的时间，研究如何将土豆转化为爱吃面包的巴黎人可能会喜欢的发酵面包。帕蒙蒂埃逐步设计出更复杂的从土豆中提取淀粉的方法，以及制作土豆酵母的方法，从而创造出完全由土豆制成的酵母面包。18世纪欧洲的作家们都盛赞帕蒙蒂埃的文章，感谢他揭示了土豆的多种性质。如今，不仅他的各项研究在帕蒙蒂埃大道得到纪念，而且法语词汇中还用含他名字的"帕蒙蒂埃烤土豆泥"（Hachis Parmentier）来命名一种顶上撒有土豆的烤牛肉馅土豆泥馅饼。毫无疑问，他会很高兴在这道日常菜中被人铭记。在许多他所发表的文章中，他反复表达了自己的信念：土豆是一种无与伦比的食物，是美洲送给人类的最伟大的礼物。[2]

当然，在帕蒙蒂埃介入之前，土豆在法国已经被人们熟知。事实上，1771年贝桑松论文比赛的所有参与者，都把食用土豆推荐为解决粮食短缺问题的最佳方案。并不只帕蒙蒂埃一人认识到它们作为小麦或燕麦补充品的潜力，也并不是只有他一人对土豆的化学成分进行分析。从18世纪30年代起，人们就开始对土豆淀粉和其他土豆加工形式进行实验。土豆面包——无论是用土豆泥、土豆淀粉还是土豆面粉制成——是欧洲各地科学家优先研究的对象。[3] 显然，把土豆引进法国并不是帕蒙蒂埃的功劳，他也不是土豆唯一的倡导者。他写的大量关于土豆的文章，不是因为土豆新奇，而是因为他想增加土豆的消费量，以此来增强法国的国力。正如他解释的那样，"食物的种类和选择对人口影响很大，所以要确保人们吃得好，再多的预防措施也不为过"。[4] 他认为，这样的事务，"值得哲学家们深思，值得政府保护"，也正是因此，在启蒙运动时期，哲学家和政府如此关注土豆。[5]

在18世纪，土豆政治知名度的日益增加，被误读为像帕蒙蒂埃这样的人物为一种当时仍被人们所排斥、受人鄙视的蔬菜做代言。而事实上，他的故事揭示的是饮食在18世纪关于国家财富和权力观念中前所未有的重要性。对帕蒙蒂埃所扮演的角色的这种误读，是一种认知的持续，即历史变化是一个自上而下的过程。更具体地说，它歪曲了普通人对农业和饮食创新的贡献。几乎可以肯定的是，农民和劳动者是第一批把土豆作为食物来源，而不是作为一种奇特植物而进行种植的欧洲人。在18世纪，他们继续掌握着精英做实验所必需的农学知识。那位伟大的土豆推广者，恰恰是从"一个叫格雷戈瓦尔（Grégoire），来自列日附近的雅莱（Jalhay）的农民"那里，第一次了解到如何种植土豆。[6]

如今，小农农业的灵活性和适应性，因其对全球粮食安全的贡献而受到了实至名归的赞扬。两公顷以下的农场种植的粮食在世界粮食产量中占比失调，因为就每公顷的产量而言，小农场通常比大农场更多产。来自世界许多地区和许多不同时期的研究表明，"不太先进"的农业实践能胜过更复杂的技术。对当地情况的详细了解，加上对有用想法的务实接受，推动了资源的有效利用，而许多小农户面临的不稳定条件，又鼓励了创新。[7]在危地马拉高地，小农户在发展方案鼓励种植的那类出口作物的农田旁边，保留着为自己的家庭种植粮食的小块菜地。这样的农户在农业实践中是不能保守的。人类学家尤莉亚·芬顿（Ioulia Evgenyevna Fenton）指出，"由于渴望任何能帮助他们从地里挤出更多钱的东西"，他们定期试验新种子、杀虫剂和化肥。[8]

另一位人类学家罗伯特·内廷，针对发展主义者观念的缺陷写了大量的文章，这种观念是"必须教会传统的集约化耕种者如何使用机器、购买的资源和科学知识进行耕作"。内廷对尼日利亚、瑞士、中国、爪哇等地的小农实践进行分析，得出的结论是，这种观念与小农农业的"土地生产力、可靠性、生态可持续性和适应性直接矛盾"。[9]正是这种不断的调整和适应，才产生了如今在安第斯山脉繁荣发展的成千上万的土豆品种，这里的农业区域利用大量实用的农艺知识，来将特定的土豆种与特定土地的土壤和环境要求相匹配。学者们把这种实用的知识称为"savoir-faire paysan"（小农智慧）或"art de lalocalité"（地方艺术）——即这类农业取得成功所需的地方性专门知识。[10] 土豆最初同样也是在农家菜园和小块田地里适应了欧洲不同的生长条件，从而迈出了成为全球主食的第一步。

承认农民对土豆历史的贡献不仅仅是一个历史正义的问题，它也关系到我们的未来。如今，生物多样性被认为是长期环境可持续性和全球粮食安全的重要组成部分。这种本质上的多样性反过来又依赖于全球小农的专业知识，他们被公认为对抗高度同质化的大规模商业性农业力量的最重要的抗衡力。小农——其中很多是妇女——种庄稼出于各种不同的目的：一些被拿到市场上去出售，一些因为它特别美味或适合做特定的菜肴，另一些被用作礼物或作为未来支出的资本被储存。不同的目的需要不同的品种，这些品种具有不同目的所需要的特定品质。在某些情况下，储存能力是最重要的，在另一些情况下，产量或味道是最重要的。这种类型的农业使得农作物的多样性具有了价值，并推动了它的发展。这类农业并不以惯性或对过去实践的本能性复制为典型特点。长期以来对自给农业漠不关心的联合国粮农组织，如今却将支撑传统土豆种植的地方知识奉为过去和现在全球粮食安全的关键。土豆的历史提醒我们，小型农业对更大的创新和变革历史的贡献是不容忽视的。

土豆的第二种贡献

跟随土豆的足迹，同样还能看到，我们对饮食的态度是如何植入一种特定的治国理念的。在18世纪前的欧洲，一个人吃什么与他本人直接相关，同样还与那些准备和供应食物的人、与他们分享食物的人，以及其他许多人也有关系。因此，一个人吃的东西形成了一张关系网，在某种程度上构成了他们的社会。饮食是，而且一直会是人类文化的核心。它是我们存在和身份的核心。不过，普通人一餐吃什么却并没有太多政治利益。

如今，这类问题反而变得如此重要，足以让整个政府计划都投入到改善我们的饮食习惯当中来。健康餐盘、膳食指南和旨在增加蔬菜消费量或减少盐摄入量的教育活动，是我们目前饮食景观中的常见特色。提供如何优化饮食的建议如今在现代治理中是被认可的组成部分。在欧洲，认为国家应该关注这些问题的观念出现在18世纪，它是国家与人民关系的更广泛的新概念的一部分。这种新概念是与商业和贸易的新思考方式同时发生的。有观点认为，个人主义是组织经济交换最有效的方式，也有观点认为，改善公共健康的途径是促进个人作出合理饮食选择，这两个观点在饮食学上是一致的。就像其他体系能够神奇地自我调节，给所有人带来最大利益一样，追求饮食幸福，也会给个人和整个政体带来更多的健康和活力。至少，人们是这样认为的。

这段历史有助于解释我们所秉承的这种观念：饮食既是个人事务，也是政府干预的合法领域。它同时也有助于解释我们对这种干预的矛盾心理。我们对个人自由的概念，极大地受惠于18世纪把个人作为"选择主体"放在首位的争论。选择吃什么，为我们提供了具体的、日常的机会来行使我们作为个人的基本自主权。正如美国饮食习惯委员会在1942年所言，"对食物的选择是成年的一种标志"。[11] 缺乏饮食选择，无论是由于外界干预还是由于贫困，都与我们从启蒙运动中继承下来的自由观念格格不入。文森特·梵高(Vincent van Gogh)有一幅作品,描绘的是农民家庭挤在一张桌子旁吃着晚餐的煮土豆,这种穷困潦倒的景象令人难过。约翰·柯文尼(John Coveney)写道:"因为选择和选择自由已经成为食物标准范畴的一部分,没有选择被视为一种需要被纠正的情况。"[12] 把穷人重新定义为选择主体的迫切要求,解释了18世纪热衷于推广土豆的人们为何会试图假装以为,那些涌向欧洲许多救济所的饥饿的穷人真的很享受慈善土豆汤(如果他们够幸运,汤里还会配上能带来幸福的拉姆福德伯爵的烤面包丁)。(参见图23)

如果,今天政府的作用是为我们提供一个有利的环境,让我们作出知情的食物选择,那么最终是我们自己要为自己的决定负起责任来。在

图23 梵高《吃土豆的人》。约翰·柯文尼说:"因为选择和选择自由已经成为食物标准范畴的一部分,没有选择被视为一种需要被纠正的情况。"行使选择权使个人能够证明对自己的健康负责,因此它是现代消费者的基本组成部分。梵高的农民家庭,吃着他们自己种植的土豆,与这种现代的观念产生了冲突,也许这就是观众发现这幅画令人难过的原因。

2002年推出的通过饮食和锻炼计划促进"更健康生活方式"的"更健康的美国"倡议中，乔治·W.布什在讲座上表示，"我经常谈到的事情之一，是真正地着手改变美国文化的必要性"。布什还说，更健康的美国"真的需要个人的责任，不是吗？它说我们要对自己的健康负责"。英国首相托尼·布莱尔在2004年关于"选择健康"的政策文件中声明，政府不能，也不应该追求让人们更健康。在他看来，"如果人们愿意，应该由他们自己作出健康的选择"。正如新西兰卫生部2018年所说，现代政府的作用，是支持我们对饮食作出"正确的选择"。[13]

应该由我们自己来决定去遵循政府和行业的指导方针，选择水煮土豆等推荐食品、避免高脂肪零食，如美国人均每年消费的76磅薯条和其他深加工土豆食品。如果我们因为吃了太多薯片而生病或超重，那是我们自己的错。2006年的一项研究指出："对肥胖者的消极态度与对少数族裔和穷人的消极态度高度相关，比如认为所有这些群体都很懒惰，缺乏自控力和意志力。"[14] 威廉·巴肯在1797年坚称，土豆将终结所有人的贫困（"挥霍无度者"除外），同样，个人责任和选择的说法就意味着，那些失去健康的人只能责怪自己。

这样的倒霉饮食者完美地代表了"多余的人"，他们被法国哲学家米歇尔·福柯认定为现代治国之道的障碍。福柯将现代政府管理民众的方法与早期政治权力的形式进行了对比。他认为，对于现代政府而言，权力不是通过定期处决罪犯，或其他断续证明政府威力的手段来间歇地进行展示。相反，它是一个促进民众中某些阶层的福祉，并对国家不愿促进其福祉的人取消一切支持的持续过程。"我们可以说，"福柯发现，在现代，"*让*人死或*让*人活的古老权利，已经被*助*人活或*不让*人死的权力取代了。"[15]

人口，有生产力的社会成员，应该得到营养、得到生活帮助、得到繁荣。正是这样的思想促使一种观念在18世纪产生：国家的力量和安全取决于人民的活力和幸福。正如福柯所述，这种个人与国家之间的新关系应包含：

> 一种循环，从国家作为对个人进行理性和有计划干预的力量开始，回到国家作为一种不断增长，或有待发展的力量……这个循环……必须成功地将国家的实力和个人的幸福联系在一起。这种幸福，对个人而言比活着更好的东西，必须在某种程度上取自并构成国家的效用：让人民的幸福成为国家的效用，让人民的幸福成为国家的实力所在。[16]

他认为，这才是西方社会历史上第一次"个人的存在和福祉与政府的干预真正产生重大关系"。正因为如此，18世纪的土豆推动者常坚称，土豆为通往个人幸福提供了一条大道。

然而，从这种现代治国之道的逻辑来看，还有一些并不属于人口一部分的人。他们只是人，并且他们碍事。政府看不到鼓励他们存在的必然理由。福柯认为，现代政治既是一个"让人活"的问题，也是一个"让人死"的问题。意大利哲学家吉奥乔·阿甘本（Giorgio Agamben）[①]曾描写过国家"让人死"的那些对象的悲惨命运，他们是边缘人，没有国家，他们的生命注定"不值得活下去"，帮助不会落在他们身上。在

[①] 吉奥乔·阿甘本（1942— ），意大利当代著名哲学家、思想家。曾于意大利马切拉塔大学、维罗纳大学、威尼斯高等建筑学院及巴黎国际哲学研究院、欧洲研究生院等多所学院和大学任教。他的研究领域广泛且影响深远，在国际学界享有极高的声誉。著述颇丰，包括《裸体》《渎神》《什么是装置》《论友爱》《教会与王国》《例外状态》《语言的圣礼》等涉及哲学、政治、文学和艺术的著作。——译注

他看来，如果我们没有功用，如果我们吃太多薯片并停止锻炼，如果我们肆意挥霍，我们都会在这个深渊的边缘摇摇欲坠，都有可能成为局外人，时时处于坐以待毙的风险中。[17]

各种类型的土豆，曾经都作为培养强壮人口的一种方式得到推广，并在个人选择和个人利益的大架构内受到理解。如今，美国约有50%的土豆作物都被加工成薯条、薯片和其他高热量、低营养的零食。我们被鼓励为了自身的健康拒绝接受这类深加工土豆，而是选择制作更简单的那些。美国政府网站支持把白煮土豆或低卡路里的填馅土豆作为更明智的饮食选择。[18]同时，不管是鼓励消费更多土豆的18世纪的广泛建议，还是当前拒绝薯条的建议，它们潜在的逻辑都是相同的，这种逻辑把我们的饮食视为既是国民幸福的重要组成部分，同时又是根本上的个人选择问题，从而也是个人责任问题。（参见美味填馅土豆食谱）

211

美味填馅土豆

美国卫生与公众服务部表示，要成为"更健康的你就是要健康地选择"食物。它的网站提供了许多官方宣传的食谱，并根据烹饪所需的时间进行了分类。它们都被描述为健康、简单和有趣的食谱。除了介绍如何准备"对你有好处"的玉米面包、约塞米蒂（Yosemite）鸡肉炖丸子之外，还有新土豆沙拉、黑色煎锅煎牛肉配青菜和红土豆，以及这个"美味填馅土豆"食谱。每半个土豆提供293毫克钾和113卡热量。如果用餐者选择吃中等大小的土豆半个以上，会对他们的健康有害，但那是他们的选择，从而也是他们的责任。

美味填馅土豆

"烤土豆填入调味的低脂农家干酪,是一道低脂、低胆固醇、低钠的宴请大菜。"

4 个中等大小的烤土豆

3/4 杯农家干酪,低脂(1%)

1/4 杯牛奶,低脂(1%)

2 汤勺软(小桶装)人造黄油

1 茶匙莳萝叶

3/4 茶匙香草调味料

4—6 滴辣椒油

2 茶匙帕尔玛干酪,切碎

1. 将土豆用叉扎破,以 220 ℃烤 60 分钟,或烤至用叉易于插入。

2. 将土豆纵向切成两半,用勺小心将土豆挖出,土豆壳内留约 1/2 英寸厚的瓤,在大碗中将土豆碾成泥。

3. 除帕尔玛干酪之外,将剩余调料用手混匀,将混合物用勺舀入土豆壳内。

4. 表面撒上 1/4 茶匙帕尔玛干酪。

5. 置于烤盘上重新放入烤箱。烤 15—20 分钟或至顶部金黄色。

*

包括鼓励吃土豆在内的活动,只有当它们被视为许多观点的一部分,而不是单个要点时,才能得到最好的理解。18 世纪土豆在整个欧洲的流行,反映出政治家和哲学家开始将个人饮食与国家富强联系起来,饮食由此在 18 世纪获得了新的政治重要性。他们把这种辩论放在

图 24　不满的臣民向腓特烈大帝扔土豆的土豆印花。仁慈的君主把土豆带给大众的故事，并没有公正对待普通人在塑造我们饮食习惯上起到的作用。正如农民兼作家温德尔·拜瑞（Wendell Berry）① 所言，吃毕竟是一种农业行为。

一种选择的语言与个人追求幸福的框架内。这种关系解释了土豆前所未有的政治知名度。日常生活、个人主义与国家之间的关系形成在 18 世纪后期，土豆的历史是其中的一部分。这种关系继续出现在如今如何平衡个人饮食自由与政体健康的讨论中。无论是集体还是个人，我们都可以通过吃来获得健康和幸福，这种诱人的承诺，依然是我们这个动荡的世界中强有力的组成部分。

注释

1. 例如，请参见 Parmentier, *Manière de Faire le Pain de Pommes de Terre sans*

---

① 温德尔·拜瑞（1934— ），美国诗人、随笔作家、农民和小说家，他出生在肯塔基的新堡，1956 年在肯塔基大学获得英语学士，1957 年获得文学硕士学位。他的著作包括《家庭经济学》《动荡的美国：文化与农业》《好土地的礼物》《性、经济、自由和社会》等。——译注

Mélange de Farine; Parmentier, *Les pommes de terre, considérées relativement à la santé & à l'économie*; Parmentier, *Traité sur la culture et les usages des pommes de terre, de la patate, et du topinambour*; Spary, *Feeding France*。

2. 他声称，"在蔬菜王国，没有一种植物能够成为比土豆更健康、更方便、更便宜的食物。我不允许自己被热情所蒙蔽，但鉴于这种植物几乎不可思议的繁殖力，谁能不赞美它呢？" Parmentier, *Les pommes de terre, considérées relativement à la santé*, 184, 195.

3. 'Farine & farineux'; *Universal Magazine* 21 (1757); *Underrättelse om potatoës eller jordpärons plantering*; Duhamel du Monceau, *Moyens de conserver la santé aux equipages des vaisseaux*, 188; Zanon, *Della coltivaziones, e dell'uso delle patate*; Young, *Farmer's Letters to the People of England*; Varenne de Béost, *La cuisine des pauvres*; Parmentier, *Les pommes de terre, considérées relativement à la santé*, 44; Parmentier, *Traité sur la culture et les usages des pommes de terre, de la patate, et du topinambour*, 18; Koerner, *Linnaeus*, 149; Spary, *Feeding France*, 61–88.

4. Parmentier, *Les pommes de terre, considérées relativement à la santé*, 133.

5. Parmentier, *Les pommes de terre, considérées relativement à la santé*, 3–4.

6. Morren, *Belgique Horticole*, III: 14.

7. 例如，请参见 Scott, *Seeing Like a State*, 189, 264; Prain et al., eds., *Biological and Cultural Diversity*; Mt. Pleasant, 'The Paradoxes of Plows and Productivity'; FAO, 'Smallholders and Family Farmers'; FAO, *The State of Food and Agriculture*; Ricciardi et al., 'How Much of the World's Food do Smallholders Produce?'

8. Ioulia Evgenyevna Fenton, personal communication, 9 Jan. 2018.

9. Netting, *Smallholders, Householders*, 8–9.

10. Lacroix, *Transformations du procès de travail agricole*, 95 (first quote); Ploeg, 'Potatoes and Knowledge', 209 (second quote).

11. Bentley, *Eating for Victory*, 63–64.

12. Pollock, 'Van Gogh and the Poor Slaves'; and Coveney, *Food, Morals and Meaning*, 92–93(quote). 梵高反而相信他的画是农民生活完好的振奋人心的证明，Vincent van Gogh to Theo van Gogh, Nuenen, 30 Apr. 1885, Vincent van Gogh: The Letters。

13. Bush, 'Remarks on the Healthier US Initiative in Dallas, Texas' (first quote); UK Department of Health, 'Choosing Health: Making Healthy Choices Easier' (second quote); Biltekoff, *Eating Right in America*, 127; New Zealand Ministry of Health, 'Eating and Activity Guidelines'.

14. Campos et al., 'The Epidemiology of Overweight and Obesity', 58. 或参见 Lang and Heasman, *Food Wars*; Guthman and Allen, 'From "Old School" to "Farm-to-School"';

Crawford, 'Health as a Meaningful Social Practice'; Jarosz, 'Defining World Hunger'; Biltekoff, *Eating Right in America*; Dowler, 'Food Banks and Food Justice in "Austerity Britain"'。

15. Foucault, *History of Sexuality*, I: 138.

16. Foucault, *Security, Territory, Population*, 327 (quote), 338–339.

17. Agamben, *Homo Sacer*, 123.

18. 例如，参见 King and Slavin, 'White Potatoes, Human Health, and Dietary Guidance'; US Department of Health and Human Services, 'A Healthier You'。

致　谢

"你应该考虑一下拉姆福德汤"——这是罗杰·库特（Roger Cooter）的建议，在我都记不清的很多年前，在克罗地亚的一个温暖惬意的傍晚，他提出了这个建议，一切便从此开始。从那时起，对于大方的同事、聪明的土豆种植者和各种乐于助人的谈话对象，"土豆计划"已经欠下了太多的开心债。我通过和他们的交流获得的与土豆相关的信息、参考资料和故事，如果完整列出表格，会填满很多页，而且会考验除了最宽容的读者之外的所有人的耐心。然而为着公道起见，我要感谢的不仅仅是罗杰·库特，还包括我亲爱的秉承福柯思想的朋友克劳迪娅·斯坦（Claudia Stein），她同样参与了创作。她对我思想的影响在本书中随处可见，虽然她总是最喜欢和我争论我的历史写法的各方面。还有一些同事也对我的思维产生了决定性的影响。在思考土豆与现代治国之道的长久关系上，雅各布·克莱因（Jakob Klein）提供了巨大的帮助，并慷慨地分享了他所掌握的中国近代史知识。基思·特赖布在经济思想史上拥有令人难以置信的智慧，他也是我感激瑞典高等研究院（Swedish Collegium for Advanced Studies）奖学金项目（以及比约恩·维特洛克［Björn Wittrock］远见卓识的领导）的众多原因之一。海伦·库里（Helen Curry）是另一个让我感激瑞典高等研究院的原因。她让我阅读了她自己正在撰写的著作，内容是科学家们为保护商业粮食作物的生物多样性所采取的不断改变的方法，她还不厌其烦地为我介绍植物育种的历史。在乌普萨拉，我还时时快乐地与大卫·坎拿代因（David Cannadine）、琳达·科利（Linda Colley）、阿南达·海迪贾迪（Anandi Hattiangadi）、尤尔根·柯卡（Jürgen Kokka）、米卡·佩雷莱（Mika Perälä）、奥托·斯鲍姆（Otto Sibaum）、雅各布·斯塔兰德（Jakob Starlander）和扬·斯滕格（Jan Stenger）交谈，最后还要感谢皮娅·坎佩贾尼以友谊和智慧给予我的日复一日的支持。在乌普萨拉的一年，我还有机会受益于莉莉-安妮·艾德曼（Lili-Anne Aldman）、杜赞奇（Prasenjit Duara）、于尔瓦·哈塞尔贝尔格（Ylva Hasselberg）、卡皮尔·拉杰（Kapil Raj）、史蒂文·沙宾（Steven Shapin）和王斌（R. Bin Wong）等专家的意见，他们帮助我了解近代早期瑞典海关记录、中国的营养治国、印度植物图解、早期现代营养学，等等。

几位非常无私的同事阅读了所有章节并提出了意见。基思·特赖布和海伦·库里阅读了数版草稿，除他们之外，什里坎特·伯特利（Shrikant Botre）也对第四章提出了极有帮助的建议。玛格丽特·亨特（Margaret Hunt）对第一章提出了意见，并为我介绍了"压弯的庄稼"。在本书撰写的早期，我得到肯尼斯·班克斯（Kenneth Banks）的肯定和他提出的宝贵意见。还有其他人与我分享了各种参考资料，"只有你才会觉得这有趣"，他们的邮件往往这样开头。我还要特别感谢他们：丽尔·巴特勒（Cheryl Butler）给我讲述了16世纪90年代土豆在南安普顿被当做礼物的故事；伯纳德·卡普（Bernard Capp）告诉了我多塞特伯爵死去的原因；莉齐·科林厄姆告诉了我约瑟夫·阿奇自传中关于土豆的内容（她还对"二战"时的粮食问题有杰出研究）；鲁迪·马特（Rudi Matthee）帮助我了解伊朗的饮食；凯瑟琳·桑特纳（Kathryn Santner）给我详细讲解了阿雷基帕修女的饮食（并为我扫描了资料）；海伦·威克利（Helen Wakely）给我提供了维康图书馆（Wellcome Library）收藏的现代早期烹饪书稿中关于土豆的资料。我还想感谢哈坎·布洛姆奎斯特（Håkan Blomqvist）为我介绍瑞典的土豆革命；尤莉亚·芬顿、利昂·西利-哈金斯（Leon Sealey-Huggins）和阿拉斯泰尔·史密斯（Alastair Smith）在小农农业方面的专业知识；蒂亚戈·萨拉维亚（Tiago Saravia）提供了关于法西斯主义者的猪和土豆的知识；孙小平（Xiaoping Sun）为我讲述 1959—1961 年中国灾荒年间土豆为帮助村民们生存所起到的作用；约纳斯·塔米拉（Joonas Tammela）告诉了我芬兰土豆英雄的故事；伊兰加·契科（Iranga Tcheko）给我讲述了她祖母在比属刚果被迫种植土豆的回忆；卡罗琳·斯蒂德曼（Carolyn Steedman）和马克·菲利普（Mark Philp）对《土豆小贩》作了文字说明。能有如此慷慨而学识渊博的朋友和同事，我深感荣幸。

由于土豆已经传播到世界如此多的角落，它在文献记录中的存在远远超过了我的语言技能。因此，我特别感谢那些帮助我通过希腊语、拉丁语、挪威语、瑞典语、芬兰语、俄语、冰岛语和汉语资料追踪其足迹的人。拉丁语和希腊语，我得到了迪热尔·阿博（Desirée Arbó）、丹尼斯·兰迪斯（Dennis Landis）、扬·斯滕格和米卡·佩雷莱的帮助；阿蒙德·佩德森（Amund Pedersen）对近代早期挪威的土豆进行了详细的研究；雅各布·斯塔兰德和汉娜·霍达克（Hanna Hodacs）指导我阅读了 8 世纪的瑞典历书和科学著作；尤西佩卡·卢克宁（Jussipekka Luukkonen）撰写了一篇关于芬兰土豆推广者的未发表论文。科勒姆·莱基（Colum Leckey）慷慨地帮助我了解俄罗斯的土豆推广，并反复查看自己的注释以找出具体细节。没有惠普娜·罗伯茨多蒂尔

（Hrefna Róbertsdóttir）的指导，我不可能了解土豆在冰岛的传播。安妮·格里森（Anne Gerritsen）、黄璐（Huang Lu）和克莱尔·唐（Claire Tang）一起翻译了一条在汉语普通话中颇具挑战性的口号，赢得了其他人的赞赏。

我一直坚信土豆值得关注，有两个人长期纵容着我的想法，那就是我的母亲丽莎·厄尔（Lisa Earle）和我的丈夫马特·韦斯顿（Matt Western）。我母亲对土豆的植物学复杂性的了解远比我希望知道的要多得多。在我脑海里还没闪现土豆这个主题的数年，甚至数十年前，她就去参观过秘鲁的国际马铃薯中心。和她谈论土豆是一件很愉快的事，她自己作为生物学家和学者的专业知识也让我受益匪浅。每当我自言自语说，嗯，这个论题现在应该清楚了，马特眼里总会闪着被逗乐的纵容的光芒。带着爱，我将本书献给他们两位。

参考文献

档案

阿雷基帕大主教档案馆，阿雷基帕
（Archivo Arzobispal de Arequipa, Arequipa）

Santa Catalina leg. 8.
Santa Teresa leg. 2.

西班牙国家图书馆，马德里
（Biblioteca Nacional de España, Madrid）

Bernardo de Cienfuegos, 'Historia de las plantas', 7 vols., c.1627–1631, Mss/3357–3363.

大英图书馆，伦敦
（British Library, London）

IOR F/4/179.
IOR H/Misc/799, 59–169.
IOR P/242/73.
IOR/F/4/379/9495.
MSS EUR/F95/2.

市政厅图书馆，伦敦
（Guildhall Library, London）

Ms 7801, box 2.

兰开夏郡记录办公室，普雷斯顿
(Lancashire Record Office, Preston)

DDB acc 6685 box 179.
DDFO 23/4.
DDM 11/61.
DDSc 9/47.
PR 718.
PR/284.

麦吉尔大学图书馆，蒙特利尔
(McGill University Library, Montreal)

Doncaster Recipes Collection, MSG 1230.

雷丁大学英国乡村博物馆特藏，雷丁
(Museum of English Rural Life Special Collections, University of Reading, Reading)

SR RASE A/I–II.
SR RASE B/I, B/II, B/VI–VII. B/X, B/XIII.

洛克菲勒基金会线上档案
(Rockefeller Foundation Archives Online)

Officers Diaries, PG 12.

皇家艺术学会档案馆，伦敦
(Royal Society of Arts Archive, London)

PR/MC/104/10/240.
RSA/PR/GE/110/26/70–71.
RSA/PR/MC/105/10/396.

拉德克利夫学院施莱辛格图书馆，马萨诸塞州坎布里奇
(Schlesinger Library, Radcliffe College, Cambridge, MA)

American Institute of Wine & Food Recipe Books, c.1690–c.1830, MC 675, box 1.
Manuscript recipe book of W. Walker, Hayes Middlesex, c.1798–1826, Recipes for Cures and Cookery, AR297.
Marion King Schlefer Recipe Collection, A/S33.
Sarah Fayerweather Cookbook, 1764, A/F283.
Sophie D. Coe Manuscript Cookbook Collection, MC 844, box 1.

国家档案馆，基尤
（The National Archives, Kew）

C 6/414/31.
E134/10Geo2/Hil3.
HO 42/35, 42/61.
WO 1/1094.

威康图书馆，伦敦
（Wellcome Library, London）

Collection of Cookery and Medical Receipts by Edward and Katherine Kidder, 1699, Ms 3107.

Collection of Cookery and Medical Receipts, c.1685–c.1725, Ms 1796.

Cookery Receipts Collected by Johnson Family of Spalding, Lincs., 1694–1831, Ms 3082.

Physical and Chyrurgicall Receipts. Cookery and Preserves. Collected by Elizabeth Jacob and Others, 1654–c.1685, Ms 3009.

Recipe Book of the Godrey-Faussett Family of Heppington, Nackington, Kent, late 17th century to mid 18th century, Ms 7998.

报纸期刊

Annals of Agriculture and Other Useful Arts, London (1795, 1797, 1799).

Appendix to the Scots Magazine, Edinburgh (1740).

Asiatic Journal and Monthly Register for British India and its Dependencies, London (1824, 1826).

Bibliothèque britannique: Littérature, Geneva (1796).

Bibliothèque britannique, ou Recueil extrait des ouvrages anglais périodiques: Sciences et Arts, Geneva (1797).

Bibliothèque physico-économique, instructive et amusant, Paris (1787, 1790).

Colonial Times and Tasmanian Advertiser, Tasmania (1826).

Correio mercantil e economico de Portugal, Lisbon (1798).

Correo mercantil de España y sus Indias, Madrid (1801).

English Review, London (1794).

European Magazine and London Review, London (1790, 1800).

Feuille de cultivateur, Paris (1799).

Friend of India, Serampore (1820).

Gentleman's Magazine and Historical Chronicle, London (1764, 1795).

Guardian, London (2006, 2013, 2014, 2018).

Journal historique & litteraire, Luxembourg (1780).

Journal historique et politique de Genève, Geneva (1779).

Journal historique et politique des principaux événemens des différentes cours de l'Europe, Geneva (1779).

L'Avantcoureur, Paris (1771).

London Gazette, London (1778, 1800).

London Review of English and Foreign Literature, London (1780).

Magazzino toscana, Florence (1775).

Massachusetts Spy, or the Worcester Gazette, Worcester, MA (1790).

Mémoires d'agriculture, d'économie rurale et domestique, Paris (1789).

Memorial literario, ó, Biblioteca periódica de ciencias y artes, Madrid (1802).

Memorial literatio, instructivo y curioso de la corte de Madrid, Madrid (1790).

Mercurio peruano de historia, literatura, y noticias públicas que da á luz la Sociedad Academica de Amantes de Lima, Lima (1792, 1794).

Monthly Magazine and British Register, London (1797).

New York Times, New York (1943, 2012).

Semanario de agricultura y artes dirigido a los párrocos, Madrid (1797, 1799, 1800, 1801, 1802, 1805, 1806).

South African Magazine, A Contribution to Colonial Literature, Cape Town (1869).

Telegraph, London (2007).

The Struggle; Devoted to the Advocacy of Free Trade and the Repeal of the Corn Laws, Preston (1843).

Times, London (1795, 1830, 1847).

Universal Magazine, London (1757).

Weekly Entertainer, London (1796).

其他

Abad-Zardoya, Carmen, 'Arquitectos en los fogones: del *theatrum machinarum* a los proyectos ilustrados para una cocina económica', *Artigrama* 26 (2011), 649–668.

Abbots, Emma-Jayne and Benjamin Coles, 'Horsemeat-Gate: The Discursive Production of a Neoliberal Food Scandal', *Food, Culture and Society* 16:4 (2013), 535–550.

Abreu, Laurinda, *The Political and Social Dynamics of Poverty, Poor Relief and Health Care in Early-Modern Portugal*, Routledge (London, 2016).

Accomplished Ladies Rich Closet of Rarities (London, 1687).

Acosta, José de, *Natural and Moral History of the Indies*, trans. Frances López-Morillas, Duke University Press (Durham, 2002 [1590]).

Act of Tonnage and Poundage, and Rates of Merchandize (Edinburgh, 1705).

Adam, James, *Practical Essays on Agriculture*, 2 vols. (London, 1789).

Adelman, Jeremy, *Republic of Capital: Buenos Aires and the Legal Transformation of the Atlantic World*, Stanford University Press (Stanford, 1999).

Africa and America Described with Anecdotes and Numerous Illustrations, 2 vols. (London, 1856).

Agamben, Georgio, *Homo Sacer: Sovereign Power and Bare Life*, trans. Daniel Heller-Roazen, Stanford University Press (Stanford, 1998).

Aguilar, Sandra, 'Cooking Modernity: Nutrition Policies, Class, and Gender in 1940s and 1950s Mexico City', *The Americas* 64:2 (2007), 177–205.

'AHR Exchange: The Question of Black Rice', *American Historical Review* 1115:1 (2010), 123–171.

Aikin, John, *A Description of the Country from Thirty to Forty Miles Round Manchester* (London, 1795).

Albala, Kenneth, *Eating Right in the Renaissance*, University of California Press (Berkeley, 2002).

Alcedo, Antonio de, *The Geographical and Historical Dictionary of America and the West Indies Containing an Entire Translation of the Spanish Work of Colonel Don Antonio de Alcedo*, ed. G. A. Thompson, 5 vols. (London, 1812).

Aldini, Tobias, *Exactissima Descriptio Rariorum Quarundam Plantarum, que continentur Rome in Horto Farnesiano* (Rome, 1625).

Aldrete, Bernardo, *Del orígen y principio de la lengua castellana o romance que oi usa en España* (Rome, 1606).

Alletz, Pons Augustin, *Agronomía, ó diccionario manual del labrador*, trans. Pedro Charro de Lorenzana (Madrid, 1817).

Alletz, Pons Augustin, *L'Agronome, ou Dictionnaire portatif du cultivateur*, 2 vols. (Paris, 1799).

Alonso de Herrera, Gabriel, *Agricultura general de Gabriel Alonso de Herrera, corregida según el texto original de la primera edición publicada en 1513 por el mismo autor y adicionada por la Real Sociedad Económica Matritense*, 2 vols. (Madrid, 1819).

Altimiras, Juan, *Nuevo arte de cocina, sacado de la escuela de la esperiencia económica* (Barcelona, 1758).

Amado Doblas, María Isabel, 'Apunte bibliográfico acerca de la batata/patata en la literatura del siglo de oro', *Isla de Arriarán* 18 (2001), 275–287.

Ambrosoli, Mauro, *The Wild and the Sown: Botany and Agriculture in Western Europe: 1350–1850*, trans. Mary McCann Salvatorelli, Cambridge University Press

(Cambridge, 1997).

Amoretti, Carlo, *Della coltivazione delle patate e loro uso. Instruzione* (Milan, 1801).

Andagoya, Pascual de, *Pascual de Andagoya, Relación y documentos*, ed. Adrian Blázquez, Historia 16 (Madrid, 1986).

Andresen, Astri and Kari Tove Elvbakken, 'From Poor Law Society to the Welfare State: School Meals in Norway, 1890–1950s', *Journal of Epidemiology & Community Health* 61:5 (2007), 374–377.

Andrews, Donna, *Philanthropy and Police: London Charity in the Eighteenth Century*, Princeton University Press (Princeton, 1989).

Angående potaters förvarande (Stockholm, 1788).

Annalen der Braunschweig Luneburgischen Churlande 9:1 (Hanover, 1795).

Appleby, Joyce, 'Ideology and Theory: The Tension between Political and Economic Liberalism in Seventeenth-Century England', *American Historical Review* 81:3 (1976), 499–515.

Arch, Joseph, *From Ploughtail to Parliament. An Autobiography*, Cresset Library (London, 1986 [c.1896]).

Argumossa y Gandara, Theodoro Ventura de, *Erudicción política despertador sobre el comercio, agricultura, y manufacturas, con avisos de buena policía, y aumento del real erario* (Madrid, 1743).

Arnold, David, 'Agriculture and "Improvement" in Early Colonial India: A Pre-History', *Journal of Agrarian Change* 5:4 (2005), 505–525.

Arnold, David, 'Hunger in the Garden of Plenty: The Bengal Famine of 1770', in *Dreadful Visitations: Confronting Natural Catastrophe in the Age of Enlightenment*, ed. Alessa Johns, Routledge (London, 1999), 81–112.

Arnold, David, 'The "Discovery" of Malnutrition and Diet in Colonial India', *Indian Economic and Social History Review* 31:1 (1994), 1–26.

Arras, Paul, ed., *Quellenbuch zur sächsischen Geschichte*, Europäisscher Geschichtsverlag (Paderborn, 2015 [1912]).

Arteta, Antonio, *Disertacion sobre la muchdumbre de niños que mueren en la infancia, y modo de remediarla, y de procurar en sus cuerpos la conformidad de sus miembros, robustez, agilidad y fuerzas competentes*, 2 vols. (Zaragoza, 1801–1802).

Astry, Diana, *Diana Astry's Recipe Book*, ed. Betty Stitt, Bedfordshire Historical Society (Streatley, 1957).

Atwater, W. O., *Principles of Nutrition and Nutritive Value of Food*, Government Printing Office (Washington, 1910).

Ausfuhrlicher Unterricht zur Bereitung der rumfortschen Spaarsuppen (Leipzig, [1805?]).

Ávila, Teresa de, *Escritos de Santa Teresa*, ed. Vicente de la Fuente, 2 vols. (Madrid, 1862).

Baldini, Filippo, *De' Pomi di terra ragionamento* (Naples, 1783).

Ballexserd, Jacques, *Dissertation sur l'éducation physique des enfans* (Paris, 1762).

Banks, Joseph, *The Indian and Pacific Correspondence of Sir Joseph Banks, 1768–1820*, ed. Neil Chambers, 5 vols., Pickering & Chatto (London, 2008).

Barber, John and Mark Harrison, *The Soviet Home Front, 1941–1945: A Social and Economic History of the USSR in World War II*, Longman (New York, 1993).

Barnard, T. C., 'Gardening, Diet and "Improvement" in Later Seventeenth Century Ireland', *Journal of Garden History* 10:1 (1990), 71–85.

Bashford, Alison and Joyce Chaplin, *The New Worlds of Thomas Robert Malthus: Rereading the Principle of Population*, Princeton University Press (Princeton, 2016).

Bass, Hans, 'The Crisis in Prussia', in *When the Potato Failed: Causes and Effects of the 'Last' European Subsistence Crisis, 1845–1850*, ed. Cormac Ó Gráda, Richard Paping and Eric Vanhaute, Brepols (Turnhout, 2007), 185–212.

Batmanglij, Najmieh, *Food of Life: Ancient Persian and Modern Iranian Cooking and Ceremonies*, Mage Publishers (Washington, 2016).

Bauhin, Gaspard, *Prodromos Theatri Botanici* (Frankfurt, 1620).

Bavetta, Sebastiano, Pietro Navarra and Dario Maimone, *Freedom and the Pursuit of Happiness: An Economic and Political Perspective*, Cambridge University Press (Cambridge, 2014).

Baxter, Richard, 'The Reverend Richard Baxter's Last Treatise', ed. Frederick J. Powicke, *Bulletin of the John Rylands Library*, 10 (1926), 163–218.

BBC News, 'Cookery to be Compulsory', 22 Jan. 2008, http://news.bbc.co.uk/1/hi/education/7200949.stm.

BBC News, 'Potatoes Spark Chile–Peru Dispute', 29 Mar. 2006, http://news.bbc.co.uk/go/pr/fr/-/2/hi/americas/4856154.stm.

BBC News, 'Tory Peer Apologises for Saying "Poor Can't Cook"', 8 Dec. 2014, www.bbc.co.uk/news/uk-politics-30379431.

BBC Radio 4, 'Woman's Hour', 17 Mar. 2014.

Beales, Derek, 'Philosophical Kingship and Enlightened Despotism', in *The Cambridge History of Eighteenth-Century Political Thought*, ed. Mark Goldie and Robert Wokler, Cambridge University Press (Cambridge, 2006), 495–524.

Becher, Johann Joachim, *Politische Discurs von den eigentlichen Ursachen deß Auffund Abnehmens der Städt, Länder und Republicken* (Frankfurt, 1688).

Beckett, Ian, 'Total War', in *War, Peace and Social Change in Twentieth-Century Europe*, ed. Clive Emsley, Arthur Warwick and Wendy Simpson, Open University Press (Milton

Keynes, 1989), 26–44.

Béguillet, Edme, *Traité des subsistances et des grains, qui servent a la nourriture de l'homme*, 6 vols. (Paris, 1780).

Beinart, William, 'African History and Environmental History', *African Affairs* 99:395 (2000), 269–302.

Beinart, William and Karen Middleton, 'Plant Transfers in Historical Perspective: A Review Article', *Environmental History* 10:1 (2004), 3–29.

Bekaert, Geert, 'Caloric Consumption in Industrializing Belgium', *Journal of Economic History* 51:3 (1991), 633–655.

Benavídez, Pedrarias de, *Secretos de chirurgia, especial de las enfermedades de morbo gallico y lamparones y mirrarchia* (Valladolid, 1567).

Bentham, Jeremy, *An Introduction to the Principles of Morals and Legislation* (London, 1789).

Bentham, Jeremy, *The Works of Jeremy Bentham*, ed. John Bowring (Edinburgh, 1843).

Bentley, Amy, *Eating for Victory: Food Rationing and the Politics of Domesticity*, University of Illinois Press (Urbana, 1998).

Benzo, Ugo, Lodovico Bertaldi and Baldassar Pisanelli, *Regole della sanitá et natura de cibi* (Turin, 1620).

Berg, Gösta, 'Die Kartoffel und die Rübe', *Ethnologia Scandinavica* (1971), 158–166.

Berg, Maxine, 'Afterword: Things in Global History', in *The Global Lives of Things: The Material Culture of Connections in the Early Modern World*, ed. Anne Gerritsen and Giorgio Riello, Routledge (London, 2016), 253–258.

Berg, Maxine and Elizabeth Eger, 'Introduction', in *Luxury in the Eighteenth Century: Debates, Desires and Delectable Goods*, ed. Maxine Berg and Elizabeth Eger, Palgrave Macmillan (Basingstoke, 2003), 1–4.

Beskrifning om jordpärons plantering (Stockholm, 1773).

Besler, Basilius, *Hortus Eystenttensis* (Nuremberg, 1640).

Bethke, Paul C., Atef M. K. Nassar, Stan Kubow, Yves N. Leclerc, Xiu-Qing Li, Muhammad Haroon, Teresa Molen, John Bamberg, Max Martin, and Danielle J. Donnelly, 'History and Origin of Russet Burbank (Netted Gem): A Sport of Burbank', *American Journal of Potato Research* 91:6 (2014), 594–609.

Bielefeldt, Heiner, 'Autonomy and Republicanism: Immanuel Kant's Philosophy of Freedom', *Political Theory* 25:4 (1997), 524–558.

Bignami, Pietro Maria, *Le patate* (Bologna, 1773).

Billingsley, John, 'On the Culture of Potatoes, and feeding Hogs with them, during Seven Years', in *Letters and Papers on Agriculture, Planting, & Selected from the*

Correspondence of the Bath and West of England Society for the Encouragement of Agriculture, Arts, Manufactures and Commerce 6 (London, 1792), 339–394.

Biltekoff, Charlotte, *Eating Right in America: The Cultural Politics of Food and Health*, Duke University Press (Durham, 2013).

Binning, Robert B. M., *Journal of Two Years' Travel in Persia Ceylon*, Etc., 2 vols. (London, 1857).

Birembaut, Arthur, 'L'École gratuite de boulangerie', in *Enseignement et diffusion des sciences en France au XVIIIe siècle*, ed. René Taton, Hermann (Paris, 1964), 493–509.

Bleichmar, Daniela, *Visible Empire: Botanical Expeditions and Visual Culture in the Hispanic Enlightenment*, University of Chicago Press (Chicago, 2012).

Bleichmar, Daniela, Paula De Vos, Kristin Huffine and Kevin Sheehan, eds., *Science in the Spanish and Portuguese Empires, 1500–1800*, Stanford University Press (Stanford, 2009).

Bligh, William, *A Voyage to the South Sea for the Purpose of Conveying the Bread-fruit Tree to the West Indies* (Cambridge, 2013 [1792]).

Blomqvist, Håkan, 'Sweden's Potato Revolution', *Solidarity*, Nov.–Dec. 2017, https://solidarity-us.org/atc/191/p5131/.

Blum, Jerome, *The End of the Old Order in Rural Europe*, Princeton University Press (Princeton, 1978).

Board of Agriculture, *Account of the Experiments tried by the Board of Agriculture in the Composition of Various Sorts of Bread* (London, 1795).

Board of Agriculture, *Communications to the Board of Agriculture on Subjects Relative to the Husbandry, and Internal Improvement, of the Country* 1: I–II (London, 1797).

Board of Agriculture, *Hints Respecting the Culture and the Use of Potatoes* (London, 1795).

Board of Agriculture, *Report of the Committee of the Board of Agriculture Appointed to Extract Information from the County Reports and other Authorities Concerning the Culture and Use of Potatoes* (London, 1795).

Bohstedt, John, 'The Myth of the Feminine Food Riot: Women as Proto-Citizens in English Community Politics, 1790–1810', in *Women and Politics in the Age of the Democratic Revolution*, ed. Harriet Applewhite and Darline Levy, University of Michigan Press (Ann Arbor, 1993), 21–60.

Bohstedt, John, *The Politics of Provisions: Food Riots, Moral Economy, and Market Transition in England, c.1550–1850*, Ashgate (Farnham, 2010).

Boissier de Sauvages, François, *Dictionnaire languedocien-françois* (Nimes, 1756).

Boll, Ernst, *Geschichte Meklenburgs mit besondere Berücksichtigung der Culturgeschichte*

(Neubrandenburg, 1856).

Bolotov, Andrei, 'O delanii iz tartofelia muki', *Trudy Vol'nago Ekonomicheskago Obshchestva* 14 (1770), 33–63.

Bonaparte, Napoleon, *Letters of Napoleon*, ed. J. M. Thompson, Dent (London, 1954).

Bonells, Jaime, *Perjuicios que acarrean al género humano y al estado las madres que rehusan criar a sus hijos* (Madrid, 1786).

Bonzon, Thierry, 'Consumption and Total Warfare in Paris (1914–1918)', in *Food and Conflict in Europe in the Age of the Two World Wars*, ed. Frank Trentmann and Flemming Just, Palgrave Macmillan (Basingstoke, 2006), 49–64.

Booth, Alan, 'Food Riots in the North-West of England, 1790–1801', *Past & Present* 77 (1977), 84–107.

Boserup, Ester, *Woman's Role in Economic Development*, intro. Nazneen Kanji, Su Fei Tan and Camilla Toulmin, Earthscan (London, 2007).

Botero, Giovanni, *The Reason of State and The Greatness of Cities*, ed. and trans. P. J. Waley and D. P. Waley, Yale University Press (New Haven, 1956).

Bowles, Guillermo, *Introducción a la historia natural y de la geografía física de España*, trans. José Nicolás de Azara (Madrid, 1775).

Boyd-Orr, John, *Food, Health and Income: Report on a Survey of Adequacy of Diet in Relation to Income*, Macmillan (London, 1936).

Brading, David John, *The First America, the Spanish Monarchy, Creole Patriots, and the Liberal State, 1492–1867*, Cambridge University Press (Cambridge, 1991).

Brady, John, 'Remedies Proposed for the Church of Ireland (1697)', *Archivium Hibernicum* 22 (1959), 163–173.

Bräker, Ulrich, *The Poor Man of Toggenburg*, 1789, trans. Margaret Clare Britton, www.keithsayers.id.au/Toggenburg/Cover.htm.

Braudel, Fernand, *Civilization and Capitalism, 15th–18th Century, vol. I: The Structure of Everyday Life*, trans. Siân Reynolds, University of California Press (Berkeley, 1982).

Braudel, Fernand, *The Mediterranean and the Mediterranean World in the Age of Philip II*, trans. Siân Reynolds, 2 vols., University of California Press (Berkeley, 1995).

Bray, Francesca, 'Chinese Literati and the Transmission of Technological Knowledge: The Case of Agriculture', in *Cultures of Knowledge: Technology in Chinese History*, ed. Dagmar Schäfer, Brill (Leiden, 2011), 299–325.

Bray, Francesca, *Technology and Gender: Fabrics of Power in Late Imperial China*, University of California Press (Berkeley, 1997).

Brennan, Thomas, *Public Drinking and Popular Culture in Eighteenth-Century Paris*, Princeton University Press (Princeton, 1988).

Broadberry, Stephen, Rainer Fremdling and Peter Solar, 'Industry', in *The Cambridge Economic History of Modern Europe*, vol. I: *1700–1850*, ed. Stephen Broadberry and K. O'Rourke, Cambridge University Press (Cambridge, 2010), 164–186.

Broadberry, Stephen and K. O'Rourke, eds., *The Cambridge Economic History of Modern Europe*, 3 vols., Cambridge University Press (Cambridge, 2010).

Brown, Stephen, 'Notes on the History of the Potato Breeding Institute, Trumpington', Jan. 2011, Trumpington Local History Group, www.trumpingtonlocalhistorygroup.org/subjects_PBIhistory.html.

Browne, Patrick, *Civil and Natural History of Jamaica* (London, 1789).

Brownell, Kelly D., Thomas Farley, Walter C. Willett, Barry M. Popkin, Frank J. Chaloupka, Joseph W. Thompson and David S. Ludwig, 'The Public Health and Economic Benefits of Taxing Sugar-Sweetened Beverages', *New England Journal of Medicine* 361 (2009), 1599–1605.

Brundage, James, 'Sumptuary Laws and Prostitution in Late Medieval Italy', *Journal of Medieval History* 13:4 (1987), 343–355.

Bruni, Luigino and Pier Luigi Porta, '*Economia civile* and *pubblica felicità* in the Italian Enlightenment', in Oeconomies in the Age of Newton, ed. Margaret Schabas and Neil de Marchi, Duke University Press (Durham, 2003), 361–385.

Brush, Stephen, *Farmer's Bounty: Locating Crop Diversity in the Contemporary World*, Yale University Press (New Haven, 2004).

Brush, Stephen, Heath Carney and Zósimo Huamán, 'Dynamics of Andean Potato Agriculture', *Economic Botany* 35:1 (1981), 70–88.

Buchan, William, *Domestic Medicine: or, a Treatise on the Prevention and Cure of Diseases by Regimen and Simple Medicines* (London, 1776).

Buchan, William, *Observations Concerning the Diet of the Common People, Recommending a Method of Living Less Expensive, and More Conducive to Health, than the Present* (London, 1797).

Buchet, Christian, *The British Navy, Economy and Society in the Seven Years War*, trans. Anita Higgie and Michael Duffy, Boydell & Brewer (Woodbridge, 2013).

Buc'hoz, Pierre-Joseph, *Manuel alimentaire des plantes* (Paris, 1771).

Burchardt, Jeremy, 'Land and the Laborer: Potato Grounds and Allotments in Nineteenth-Century Southern England', *Agricultural History* 74:3 (2000), 667–684.

Burke, Edmund, *Thoughts and Details on Scarcity Originally Presented to the Right Hon. William Pitt, in the Month of November 1795* (London, 1800).

Burnett, John, *Plenty and Want: A Social History of Food in England from 1815 to the Present Day*, Routledge (London, 1989).

Burnett, John and Derek Oddy, eds., *The Origins and Development of Food Policies in Europe*, Leicester University Press (London, 1994).

Bush, George W., 'Remarks on the Healthier US Initiative in Dallas, Texas', 18 July 2003, *Public Papers of the Presidents of the United States* III: 893, The US national Archives and Records Administration, www.gpo.gov/fdsys/pkg/PPP2003-book2/html/PPP-2003-book2-doc-pg889-2.htm.

Business in the Community, *Healthy People=Healthy Profits* (London, 2009).

Butler, Cheryl, ed., *The Book of Fines: The Annual Accounts of the Mayors of Southampton,* vol. III: *1572–1594*, Southampton Records Series 44 (Southampton, 2010).

[Butler, Nathaniel], *The History of the Bermudaes or Summer Islands*, ed. John Henry Lefroy (London, 1882).

Cabarrús, Francisco, 'Discurso sobre la libertad de comercio', 1778, reproduced in Santos Manuel Coronas González, 'Espíritu ilustrado y liberación del tráfico comercial con Indias', *Anuario de historia del derecho español* (1992), 102–111.

Cadet de Vaux, Antoine-Alexis, *L'Ami de l'économie aux amis de l'humanité sur les pains divers dans la composition desquels entre le pomme-de-terre* (Paris, 1816).

Cadet de Vaux, Antoine-Alexis, 'Variétés', *La Décade philosophique, littéraire et politique* 24 (Paris, 1800), 367–371.

Cadet de Vaux, Antoine-Alexis, [Augustin Pyramus] Decandolle, [Benjamin-Jules-Paul] Delessert, Money and Antoine Augustin Parmentier, *Recueil de rapports, de mémoires et d'expériences sur les soupes économiques et les fourneaux a la Rumford* (Paris, 1801).

Caird, James, *Our Daily Food, Its Price, and Sources of Supply* (London, 1868).

Campbell, Jodi, *At the First Table: Food and Social Identity in Early Modern Spain*, University of Nebraska Press (Lincoln, 2017).

Campillo y Cosío, José del, *Nuevo sistema de gobierno económico para la América, con los males y daños que le causa el que hoy tiene, de los que participa copiosamente España y remedios universales para que la primera tenga considerables ventajas, y al segunda mayores intereses* (Madrid, 1789).

Campini, Antonio, *Saggi d'Agricoltura del medico Antonio Campini* (Turin, 1774).

Camporesi, Piero, *Bread of Dreams: Food and Fantasy in Early Modern Europe*, trans. David Gentilcore, Polity Press (Cambridge, 1989).

Camporesi, Piero, *The Magic Harvest: Food, Folklore and Society*, trans. Joan Hall, Polity Press (Cambridge, 1993).

Campos, Paul, Abigail Saguy, Paul Ernsberger, Eric Oliver and Glenn Gaesser, 'The Epidemiology of Overweight and Obesity: Public Health Crisis or Moral Panic?',

International Journal of Epidemiology 35:1 (2006), 55–60.

Cañizares-Esguerra, Jorge, *How to Write the History of the New World: Histories, Epistomologies and Identities in the Eighteenth-Century Atlantic World*, Stanford University Press (Stanford, 2001).

Cañizares-Esguerra, Jorge, *Nature, Empire and Nation: Explorations of the History of Science in the Iberian World*, Stanford University Press (Stanford, 2006).

Capatti, Alberto and Massimo Montanari, *Italian Cuisine: A Cultural History*, trans. Aine O'Healy, Columbia University Press (New York, 2003).

Caradonna, Jeremy L., *The Enlightenment in Practice: Academic Prize Contests and Intellectual Culture in France, 1670–1794*, Cornell University Press (Ithaca, 2012).

Cardano, Geronimo, *De Rerum Varietate* (Basle, 1557).

Carneiro Dias, Camila and Maria Conceição da Costa, 'Indigenous Claims to Native Crops and Plant Genebanks: A Case Study from Peru', unpublished conference paper, IV Globelics Conference, Mexico City, 2008.

Carney, Judith, *Black Rice: The African Origins of Rice Cultivation in the Americas*, Harvard University Press (Cambridge, MA, 2001).

Carney, Judith and Richard Rosomoff, *In the Shadow of Slavery: Africa's Botanical Legacy in the Atlantic World*, University of California Press (Berkeley, 2009).

Carpenter, Kenneth, *Protein and Energy: A Study of Changing Ideas in Nutrition*, Cambridge University Press (Cambridge, 1994).

Carr, Edward, 'Postmodern Conceptualizations, Modernist Applications: Rethinking the Role of Society in Food Security', *Food Policy* 31 (2006), 14–29.

Carrió de la Vandera, Alonso, *El lazarillo de ciegos caminantes desde Buenos Aires, hasta Lima*, 1773, Biblioteca Virtual Miguel de Cervantes, www.cervantesvirtual.com.

Casado Soto and José Luis, 'Notas sobre la implantación del maíz en Cantabria y la sustitución de otros cultivos', in *Población y sociedad en la España cantábrica durante el siglo XVII*, Institución Cultural de Cantabria (Santander, 1985), 159–173.

Cassidy, Martin, 'Comber Spuds get European Protected Status', 25 Jan. 2012, BBC News, www.bbc.com/news/uk-northern-ireland-16727411.

Casteau, Lancelot de, *Ouverture de cuisine* (Liège 1604).

Castro-Gómez, Santiago, *La hybris del punto cero: ciencia, raza e ilustración en la Nueva Granada (1750–1816)*, Pontificia Universidad Javeriana (Bogotá, 2005).

Centro Internacional de la Papa, *Informe anual 1975*, Centro Internacional de la Papa (Lima, 1975).

Centro Internacional de la Papa, *Informe anual 1979*, Centro Internacional de la Papa (Lima, 1979).

Centro Internacional de la Papa, 'Potato', https://cipotato.org/crops/potato/.

Chakrabarty, Dipesh, *Provincializing Europe: Postcolonial Thought and Historical Difference*, Princeton University Press (Princeton, 2000).

Chamberlayne, Edward, *Angliae Notitia: or the Present State of England* (London, 1702).

Chambers, Sarah, 'Little Middle Ground: The Instability of a Mestizo Identity in the Andes, Eighteenth and Nineteenth Centuries', in *Race and Nation in Modern Latin America*, ed. Nancy Appelbaum, Anne Macpherson and Karin Alejandra Rosemblatt, University of North Carolina Press (Chapel Hill, 2003), 32–55.

Charbit, Yves, *The Classical Foundations of Population Thought from Plato to Quesnay*, Springer (London, 2011).

Chen, Te-Ping, 'Pushing the Potato: China Wants People to Eat More "Earth Beans"', *Wall Street Journal Online*, 11 Jan. 2015, https://blogs.wsj.com/chinarealtime/2015/01/09/pushing-the-potato-china-wants-people-to-eat-moreearth-beans/.

China News Service, 'China to Boost Potato Cultivation', 6 Jan. 2015, www.chinanews.co.jp/node?page=1433.

Christiansen González, Jorge, *El cultivo de la papa en el Perú*, Editorial Jurídica (Lima, 1967).

Cieza de León, Pedro de, *Parte primera de la Chronica del Perú* (Antwerp, 1554).

Clarkson, L. A. and E. Margaret Crawford, *Feast and Famine: Food and Nutrition in Ireland 1500–1920*, Oxford University Press (Oxford, 2001).

Clericuzio, Antonio, 'Chemical and Mechanical Theories of Digestion in Early Modern Medicine', *Studies in History and Philosophy of Biological and Biomedical Sciences* 43 (2012), 329–337.

Clusius, Carolus (Charles de L'Écluse), *Clusius Correspondence: A Digital Edition-in-Progress*, ed. Esther van Gelder, http://clusiuscorrespondence.huygens.knaw.nl.

Clusius, Carolus, *Rariorum Plantarum Historia* (Antwerp, 1601).

Coates, Colin, *The Metamorphosis of Landscape and Community in Early Quebec*, McGill-Queen's University Press (Montreal and Kingston, 2000).

Cobbett, William, *Cobbett in Ireland: A Warning to England*, ed. Denis Knight, Lawrence & Wishart (London, 1984).

Cobbett, William, *Cobbett's Two-Penny Trash or, Politics for the Poor* 1 (London, 1831).

Cobbett, William, *Cobbett's Weekly Political Register* (London, 1822–1832).

Cobbett, William, *Cottage Economy* (London, 1826).

Coe, Sophie, *America's First Cuisines*, University of Texas Press (Austin, 1994).

Cohen, Patricia Cline, *A Calculating People: The Spread of Numeracy in Early America*, University of Chicago Press (Chicago, 1982).

Cole, Joshua, *The Power of Large Numbers: Population, Politics, and Gender in Nineteenth-Century France*, Cornell University Press (Ithaca, 2000).

Cole, Mark, 'Feeding the Volk: Food, Culture and the Politics of Nazi Consumption, 1933–1945', DPhil, University of Florida, 2011.

Coles, William, *Adam in Eden: or, Natures Paradise. The History of Plants, Fruits, Herbs and Flowers* (London, 1657).

Collection of Receipts in Cookery, Physick and Surgery (London, 1746).

Collingham, Lizzie, *Curry: A Biography*, Chatto & Windus (London, 2005).

Collingham, Lizzie, *The Taste of War: World War Two and the Battle for Food*, Penguin (London, 2011).

Collins, David John, *Practical Rules for the Management and Medical Treatment of the Negro Slaves in the Sugar Colonies* (London, 1803).

[Colquhoun, Patrick], *An Account of a Meat and Soup Charity* (London, 1797).

Compleat Planter and Cyderist (London, 1690).

Congreso de la República, Resolución Suprema N. 009–2005-AG, 23 Feb. 2005, *Normas legales: El peruano: Diario Oficial* 22: 9018 (Lima, 2005).

Conrad, Sebastian, 'The Enlightenment in Global History: A Historiographical Critique', *American Historical Review* 117:4 (2012), 999–1027.

Conrad, Sebastian, *What is Global History?*, Princeton University Press (Princeton, 2016).

Cooper, Joseph, *The Art of Cookery Refin'd and Augmented* (London, 1654).

Corrado, Vincenzo, *Il credenziere di bueon gusto* (Naples, 1820 [1778]).

Coulson, Andrew, *Tanzania: A Political Economy*, Oxford University Press (Oxford, 2013).

Coup d'oeil sur les quatre concours (Lyon, 1791).

Covarrubias, José Enrique, *En busca del hombre útil: Un estudio comparativo del utilitarianismo neomercantilista en México y Europa, 1748–1833*, Universidad Nacional Autónoma de México (Mexico City, 2005).

Coveney, John, *Food, Morals and Meaning: The Pleasure and Anxiety of Eating*, Routledge (London, 2006).

Crawford, Robert, 'Health as a Meaningful Social Practice', *Health* 10:4 (2006), 401–420.

Crosby, Alfred, *The Columbian Exchange: Biological and Cultural Consequences of 1492*, Greenwood Press (Westport, 1972).

Crosby, Alfred, *Ecological Imperialism: The Biological Expansion of Europe, 900–1900*, Canto (Cambridge, 1986).

Crowe, Hilary, 'Profitable Ploughing of the Uplands? The Food Production Campaign in the First World War', *Agricultural History Review* 55:2 (2007), 205–228.

Cuisinière républicaine, qui enseigne la manière simple d'accomoder les Pommes de terre avec quelques avis fur les soins nécessaires pour les conserver (Paris, year III [1794/5]), facsimile edition of Daniel Morcrette (Luzarches, 1976).

Cullather, Nick, 'Development? It's History', *Diplomatic History* 24:4 (2000), 641–653.

Cullather, Nick, 'The Foreign Policy of the Calorie', *American Historical Review* 112:2 (2007), 337–364.

Cullather, Nick, *The Hungry World: America's Cold War Battle Against Poverty in Asia*, Harvard University Press (Cambridge, MA, 2010).

Cunningham, James, 'Part of Two Letters to the Publisher from Mr James Cunningham, F.R.S. and Physician to the English at Chusan in China, Giving an Account of his Voyage Thither', *Philosophical Transactions* 23 (1703), 1201–1209.

Die Curieuse ... Köchin (Nürnberg, 1706).

Cwiertka, Katarzyna, 'Militarization of Nutrition in Wartime Japan', *ILAS Newsletter* 38 (2005), 15.

Cwiertka, Katarzyna, 'Popularizing a Military Diet in Wartime and Postwar Japan', *Asian Anthropology* 1:1 (2002), 1–30.

Dalrymple, Alexander, *The Poor Man's Friend* (London, 1795).

Daly, Mary, 'Something Old and Something New: Recent Research on the Great Irish Famine', in *When the Potato Failed: Causes and Effects of the 'Last' European Subsistence Crisis, 1845–1850*, ed. Cormac Ó Gráda, Richard Paping and Eric Vanhaute, Brepols (Turnhout, 2007), 59–78.

Daunton, Martin, *Wealth and Welfare: An Economic and Social History of Britain 1851–1951*, Oxford University Press (Oxford, 2007).

Davalos, José Manuel, *De morbis nonnullis Limae, grassantibus ipsorumque therapeia* (Montpellier, 1787).

Davies, H. R., *Yün-nan: The Link between Indian and the Yangtze*, Cambridge University Press (Cambridge, 1909 [2010 digital edition]).

Davis, Belinda, *Home Fires Burning: Food, Politics, and Everyday Life in World War I Berlin*, University of North Carolina Press (Chapel Hill, 2000).

Davis, Mike, *Late Victorian Holocausts: El Niño and the Making of the Third World*, Verso (London, 2001).

Davis, Natalie Zemon, 'Poor Relief, Humanism and Heresy', in *Society and Culture in Early Modern France*, Stanford University Press (Stanford, 1975), 17–64.

Dawson, Thomas, *Good Huswife's Jewell* (London, 1596).

De Jong, Hielke, Joseph B. Sieczka and Walter de Jong, *The Complete Book of Potatoes: What Every Grower and Gardener Needs to Know*, Timber Press (Portland, 2011).

De Nijs, Thimo, 'Food Provision and Food Retailing in The Hague, 1914–1930', in *Food and Conflict in Europe in the Age of the Two World Wars*, ed. Frank Trentmann and Flemming Just, Palgrave Macmillan (Basingstoke, 2006), 65–87.

Dean, Mitchell, *The Constitution of Poverty: Toward a Genealogy of Liberal Governance*, Routledge (London, 1991).

Dechambre, Amédée, ed., *Dictionnaire encyclopédique des sciences medicales* (Paris, 1877), fourth series, FAA-FET.

Delle Piane, Niccolò, *De'pomi di terra, ossia patate* (Genoa, 1793).

DeLoughrey, Elizabeth, 'Globalizing the Routes of Breadfruit and Other Bounties', *Journal of Colonialism and Colonial History* 8:3 (2007).

Demerson, Paula, 'La distribución de sopas económicas por la Real Sociedad Matritense en 1803–1804', *Boletín de la Real Academia de Historia* 164 (1969), 119–135.

Dening, Greg, *Mr Bligh's Bad Language: Passion, Power and Theatre on the Bounty*, Cambridge University Press (Cambridge, 1992).

Des-Essartz, Jean-Charles, *Traité de l'éducation corporelle des enfants en bas âge, ou réflexions-pratiques sur les moyens de procurer une meilleure constitution aux citoyens* (Paris, 1760).

DeSoucey, Michaela, 'Gastronationalism: Food Traditions and Authenticity Politics in the European Union', *American Sociological Review* 75:3 (2010), 432–455.

Dewey, P. E., *British Agriculture in the First World War*, Routledge (London, 1989).

Dickler, Robert, 'Organization and Change in Productivity in Eastern Prussia', in *European Peasants and Their Markets: Essays in Agrarian Economic History*, ed.William Parker and Eric Jones, Princeton University Press (Princeton, 1975), 269–292.

Dillon, John Talbot, *Travels Through Spain, with a View to Illustrate the Natural History and Physical Geography of that Kingdom* (London, 1782).

Dombey, Joseph, *Joseph Dombey: Médecin, naturaliste, archéologue, exploratuer du Pérou, du chili et du Brésil (1778–1785): Sa vie, son oeuvre, sa correspondence*, ed. Ernest-Théodore Hamy (Paris, 1905).

Dowler, Elizabeth, 'Food Banks and Food Justice in "Austerity Britain"', in *First World Hunger Revisited: Food Charity or the Right to Food*, ed. Graham Riches and Tiina Silvasti, Palgrave Macmillan (London, 2014), 160–175.

Doyle, Enrique, *Instrucción formada de orden del Consejo por D. Enrique Doyle, para el cultivo y uso de las patatas* (Madrid, 1785).

Doyle, Enrique, *Tratado sobre el cultivo, uso y utilidades de las patatas o papas, corregido y considerablemente aumentado* (Madrid, 1804).

Doyle, Enrique, *Tratado sobre el cultivo, uso y utilidades de las patatas ó papas, é instrucción para su mejor propagación* (Madrid, 1797).

Doyle, Enrique, *Tratado sobre la cría y propagación de pastos y ganados*, 2 vols. (Madrid, 1799).

Drake, Michael, *Population and Society in Norway, 1735–1865*, Cambridge University Press (Cambridge, 1969).

Drayton, Richard, *Nature's Government: Science, Imperial Britain, and the 'Improvement' of the World*, Yale University Press (New Haven, 2000).

Drinot, Paulo, 'Food, Race and Working-Class Identity: Restaurantes Populares and Populism in 1930s Peru', *The Americas* 62:2 (2005), 425–470.

Dubroca, Jean François, *Conversaciones de un padre con sus hijos sobre la historia natural*, trad. Manuel Maria de Ascargorta y Ramírez (Madrid, 1802).

Duhamel du Monceau, Henri Louis, *Moyens de conserver la santé aux equipages des vaisseaux* (Paris, 1759).

Dusinberre, William, *Them Dark Days: Slavery in the American Rice Swamps*, Oxford University Press (Oxford, 1996).

Dyck, Ian, *William Cobbett and Rural Popular Culture*, Cambridge University Press (Cambridge, 1992).

Dyck, Ian, 'William Cobbett and the Rural Radical Platform', *Social History* 18:2 (1993), 185–204.

Dyer, Hugh Mc Neile, *The West Coast of Africa as Seen from the Deck of a Man-of-War* (London, 1876).

Eagle, F. K. and Edward Younge, eds., *Collection of the Reports of Cases, the Statutes, and Ecclesiastical Laws Relating to Tithes*, 4 vols. (London, 1826).

Earle, Rebecca, *The Body of the Conquistador: Food, Race and the Colonial Experience in Spanish America, 1492–1700*, Cambridge University Press (Cambridge, 2012).

Earle, Rebecca, 'Food, Colonialism and the Quantum of Happiness', *History Workshop Journal* 84 (2017), 170–193.

Earle, Rebecca, 'The Political Economy of Nutrition in the Eighteenth Century', *Past & Present* 242:1 (2019), 79–117.

Eastwick, Edward, *Journal of a Diplomate's Three Years' Residence in Persia* (London, 1864), 79–117.

Eden, Frederick Morton, *The State of the Poor, or a History of the Labouring Classes in England, from the Conquest to the Present Period*, 3 vols., Cambridge University Press (Cambridge, 2011 [1797]).

Edwards, Bryan, The History, *Civil and Commercial, of the British Colonies in the West Indies*, 2 vols. (London, 1793).

Eichler, Helga, 'Die Leipziger Ökonomische Sozietät im 18. Jahrhundert', *Jahrbuch für*

Geschichte des Feudalismus 2 (1978), 357–386.

El Excmo. Sr. Secretario de Estado y del Despacho de Hacienda dice a esta Junta Superior con fecha del de 19 de marzo próximo pasado lo que sigue (La Coruña, 27 April 1812).

Elina, Olga, Susanne Heim and Nils Roll-Hansen, 'Plant Breeding on the Front: Imperialism, War, and Exploitation', *Osiris* 20 (2005), 161–179.

Elliot, Gill, 'Cookery Examined – 1937–2007: Evidence from Examination Questions of the Development of a Subject over Time', *Research Matters* 6 (2008), 24–30.

Ellis, John, *A Description of the Mangostan and the Bread-Fruit: the First, Esteemed one of the Most Delicious, the Other, the Most Useful of all the Fruits of the Eat Indies, to which are added Directions to Voyagers, or Bringing over these and other Vegetable Productions, which would be extremely beneficial to the Inhabitants of our West India Islands* (London, 1775).

Elsholtz, Johann Sigismund, *Diaeteticon* (Cölln an der Spree, 1682).

Eltis, David John, Philip Morgan and David Richardson, 'Agency and Diaspora in Atlantic History: Reassessing the African Contribution to Rice Cultivation in the Americas', *American Historical Review* 112:5 (2007), 1329–1358.

Encyclopédie méthodique: Arts et métiers mécaniques 2 (Paris, 1783).

Engel, Samuel, *Traité de la nature, de la culture et de l'utililé des pommes de terre par un Ami des Hommes* (Lausanne, 1771).

Ensayos de comidas económicas á la Rumford hechos por una comisión nombrada á este fin por la Real Sociedad Económica Matritense (Oviedo, 1803).

Escobar, Arturo, 'Power and Visibility: Development and the Invention and Management of the Third World', *Cultural Anthropology* 3:4 (1988), 428–443.

Essay on Modern Luxuries: An Essay on Tea, Sugar, White Bread and Butter, Country Alehouses, Strong Beer and Geneva, and other Modern Luxuries (Salisbury, 1777).

Essemyr, Mats, 'Food Policies in Sweden during the World Wars', in *The Origins and Development of Food Policies in Europe*, ed. John Burnett and Derek Oddy, Leicester University Press (London, 1994), 161–177.

Evans, E. J., 'Tithing Customs and Disputes: The Evidence of Glebe Terriers', *Agricultural History Review* 18:1 (1970), 17–35.

Evans, Eric, *The Contentious Tithe: The Tithe Problem and English Agriculture, 1750–1850*, Routledge (London, 1976).

Evans, Eric, 'Some Reasons for the Growth of English Anti-Clericalism, c. 1750–c.1830', *Past & Present* 66 (1975), 84–109.

Evelyn, John, *Kalendarium Hortense* (London, 1666).

Extractos de las Juntas Generales celebradas por la Real Sociedad Bascongada de los

Amigos del País en la ciudad de Vitoria por julio de 1786 (Vitoria, 1786).

Extractos de las Juntas Generales celebradas por la Real Sociedad Bascongada de los Amigos del País en la villa de Vergara por setiembre de 1779 (Vitoria, [1779]).

FAO, *Approaches to World Food Security: Selected Working Papers of the Commodities and Trade Division*, United Nations (Rome, 1983).

FAO, 'Background', International Year of the Potato, www.fao.org/potato-2008/en/aboutiyp/background.html#fao.

FAO, 'Constitution of the United Nations Food and Agriculture Organization', Québec, 16 Oct. 1945, University of Oslo, www.jus.uio.no/english/services/library/treaties/14/14–01/food-organization.xml.

FAO, 'FAO Overview: Major Weaknesses in Present Agricultural Production and Related Research Efforts in Developing Countries', Rome, 12 Jan. 1971, 6, AGR-RES/71–5, CGSpace, https://cgspace.cgiar.org/bitstream/handle/10947/519/cg7101b.pdf?sequence=1&is Allowed=y.

FAO, 'New Light on a Hidden Treasure. An End-of-Year Review', *International Year of the Potato 2008* (Rome, 2008).

FAO, 'Potato and Biodiversity', International Year of the Potato, www.fao.org/potato2008/en/potato/biodiversity.html.

FAO, 'Potato and Gender', International Year of the Potato, www.fao.org/potato2008/en/potato/gender.html.

FAO, 'Potato and Water Resources', International Year of the Potato, www.fao.org/potato-2008/en/potato/water.html.

FAO, 'Potato World: Africa', International Year of the Potato, www.fao.org/potato-2008/en/world/africa.html.

FAO, *Report of the Council of FAO*, 64th Session, Rome, 18–29 Nov. 1974, www.fao.org/docrep/meeting/007/F5340E/F5340E03.htm#ch2.4.

FAO, Rome Declaration on World Food Security, Rome, 13 Nov. 1996.

FAO, *Roots, Tubers, Plantains and Bananas in Human Nutrition* (Rome, 1990).

FAO, 'Smallholders and Family Farmers' (2012), www.fao.org/fileadmin/templates/nr/sustainabilitypathways/docs/Factsheet_SMALLHOLDERS.pdf.

FAO, *The State of Food and Agriculture: Innovation in Food and Agriculture* (Rome, 2014).

FAO, 'Why Potato?', International Year of the Potato, www.fao.org/potato-2008/en/aboutiyp/index.html.

FAO, FAOSTAT, www.fao.org/faostat/en/#data.

'Farine & farineux', 1756, *Encyclopédie, ou dictionnaire raisonné des sciences, des arts et des métiers, etc.*, ed. Denis Diderot and Jean le Rond d'Alembert, University of Chicago: ARTFL Encyclopédie Project (Spring 2013 Edition), ed. Robert Morrissey, http://encyclopedie.uchicago.edu/.

Farnham, Mrs. Frank, Mrs. Francis Harding, Mrs. Roland Hopkins, Mrs. Henry Prout and Mrs. Howard Rogers, eds., *Choice War Time Recipes* (Chestnut Hill, 1918).

Faulhaber, Lilian, 'Cured Meat and Idaho Potatoes: A Comparative Analysis of European and American Protection and Enforcement of Geographic Indications of Foodstuffs', *Columbia Journal of European Law* 11 (2005), 623–664.

Fenzi, Marianna and Christophe Bonneuil, 'From "Genetic Resources" to "Ecosystems Services": A Century of Science and Global Policies for Crop Diversity Conservation', *Culture, Agriculture, Food and Environment* 38:2 (2016), 72–83.

Ferrières, Madeleine, 'Le cas de la pomme de terre dans le Midi', in *L'échec a-t-il des vertus économiques?*, ed. Natacha Coquery and Matthieu de Oliveira, OpenEdition (Paris, 2015), 205–215.

Fielding, Henry, 'An Enquiry into the Causes of the Late Increase of Robbers', 1751, in *An Enquiry into the Causes of the Late Increase of Robbers and Related Writings*, ed. Malvin Zirker, Oxford University Press (Oxford, 1988), 75–174.

Fiori, Stefano, 'Individuals and Self-Interest in Adam Smith's Wealth of Nations', *Cahiers d'économie politique* 49 (2005), 19–31.

Fleischacker, Samuel, *A Third Concept of Liberty: Judgment and Freedom in Kant and Adam Smith*, Princeton University Press (Princeton, 1999).

Foreign Essays on Agriculture and Arts consisting chiefly of the Most Curious Discoveries made in the Several Provinces of France, Flanders, Germany, Switzerland, etc. and Communicated to the Learned in those Countries for the Improvement of British Husbandry (London, 1766).

Forster, John, *Englands Happiness Increased, or, A Sure and Easie Remedy against all Succeeding Dear Years* (London, 1664).

Forster, Robert, 'The Noble as Landlord in the Region of Toulouse at the End of the Old Regime', *Journal of Economic History* 17 (1957), 224–244.

Foster, Nelson and Linda S. Cordell, eds., *Chillies to Chocolate: Food the Americas Gave the World*, University of Arizona Press (Tucson, 1992).

Foucault, Michel, *Discipline and Punish: The Birth of the Prison*, trans. Alan Sheridan, Penguin (London, 1991).

Foucault, Michel, *History of Sexuality*, trans. Robert Hurley, Random House (New York, 1978).

Foucault, Michel, Security, *Territory, Population: Lectures at the Collège de France,*

1977–1978, ed. Michel Senellart, trans. Graham Burchell, Palgrave Macmillan (New York, 2009).

Foucault, Michel, 'The Subject and Power', *Critical Inquiry* 8:4 (1982), 777–795.

Fox, Adam, 'Food, Drink and Social Distinction in Early Modern England', in *Remaking English Society: Social Relations and Social Change in Early Modern England*, ed. S. Hindle, A. Shepard and J. Walter, Boydell & Brewer (Woodbridge, 2013), 165–187.

Frangsmyr, Tore, J. L. Heilbron and Robin Rider, eds., *The Quantifying Spirit in the Eighteenth Century*, University of California Press (Berkeley, 1990).

Frauenzimmer-Lexikon (Leipzig, 1715).

Frost, Alan, 'The Antipodean Exchange: European Horticulture and Imperial Designs', in *Visions of Empire: Voyages, Botany, and Representations of Nature*, ed. David Philip Miller and Peter Hanns Reill, Cambridge University Press (Cambridge, 1996), 58–79.

Frost, John, *The Book of Travels in Africa, from the Earliest Ages to the Present Time* (New York, 1848).

Fryer, John, *A New Account of East India and Persia being Nine Years' Travels, 1672–1681*, ed. W. Crooke, 3 vols., Hakluyt Society (London, 1909–1915).

Gadd, Carl-Johan, 'On the Edge of a Crisis: Sweden in the 1840s', in *When the Potato Failed: Causes and Effects of the 'Last' European Subsistence Crisis, 1845–1850*, ed.Cormac Ó Gráda, Richard Paping and Eric Vanhaute, Brepols (Turnhout, 2007), 313–342.

Gallagher, Catherine and Stephen Greenblatt, 'The Potato in the Materialist Imagination', in *Practicing New Historicism*, University of Chicago Press (Chicago, 2000), 110–135.

Gallego, José Andrés, *El motín de Esquilache, América y Europa*, Consejo Superior de Investigaciones Científicas (Madrid, 2003).

Galli, Marika, *La conquête alimentaire du Nouveau Monde. Pratiques et representations franco-italiennes des nouveaux produits du XVIe au XVIIIe siècle*, L'Harmattan (Paris, 2016).

Galloway, J. H., 'Agricultural Reform and the Enlightenment in Late Colonial Brazil', *Agricultural History* 53:4 (1979), 763–779.

Gandhi, Leela, *Affective Communities: Anticolonial Thought, Fin-de-Siècle Radicalism, and the Politics of Friendship*, Duke University Press (Durham, 2005).

García, Gregorio, *Origen de los indios del Nuevo Mundo*, ed. Franklin Pease, Fondo de Cultura Economica (Mexico City, 1981 [1607]).

Garnett, T. and A. Wilkes, *Appetite for Change: Social, Economic and Environmental Transformations in China's Food System*, Food Climate Research Network,

Universityof Oxford (Oxford, 2014), www.fcrn.org.uk/sites/default/files/fcrn_china_mapping_study_final_pdf_2014.pdf.

Gartner, Christian, *Horticultura* (Trondheim, 1694).

Gasparini, Danilo, *Polenta e formenton. Il mais nelle campagne venete tra XVI e XX secolo*, Cierre (Verona, 2002).

General Report of the Committee of Subscribers, to a Fund for the Relief of the Industrious Poor, Resident in the Cities of London and Westminster, the Borough of Southwark, and the Several Out Parishes of the Metropolis (London, 1800).

'Genève (commune)', *Dictionnaire Historique de la Suisse*, www.hls-dhs-dss.ch/textes/f/F2903.php.

Gentilcore, David, *Food and Health in Early Modern Europe: Diet, Medicine, and Society, 1450–1800*, Bloomsbury (London, 2016).

Gentilcore, David, *Italy and the Potato: A History, 1550–2000*, Bloomsbury (London, 2012).

George, Susan, *How the Other Half Dies: The Real Reasons for World Hunger*, Penguin (Harmondsworth, 1986 [1976]).

Gerarde, John, *Herball, or General History of Plantes* (London, 1597).

Gerarde, John, *The Herbal or General History of Plants. The Complete 1633 Edition as Revised and Enlarged by Thomas Johnson*, Dover (New York, 1975).

Gerbi, Antonello, *The Dispute of the New World: The History of a Polemic, 1750–1900*, trans. Jeremy Moyle, University of Pittsburgh Press (Pittsburgh, 1973 [1955]).

Gibson, Mark, *The Feeding of Nations: Redefining Food Security for the 21st Century*, CRC Press (Boca Raton, 2012).

Girdler, J. S., *Observations on the Pernicious Consequences of Forestalling, Regrating, and Ingrossing* (London, 1800).

Gitomer, Charles, *Potato and Sweetpotato in China: Systems, Constraints, and Potential*, International Potato Center (Lima, 1996).

Glaiser, Benjamin, 'French and Indian War Diary of Benjamin Glaiser of Ipswich, 1758–1760', *Essex Institute Historical Collections* 86 (1950), 65–92.

Goldie, Mark and Robert Wokler, eds., *The Cambridge History of Eighteenth-Century Political Thought*, Cambridge University Press (Cambridge, 2006).

Gómez de Ortega, Casimiro, *Elementos teóricos-prácticos de agricultura ... traducidos del francés del célebre Mr. Duhamel de Monceau*, 2 vols. (Madrid, 1805).

Gonnella, Anna, 'L'assistenza pubblica a Trieste: l'alimentazione nell'Istituto dei Poveri (1818–1918)', in *Archivi per la stories dell'alimentazione*, ed. Paola Carucci, 3 vols. (Rome, 1995), III, 1590–1608.

González, Pedro María, *Tratado de las enfermedades de la gente de mar, en que se exponen sus causas, y los medios de precaverlas* (Madrid, 1805).

Good, John Mason, *Dissertation on the Best Means of Maintaining and Employing the Poor in Parish Work-Houses* (London, 1798).

Graves, Christine, *The Potato Treasure of the Andes: From Agriculture to Culture*, International Potato Center (Lima, 2001).

Gray, John, *Liberalism*, Open University Press (Milton Keynes, 1986).

Gray, Peter, 'The European Food Crisis and the Relief of Irish Famine, 1845–1850', in *When the Potato Failed: Causes and Effects of the 'Last' European Subsistence Crisis, 1845–1850*, ed. Cormac Ó Gráda, Richard Paping and Eric Vanhaute, Brepols (Turnhout, 2007), 95–107.

Gray, Peter, *Famine, Land and Politics: British Government and Irish Society, 1843–50*, Irish Academic Press (Dublin, 1999).

Grey, Elizabeth, Countess of Kent, *A Choice Manual* (London, 1687).

Grieco, Allen, 'The Social Politics of Pre-Linnaean Botanical Classification', *I Tatti Studies* 4 (1991), 131–149.

Grimaux, Éduard, *Lavoisier, 1743–1794* (Paris, 1888).

Grote, John, *An Examination of the Utilitarian Philosophy*, ed. Joseph Bickersteth Mayor (Cambridge, 1870).

Grove, Richard, 'The Great El Niño of 1789–93 and Its Global Consequences: Reconstructing an Extreme Climate Event in World Environmental History', *Medieval History Journal* 10:1–2 (2007), 75–98.

Guaman Poma de Ayala, Felipe, 'El primer nueva corónica y buen gobierno', 1615–1616, The Guaman Poma Website, Det Kongelige Bibliotek, Copenhagen, www.kb.dk/permalink/2006/poma/info/en/frontpage.htm.

Guðmundsdóttir, Jóhanna Þ., 'Viðreisn garðræktar á síðari hluta 18. aldar: viðbrögð og viðhorf almennings á Íslandi', *Saga* 52:1 (2014), 9–41.

Guilding, Landsdown, *An Account of the Botanic Garden in the Island of St. Vincent from its Establishment to the Present Time* (Glasgow, 1825).

Guimarães Sá, Isabel dos, 'Circulation of Children in Eighteenth-Century Portugal', in *Abandoned Children*, ed. Catherine Panter-Brick and Malcolm Smith, Cambridge University Press (Cambridge, 2000), 27–40.

Gurney, Peter, '"Rejoicing in Potatoes": The Politics of Consumption in England during the "Hungry Forties"', *Past & Present* 203 (2009), 99–136.

Gutaker, Rafael, Clemens Weiß, David Ellis, Noelle Anglin, Sandra Knapp, José Luis Fernández-Alonso, Salomé Prat and Hernán Burbano, 'The Origins and Adaptation of European Potatoes Reconstructed from Historical Genomes', *Nature, Ecology & Evolution* 3 (2019), 1093–1101.

Guthman, Julie and Patricia Allen, 'From "Old School" to "Farm-to-School": Neoliberalism from the Ground Up', *Agriculture and Human Values* 23 (2006), 401–415.

Haitin, Marcel, 'Prices, the Lima Market, and the Cultural Crisis of the Late Eighteenth Century in Peru', *Jahrbuch für Geschichte von Staat, Wirtschaft und Gesellschaft Latein-amerikas* 22 (1985), 167–199.

Hall, T., *The Queen's Royal Cookery* (London, 1709).

Halldórsson, Björn, *Korte Beretninger om nogle Forsög til Landvæsenets og i sær Havedyrkningens Forbedring i Island* (Copenhagen, 1765).

Hamilton, Alexander, *A New Account of the East Indies*, 2 vols. (London, 1744).

[Hamilton, John], *The Country-Man's Rudiments, or an Advice to the Farmers of East-Lothian* (Edinburgh, 1713 [1699]).

Hammer, Christopher, *Afhandling om Patatos: med endeel Tanker i Land Huusholdningen* (Christiania, 1766).

Hammond, J. L. and Barbara Hammond, *The Village Labourer: A Studyin the Government of England before the Reform Bill*, Longmans, Green and Co. (London, 1920).

Hammond, R. J., *Food and Agriculture in Britain, 1939–45: Aspects of Wartime Control*, Stanford University Press (Stanford, 1954).

Hancock, Simon, 'The Social Impact of the First World War in Pembrokeshire', DPhil, Cardiff University, 2015.

Hanway, Jonas, *A Candid Historical Account of the Hospital for the Reception of Exposed and Deserted Young Children* (London, 1759).

Hanway, Jonas, *The Great Advantage of Eating Pure and Genuine Bread, Comprehending the Heart of the Wheat* (London, 1773).

Harcout, Bernard, *The Illusion of Free Markets: Punishment and the Myth of Natural Order*, Harvard University Press (Cambridge, MA, 2011).

Harrison, Regina, *Signs, Songs and Memory in the Andes*, University of Texas Press (Austin, 1989).

Harsanyi, David John, *Nanny State: How Food Fascists, Teetotaling Do-Gooders, Priggish Moralists, and Other Boneheaded Bureaucrats are Turning America into a Nation of Children*, Broadway Books (New York, 2007).

Hartley, E. R., *How to Feed the Children* (Bradford, 1908).

Harvard School of Public Health, 'Public Health and the US Economy', 2012, www.hsph.harvard.edu/news/magazine/public-health-economy-election/.

Hawkes, J. G., 'N.I. Vavilov — The Man and his Work', *Biological Journal of the Linnaean*

Society 39 (1990), 3–6.

Hawkes, J. G., *Potato Collecting Expeditions in Mexico and South America*, Imperial Bureau of Plant Breeding and Genetics, School of Agriculture (Cambridge, 1941).

Hawkes, J. G. and J. Francisco-Ortega, 'The Early History of the Potato in Europe', *Euphytica* 70 (1993), 1–7.

Hawkes, J. G. and J. Francisco-Ortega, 'The Potato in Spain during the Late Sixteenth Century,' *Economic Botany* 46:1 (1992), 86–97.

Helgi Library, 'Potato Consumption Per Capita in the World', 2011, www.helgilibrary.com/indicators/potato-consumption-per-capita/world/.

Helmhardt von Hohberg, Wolf, *Georgica Curiosa Aucta. Das ist, Umständlicher Bericht und klarer Unterricht von dem vermehrten und Verbesserten Adelichen Land-und Feld-Leben*, c.1688 (Nürnberg, 1716).

Helmhardt von Hohberg, Wolf, *Herrn von Hohbergs Georgica Curiosa Aucta. Oder: Adelichen Land und Feld-Lebens*, c.1688 (Nürnberg, 1715).

Helstocky, Carol, 'Recipe for the Nation: Reading Italian History through *La scienza in cucina and La cucina futurista*', *Food and Foodways* 11 (2003), 113–140.

Helstocky, Carol, 'The State, Health, and Nutrition', in *The Cambridge World History of Food*, ed. Kenneth Kiple and Kriemhild Coneè Ornelas, 2 vols. Cambridge University Press (Cambridge, 2001), II, 1577–1584.

Henriksen, Ingrid, 'A Disaster Seen from the Periphery: The Case of Denmark', in *When the Potato Failed: Causes and Effects of the 'Last' European Subsistence Crisis, 1845–1850*, ed. Cormac Ó Gráda, Richard Paping and Eric Vanhaute, Brepols (Turnhout, 2007), 293–312.

Henry, David John, *The Complete English Farmer, or, A Practical System of Husbandry* (London, 1771).

Hertzberg, Peter Harboe, *Underretning for Bønder i Norge om den meget nyttige Jord-Frukt Potatos: at plante og bruge* (Bergen, 1774).

Hildesheim, Wilhelm, *Die Normal-Diät. Physiologisch-chemischer Versuch zu Ermittlung des normalen Nahrungsbedürfnisse der Menschen, behufs Aufstellung einer Normal-Diät, mit besonderer Rücksicht auf das Diät-Regulative des neuen Reglements für die Friedens-Garnison-Lazarethe, und die Natural-Verpflegung des Soldaten sowie auf die Verpflegung der Armen* (Berlin, 1856).

Hiler, David, 'La pomme de terre révolutionnaire', in *Regards sur la Révolution genevoise, 1792–1798*, ed. Louis Binz, Société d'Histoire et d'Archéologie de Genève (Geneva, 1992), 91–117.

Hirschman, Albert O., *The Passions and the Interests: Political Arguments for Capitalism before Its Triumph*, Princeton University Press (Princeton, 1997).

Historic Royal Palaces Blog, 'History of the Sweet Potato', 26 Nov. 2015, http://blog.hrp.org.uk/gardeners/history-of-sweet-potato.

Ho, Ping-Ti, 'The Introduction of American Food Plants into China', *American Anthropologist* 57:2 (1955), 191–201.

Hobbes, Thomas, *Leviathan, or, The Matter, Forme, and Power of a Common Wealth, Ecclesiasticall and Civil* (London, 1651).

Hobsbawm, E. J. and George Rudé, *Captain Swing*, Lawrence & Wishart (Woking, 1969).

Hochstrasser, T. J., 'Physiocracy and the Politics of Laissez-Faire', in *The Cambridge History of Eighteenth-Century Political Thought*, ed. Mark Goldie and Robert Wokler, Cambridge University Press (Cambridge, 2006), 419–442.

Hoffman, Philip T., *Growth in a Traditional Society: The French Countryside 1450–1815*, Princeton University Press (Princeton, 1996).

Holland, Michael, 'Swing Revisited: The Swing Project', *Family & Community History* 7:2 (2004), 87–100.

Hont, Istvan, *Jealousy of Trade: International Competition and the Nation-State in Historical Perspective*, Harvard University Press (Cambridge, MA, 2005).

Horrell, Sara and Deborah Oxley, 'Hasty Pudding versus Tasty Bread: Regional Variations in Diet and Nutrition during the Industrial Revolution', *Local Population Studies* 88:1 (2012), 9–30.

House of Commons Sessional Papers of the Eighteenth Century 1715–1800, House of Commons Parliamentary Papers, Parlipapers.chadwyck.co.uk.

Housekeeper's Receipt Book, or, Repository of Domestic Knowledge (London, 1813).

Howard, Henry, *England's Newest Way in All Sorts of Cookery* (London, 1708).

Huarte de San Juan, Juan, *Examen de Ingenios, or The Examination of Mens Wits* (London, 1594).

Hughes, William, *The American Physitian, or, A Treatise of the Roots, Plants, Trees, Shrubs, Fruit, Herbs, &c. Growing in the English Plantations in America* (London, 1672).

Hunter, Alexander, *Culina Famulatrix Medicinae: Or, Receipts in Cookery, Worthy the Notice of those Medical Practitioners, who Ride in their Chariots with a Footman Behind and who Receive Two-guinea Fees from their Rich and Luxurious Patients* (York, 1804).

Hutcheson, Francis, *Inquiry into the Original of Our Ideas of Beauty and Virtue* (London, 1725).

Hutchison, E. P., 'Swedish Population Thought in the Eighteenth Century', *Population*

Studies 12:1 (1959), 81–102.

Ibáñez Rodríguez, Santiago, 'El diezmo en la Rioja (XVI–XVIII)', *Brocar* 18 (1994), 189–222.

Ingebretson, Britta, 'The *Tuhao* and the Bureaucrat: The Qualia of "Quality" in Rural China', *Signs and Society* 5:2 (2017), 243–268.

Instrucção sobre a cultura das batatas, traduzida do inglez por ordem superior (Lisbon, 1800).

Jaffe, Catherine, '"Noticia de la vida y obras del Conde de Rumford" (1802) by María Lorenza de los Ríos, Marquesa de Fuerte-Híjar: Authorizing a Space for Female Charity', *Studies in Eighteenth Century Culture* 38 (2009), 91–115.

Jansky, S. H., L. P. Jin, K. Y. Xie, C. H. Xie and D. M. Spooner, 'Potato Production and Breeding in China', *Potato Research* 52:57 (2009), 57–65.

Jarosz, Lucy, 'Defining World Hunger: Scale and Neoliberal Ideology in International Food Security Policy Discourse', *Food, Culture & Society* 14:1 (2011), 117–139.

Jefferson, Thomas, 'Summary of Public Service', after 2 Sept. 1800, *The Papers of Thomas Jefferson Digital Edition 32: 1 June 1800–16 February 1801*, ed. Barbara Obergand J. Jefferson Looney, University of Virginia Press (Charlottesville, 2008–2016), 124, http://rotunda.upress.virginia.edu/founders/TSJN-01-32-02-0080.

Jia, Ruixue, 'Weather Shocks, Sweet Potatoes and Peasant Revolts in Historical China', *Economic Journal* 124:575 (2013), 92–118.

Jiménez de la Espada, Marcos, ed., *Relaciones geográficas de las Indias: Perú*, 3 vols., Biblioteca de Autores Españoles (Madrid, 1965).

Jing, Jun, 'Introduction: Food, Children, and Social Change in Contemporary China', in *Feeding China's Little Emperors: Food, Children, and Social Change*, ed. Jun Jing, Stanford University Press (Stanford, 2000), 1–26.

Jones-Brydges, Harford, *Account of the Transactions of his Majesty's Mission to the Court of Persia in the Years 1807–11* (London, 1834).

Jones, Robert, *Provincial Development in Russia: Catherine II and Jacob Sievers*, Rutgers University Press (New Brunswick, 1984).

Jonsson, Fredrik, *Enlightenment's Frontier: The Scottish Highlands and the Origins of Environmentalism*, Yale University Press (New Haven, 2013).

Junta Pública de la Real Sociedad Económica de Amigos del País de Valencia (Valencia, 1801).

Jütte, Robert, *Poverty and Deviance in Early Modern Europe*, Cambridge University Press (Cambridge, 1994).

Kaldy, M. S., 'Protein Yield of Various Crops as Related to Protein Value', *Economic*

Botany 26:2 (1972), 142–144.

Kamminga, Harmke, 'Nutrition for the People, or the Fate of Jacob Moleschott's Contest for a Humanist Science', in *The Science and Culture of Nutrition, 1840–1940*, ed. Harmke Kamminga and Andrew Cunningham, Rodopi (Amsterdam, 1995), 15–47.

Kamminga, Harmke and Andrew Cunningham, eds., *The Science and Culture of Nutrition, 1840–1940*, Rodopi (Amsterdam, 1995).

Kant, Immanuel, *Metaphysics of Morals*, trans. and ed. Mary Gregor, intro. Roger Sullivan, Cambridge University Press (Cambridge, 1996).

Kant, Immanuel, 'What Does it Mean to Orient Oneself in Thinking?', 1786, in *Religion and Rational Theology, The Cambridge Edition of the Works of Immanuel Kant*, ed. A. Wood and G. di Giovanni, Cambridge University Press (Cambridge, 1996), 1–18.

Kant, Immanuel, 'What is Enlightenment?', 1784, German History in Documents and Images, http://ghdi.ghi-dc.org/sub_document.cfm?document_id=3589.

Kaplan, Steven, *Bread, Politics and Political Economy in the Reign of Louis XV*, 2 vols., Springer (The Hague, 1976).

Kaplan, Steven, *The Stakes of Regulation: Perspectives on Bread, Politics and Political Economy Forty Years Later*, Anthem Press (London, 2015).

Kåre, Lunden, ed., *Norges landbrukshistorie II: 1350–1814: Frå svartedauden til 17.Mai*, Samlaget (Oslo, 2002).

Kaufmann, Franz-Xavier, *Variations of the Welfare State: Great Britain, Sweden, Franceand Germany between Capitalism and Socialism*, trans. Thomas Dunlap, Springer (Heidelberg, 2013).

Kaye, John William, *The Life and Correspondence of Major-General Sir John Malcolm*, 2 vols. (London, 1856).

Keith, George Skene, *A General View of the Agriculture of Aberdeenshire* (Aberdeen, 1811).

Khodnev, A. I., *Istoriia Imperatorskago Vol'nago Ekonomicheskago Obshchestva s 1765 do 1865* (St. Petersburg, 1865).

Khondker, Habibul, 'Famine Policies in Pre-British India and the Questionof Moral Economy', *South Asia: Journal of South Asian Studies* 9:1 (1986), 25–40.

King, Janet and Joanne Slavin, 'White Potatoes, Human Health, and Dietary Guidance', *Advances in Nutrition* 4:3 (2013), 393S–401S.

Kiple, Kenneth, *A Moveable Feast: Ten Millennia of Food Globalization*, Cambridge University Press (Cambridge, 2007).

Kiple, Kenneth and Kriemhild Coneè Ornelas, eds., *The Cambridge World History of Food*, 2 vols., Cambridge University Press (Cambridge, 2001).

Kisbán, Eszter, 'The Beginnings of Potato Cultivation in Transylvania and Hungary: Government Policy and Spontaneous Process', in *The Origins and Development of Food Policies in Europe*, ed. John Burnett and Derek Oddy, Leicester University Press (London, 1994), 178–193.

Klein, Jakob, 'Connecting with the Countryside? "Alternative" Food Movements with Chinese Characteristics', in *Ethical Eating in the Postsocialist and Socialist World*, ed. Yuson Jung, Jakob Klein and Melissa Caldwell, University of California Press (Berkeley, 2014), 116–143.

Knight, Roger and Martin Wilcox, *Sustaining the Fleet, 1793–1815: War, the British Navy and the Contractor State*, Boydell & Brewer (Woodbridge, 2010).

Knight, Thomas Andrew, 'On Raising New and Early Varieties of the Potatoe (*Solanum Tuberosum*)', 6 Jan. 1807, *Transactions of the Horticultural Society of London* (London, 1812), 57–59.

Koenker, Roger, 'Was Bread Giffen? The Demand for Food in England Circa 1790', *Review of Economics and Statistics* 59:2 (1977), 225–229.

Koerner, Lisbet, *Linnaeus: Nature and Nation*, Harvard University Press (Cambridge, MA, 1999).

Komlos, John, 'The New World's Contribution to Food Consumption during the Industrial Revolution', *Journal of European Economic History* 27:1 (1998), 67–82.

Kreuger, Rita, 'Mediating Progress in the Provinces: Central Authority, Local Elites, and Agrarian Societies in Bohemia and Moravia', *Austrian History Yearbook* 35 (2004), 49–79.

Kümin, Beat, *Drinking Matters: Public Houses and Social Exchange in Early Modern Central Europe*, Palgrave Macmillan (Basingstoke, 2007).

Labat, Jean-Baptiste, *Nouveau voyage aux îles de l'Amerique* (Paris, 1722).

Lacroix, Anne, *Transformations du procès de travail agricole: incidences de l'industrialisation sur les conditions de travail paysannes*, Institut de Recherche Économique et de Planification (Grenoble, 1981).

Lafuente, Antonio, 'Enlightenment in an Imperial Context: Local Science in the Late Eighteenth-Century Hispanic World', *Osiris* 15 (2000), 155–173.

Lagrange, Joseph-Louis, 'Essai d'arithmétique politique sur les premiers besoins de l'intérieur de la république', 1796, in *Oeuvres de Lagrange*, ed. J.-A. Serret, 14 vols. (Paris, 1867–1892), VII, 571–582.

Lang, James, *Notes of a Potato Watcher*, Texas A&M University Press (College Station, 2001).

Lang, Tim and Michael Heasman, *Food Wars: The Global Battle for Mouths, Minds and*

Markets, Earthscan (London, 2004).

Langer, William, 'American Foods and Europe's Population Growth 1750–1850', *Journal of Social History* 8:2 (1975), 51–66.

Langworthy, C. F., *Potatoes and Other Root Crops as Food*, Government Printing Office (Washington, 1907).

Larriba, Elisabel, 'Un intento de reforma agraria por y para las clases productoras: el *Semanario de Agricultura y Artes Dirigido a los Párrocos* (1797–1808)', *Brocar* 23 (1999), 87–117.

Larumbe, Josef María, *Epítome cristiano de agricultura* (Pamplona, 1800).

Las Casas, Bartolomé de, *Apologética historia sumaria, c.1552, in Obras escogidas*, ed. Juan Pérez de Tudela Bueso, 5 vols., Biblioteca de Autores Españoles (Madrid, 1958), III.

Laufer, Berthold, *American Plant Migration, part 1: The Potato*, Field Museum of Natural History Anthropological Series Publication 418 (Chicago, 1938).

Laurell, Axel, *Lyhykäinen kirjoitus potatesten eli maan-päronain wiljelemisestä*, Säilyttämisestä ja hyödytyxestä huonen hallituxesa (Turku, 1773).

Laurence, Edward, *The Duty of a Steward to his Lord* (London, 1727).

Lavery, Brian, ed., *Shipboard Life and Organisation, 1731–1815*, Ashgate (Aldershot, 1998).

Lavoisier, Antoine Laurent, *Oeuvres de Lavoisier*, 6 vols. (Paris, 1864–1893).

Lebeschu de la Bastays, M., *L'Ami des navigateurs, ou Instruction destinée à préserver les gens de mer des maladies qui sont propres à leur état, & à les faire jouir d'une aussi bonne santé sur les vaisseaux qu'à terre* (Nantes, 1787).

Lee, Seung-joon, 'Taste in Numbers: Science and the Food Problem in Republican Guangzhou, 1927–1937', *Twentieth-Century China* 35:2 (2010), 81–103.

Legrand d'Aussy, Pierre-Jean-Baptiste, *Histoire de la vie privée des Français*, 3 vols. (Paris, 1782).

Leisering, Lutz, 'Nation State and Social Policy: An Ideational and Political History', introductory chapter to Franz-Xavier Kaufmann, *Variations of the Welfare State: Great Britain, Sweden, France and Germany between Capitalism and Socialism*, trans. Thomas Dunlap, Springer (Heidelberg, 2013), 1–22.

Lémery, Louis, *Traité des aliments, où l'on trouve la différence, & le choix qu'on en doit faire*, ed. Jacques Jean Bruhier, third edition, 2 vols. (Paris, 1755).

Lemire, Beverly, '"Men of the World": British Mariners, Consumer Practice, and Material Culture in the Era of Global Trade, c. 1660–1800', *Journal of British Studies* 54:2 (2015), 288–319.

Lemnius, Levinus, *The Touchstone of Complexions*, trans. T.N. (London, 1633).

Leon Pinelo, Antonio, *Question moral si el chocolate quebranta el ayuno elesiástico* (Madrid, 1638).

Letters and Papers on Agriculture, Planting, &c. Selected from the Correspondence-Book of the Society Instituted at Bath for the Encouragement of Agriculture, Arts, Manufactures, and Commerce (Bath and London, 1783–1788).

Lindroth. Sten, Kungl. *Svenska vetenskapsakademiens historia 1739–1818*, part I, vol.I: *Tiden intill Wargentins död* (1783), Almquist & Wiksell (Stockholm, 1967).

Linnaeus, Carl, *Dissertatio Academicum de Pane Diaetetico* (Uppsala, 1757).

Linnaeus, Carl, *Skånska resa år 1749* (Stockholm, 1751).

Linschoten, J. H., *The Voyage of John Huyghen van Linschoten to the East Indies. From the Old English Translation of 1598*, ed. A. C. Burnell and P. A. Tiele, 2 vols., Hakluyt Society (London, 1885).

Liu, Jiancheng, Hu Lianquan, Yang Jingwu and Shu Xiaojun, *Sichuan Cuisine for the Masses* [大众川菜 *Dazhong Chuancai*], Sichuan Science and Technology Publishing House (Chengdu, 1995).

Lloyd, David, 'The Political Economy of the Potato', *Nineteenth-Century Contexts* 29:2–3 (2007), 311–335.

Lo, Vivienne and Penelope Barrett, 'Cooking up Fine Remedies: On the Culinary Aesthetic in a Sixteenth-Century Chinese Materia Medica', *Medical History* 49:4 (2005), 395–422.

[Lobb, Theophilus], *Primitive Cookery: or the Kitchen Garden Display'd* (London, 1767).

Locke, John, *Second Treatise of Government*, ed. Joseph Carrig, Barnes & Noble (New York, 2004 [1690]).

Long, Janet, ed., *Conquista y comida: consecuencias del encuentro de dos mundos*, Universidad Nacional Autónoma de México (Mexico City, 1997).

Lorry, Anne-Charles, *Essai sur les alimens, pour servir de commentaire aux livres diététiques d'Hippocrate* (Paris, 1757).

Louderback, Lisbeth and Bruce Pavlik, 'Starch Granule Evidence for the Earliest Potato Use in North America', *Proceedings of the National Academy of Sciences* 114:29 (2017), 7606–7610.

Lowood, Henry, *Patriotism, Profit and the Promotion of Science in the German Enlightenment: The Economic and Scientific Societies, 1760–1815*, Garland (New York, 1991).

Lunan, John, *Hortus Jamaicensis, or a Botanical Description, (According to the Linnean System) and an Account of the Virtues, &c. of its Indigenous Plants Hitherto Known, as also of the Most Useful Exotics*, 2 vols. (Jamaica, 1814).

Macdonald, Janet, *British Navy's Victualling Board, 1793–1815: Management Competence and Incompetence*, Boydell & Brewer (Woodbridge, 2010).

Macdonald, Janet, *Feeding Nelson's Navy: The True Story of Food at Sea in the Georgian Era*, Frontline (London, 2014).

Machiavelli, Niccolò, *The Prince*, trans. and ed. Peter Bondanella, intro. Maurizio Viroli, Oxford University Press (Oxford, 2005 [1532]).

Mackay, David, *In the Wake of Cook: Exploration, Science and Empire, 1780–1801*, St. Martin's Press (New York, 1985).

Mackenzie, Fiona, 'Contested Ground: Colonial Narratives and the Kenyan Environment, 1920–1945', *Journal of Southern African Studies* 26:4 (2000), 697–718.

Mackenzie, Henry, *Prize Essays and Transactions of the Highland Society of Scotland* 1 (Edinburgh, 1799).

Magazzini, Vitale, *Coltivazione toscana del molto rever. P.D. Vitale Magazzini Monaco Vallombrosano*, third edition (Florence, 1669).

Mahlerwein, Gunter, 'The Consequences of the Potato Blight in Southern Germany', in When the Potato Failed: *Causes and Effects of the 'Last' European Subsistence Crisis, 1845–1850*, ed. Cormac Ó Gráda, Richard Paping and Eric Vanhaute, Brepols (Turnhout, 2007), 213–221.

Malanima, Paolo, 'Urbanization', in *The Cambridge Economic History of Modern Europe, vol. I: 1700–1850*, ed. Stephen Broadberry and K. O'Rourke, Cambridge University Press (Cambridge, 2010), 235–263.

Malte-Brun, Conrad, *Universal Geography, or a Description of all the Parts of the World, on a New Plan, According to the Great Natural Divisions of the Globe*, 6 vols. (Edinburgh, 1822).

Malthus, Thomas Robert, *An Essay on the Principle of Population*, 1798, ed. Philip Appleman, Norton (New York, 1974).

Malthus, Thomas Robert, *An Essay on the Principle of Population. The 1803 Edition*, ed. Shannon Stimson, Yale University Press (New Haven, 2018).

Malthus, Thomas Robert, 'Newenham on the State of Ireland', *Edinburgh Review or Critical Journal* 14 (Edinburgh, 1809).

Mandeville, Bernard, *The Fable of the Bees: Or, Private Vices, Publick Benefits* (London, 1714).

Manetti, Saverio, *Delle specie diverse di frumento e di pane siccome della panizzazione* (Florence, 1765).

Markham, Gervase, *The Husbandman's Jewel* (London, 1695).

Marquis de Chastellux, *De la félicité publique, ou Considérations sur le sort des hommes*, 2 vols. (Amsterdam, 1772).

Marquis de Langle (Jean Marie Jérome Fleuriot), *Voyage de Figaro en Espagne* (Seville, 1785).

Marshall, Woodville, 'Provision Ground and Plantation Labor in Four Windward Islands: Competition for Resources during Slavery', in *Cultivation and Culture: Labor and the Shaping of Slave Life in the Americas*, ed. Ira Berlin and Philip Morgan, University Press of Virginia (Charlottesville, 1993), 203–220.

Martin, John, *The Development of Modern Agriculture: British Farming since 1931*, Palgrave Macmillan (Basingstoke, 2000).

Marx, Karl, 'The Eighteenth Brumaire of Louis Napoleon', 1852, in *Marx's Eighteenth Brumaire: (Post)modern Interpretations*, ed. Mark Cowling and James Martin, trans. Terrell Carver, Pluto Press (London, 2002), 19–110.

Mason, Charlotte, *The Lady's Assistant for Regulating and Supplying the Table: Being a Complete System of Cookery* (London, 1787).

Matta, Raúl, 'Valuing Native Eating: The Modern Roots of Peruvian Food Heritage', *Anthropology of Food* (2013), https://journals.openedition.org/aof/7361.

Matthee, Rudolph, 'Patterns of Food Consumption in Early Modern Iran' (2016), *Oxford Handbooks Online*, www.oxfordhandbooks.com/view/10.1093/oxfordhb/9780199935369.001.0001/oxfordhb-9780199935369-e-13.

Maxwell, Simon, 'Food Security: A Post-Modern Perspective', *Food Policy* 21:2 (1996), 155–170.

Mayer, E., *The Articulated Peasant*, Westview (Boulder, 2002).

Mazumdar, Sucheta, 'The Impact of New World Food Crops on the Diet and Economy of China and India, 1600–1900', in *Food in Global History*, ed. Raymond Grew, Westview (Boulder, 1999), 58–78.

McCann, James, *Maize and Grace: Africa's Encounter with a New World Crop, 1500–2000*, Harvard University Press (Cambridge, MA, 2005).

McCann, James, *Stirring the Pot: A History of African Cuisine*, Ohio State University Press (Athens, 2009).

McClellan III, James, *Colonialism and Science: Saint Domingue in the Old Regime*, University of Chicago Press (Chicago, 2010).

McClellan III, James and François Regourd, 'The Colonial Machine: French Science and Colonization in the Ancien Regime', *Osiris* 15 (2000), 31–50.

McClure, Norman, ed., *The Letters of John Chamberlain*, 2 vols., American Philosophical Society (Philadelphia, 1939).

McCook, Stuart, *States of Nature: Science, Agriculture, and Environment in the Spanish Caribbean, 1760–1940*, University of Texas Press (Austin, 2002).

McCormick, Ted, 'Population: Modes of Seventeenth-Century Demographic Thought', in *Mercantalism Reimagined: Political Economy in Early Modern Britain and its Empire*, ed. Philip Stern and Carl Wennerlind, Oxford University Press (Oxford, 2014), 25–45.

McCulloch, John Ramsay, 'Cottage System', *Supplement to the Fourth, Fifth, and Sixth Editions of the Encyclopaedia Britannica* (Edinburgh, 1824), III, 378–387.

McMahon, Darrin, *Happiness: A History*, Atlantic Monthly (New York, 2006).

McNeil, William, 'How the Potato Changed the World's History', *Social Research* 66:1 (1999), 67–83.

Mehta, Pratap Bhanu, 'Self-Interest and Other Interests', in *The Cambridge Companion to Adam Smith*, ed. Knud Haakonssen, Cambridge University Press (Cambridge, 2006), 246–269.

Meinert, C. A., *Armee- und Volks-Ernährung. Ein Versuch Professor C. von Voit's Ernährungstheorie für die Praxis zu verwerthen*, 2 vols. (Berlin, 1880).

Meléndez, Mariselle, *Deviant and Useful Citizens: The Cultural Production of the Female Body in Eighteenth-Century Peru*, Vanderbilt University Press (Nashville, 2011).

Melon, Jean-François, *Essai politique sur le commerce* (n.p., 1736).

Meneghello, Laura, *Jacob Moleschott: A Transnational Biography — Science, Politics, and Popularization in Nineteenth-Century Europe*, Transcript (Bielefeld, 2017).

Meredith, David and Deborah Oxley, 'Food and Fodder: Feeding England, 1700–1900', *Past & Present* 222:1 (2013), 163–214.

Messer, Ellen, 'Three Centuries of Changing European Tastes for the Potato', in *Food Preferences and Tastes: Continuity and Change*, ed. Helen Mac Beth, Berghahn Books (New York, 1997), 101–114.

Metodo facile, e sperimentato per coltivare le patate (Florence, 1801).

Michiel, Pietro Antonio, *I cinque libri di piante*, Real Instituto Veneto di Scienze, Lettere ed Arti (Venice, 1940 [c.1570]).

Mill, James, 'Colony', *Supplement to the Encyclopaedia Britannica* (London, 1825), 257–273.

Miller, David Philip and Peter Hanns Reill, eds., *Visions of Empire: Voyages, Botany, and Representations of Nature*, Cambridge University Press (Cambridge, 1996).

Miller, Judith, *Mastering the Market: The State and the Grain Trade in Northern France, 1700–1860*, Cambridge University Press (Cambridge, 1999).

Miller, Philip, *The Gardeners Dictionary*, 3 vols. (London, 1754).

Ministry of Agriculture and Rural Affairs of the People's Republic of China, 'China to Position Potato as Staple Food', 8 Jan. 2015, http://english.agri.gov.cn/news/dqnf/201501/t20150109_24781.htm.

Ministry of National Service, 1917–1919, Report upon the Physical Examination of Men of Military Age by National Service Medical Boards from November 1st 1917–October 31st 1918, *Parliamentary Papers*, 1919, 26:307, Cmd. 504.

Miodunka, Piotr, 'L'essor de la culture de la pomme de terre au sud de la Pologne jusqu'au mileau du XIXe siècle', *Histoire & Sociétés Rurales* 41:2 (2014), 67–84.

Mitchison, Rosalind, 'The Old Board of Agriculture (1793–1822)', *English Historical Review* 74:290 (1959), 41–69.

Moffet, Thomas, *Health's Improvement, or, Rule Comprizing and Discovering the Nature, Method, and Manner of Preparing all sorts of Food Used in this Nation* (London, 1655).

Moheau, Jean-Baptiste, *Recherches et considérations sur la population de la France* (Paris, 1778).

Moleschott, Jacob, *Der Kreislauf des Lebens*, 2 vols. (Gießen, 1887).

Moleschott, Jacob, *Lehre der Nahrungsmittel: Für das Volk* (Erlangen, 1858).

Molina, Cristóbal de, *Relación de las fábulas y ritos de los incas*, ed. Paloma Jiménez del Campo, Verveurt (Madrid, 2010 [c.1573]).

Molina, Juan Ignacio, *Compendio de la historia civil del reyno de Chile* (Madrid, 1795).

Molinier, Alain, *Stagnations et croissance: Le Vivarais aux XVIIe–SVIIIe siècles*, L'École des Hautes Études en Sciences Sociales (Paris, 1985).

Molokhovets, Elena, *Classic Russian Cooking: Elena Molokhovets' A Gift to Young Housewives*, ed. and trans. Joyce Toomre, Indiana University Press (Bloomington, 1992).

Monardes, Nicolás, *Joyfull News out of the New-found Worlde* (London, 1596).

Montesquieu (Charles-Louis de Secondat), *De l'Esprit des loix*, 1748, in *Oeuvres complètes de Montesquieu*, ed. Édouard Laboulaye (Paris, 1876), III.

More, Hannah, *The Cottage Cook or, Mrs. Jones's Cheap Dishes: Shewing the Way to do Much Good with Little Money* (London, [1795]).

More, Hannah, *The Way to Plenty, Or, the Second Part of Tom White* (London, 1796).

Morel, Marie-France, 'Children', in *Encyclopedia of the Enlightenment*, ed. Michel Delon and Philip Stewart, trans. Gwen Wells, Routledge (London, 2001), 243–247.

Morel, Marie-France, 'Théories et pratiques de l'allaitement en France au XVIIIe siècle', *Annales de démographie historique* 1 (1976), 393–427.

Morell, Mats L. W., 'Diet in Sweden during Industrialization, 1870–1939: Changing Trends and the Emergence of Food Policy', in *The Origins and Development of Food Policies in Europe*, ed. John Burnett and Derek Oddy, Leicester University Press (London, 1994), 232–248.

Morineau, Michel, 'The Potato in the Eighteenth Century', in *Food and Drink in History*, ed. Robert Forster and Orest Ranum, trans. Elborg Forster and Patricia Ranum, Johns Hopkins University Press (Baltimore, 1979), 17–36.

Moris, Jon, Extension *Alternatives in Tropical Africa*, Overseas Development Institute (London, 1991).

Morren, Charles, *Belgique Horticole*, 35 vols. (Liège, 1853).

Moskoff, William, *The Bread of Affliction: The Food Supply in the USSR during World War II*, Cambridge University Press (Cambridge, 1990).

Moyer, Johanna, '"The Food Police": Sumptuary Prohibitions on Food in the Reformation', in *Food and Faith in Christian Culture*, ed. Ken Albala and Trudy Eden, Columbia University Press (New York, 2011), 59–81.

Mt. Pleasant, Jane, 'The Paradoxes of Plows and Productivity: An Agronomic Comparison of Cereal Grain Production under Iroquois Hoe Culture and European Plow Culture in the Seventeenth and Eighteenth Centuries', *Agricultural History* 85:4 (2011), 460–492.

Muldrew, Craig, *Food, Energy and the Creation of Industriousness: Work and Material Culture in Agrarian England, 1550–1780*, Cambridge University Press (Cambridge, 2011).

Murphey, Rhoads, 'Provisioning Istanbul: The State and Subsistence in the Early Modern Middle East', *Food and Foodways*, 2:1 (1988), 217–263.

Murphy, Rose and Lee Kass, *Evolution of Plant Breeding at Cornell University*, Internet-First University Press (Ithaca, 2007).

Murra, John V., *The Economic Organization of the Inka State, Research in Economic Anthropology, Supplement* 1, JAI Press (Greenwich, CT, 1980).

Murra, John V., 'Rite and Crop in the Inca State', in *Culture in History: Essays in Honor of Paul Radin*, ed. Stanley Diamond, Octagon (New York, 1960), 393–407.

Murray, Laura May Kaplan, 'New World Food Crops in China: Farms, Food, and Families in the Wei River Valley, 1650–1910', DPhil., University of Pennsylvania, 1985.

Murrell, John, *A New Book of Cookerie* (London, 1617).

Nadeau, Carolyn, *Food Matters: Alonso Quijano's Diet and the Discourse of Food in Early Modern Spain*, University of Toronto Press (Toronto, 2016).

Nally, David, 'The Biopolitics of Food Provisioning', *Transactions of the Institute of British Geographers* 36 (2011), 37–53.

Nanny State Index, http://nannystateindex.org/.

Naranjo Vargas, Plutarco, 'La comida andina antes del encuentro', in *Conquista y comida: consecuencias del encuentro de dos mundos*, ed. Janet Long, Universidad Nacional Autónoma de México (Mexico City, 1997), 31–43.

Narayanan, Divya, 'Cultures of Food and Gastronomy in Mughal and Post-Mughal India', Inaugural dissertation, Ruprecht-Karls-Universität Heidelberg, 2015.

National Cancer Institute, 'Acrylamide and Cancer Risk', 2017, www.cancer.gov/about-cancer/causes-prevention/risk/diet/acrylamide-fact-sheet.

Netting, Robert McC., *Balancing on an Alp: Ecological Change & Continuity in a Swiss Mountain Community*, Cambridge University Press (Cambridge, 1981).

Netting, Robert McC., *Smallholders, Householders: Farm Families and the Ecology of Intensive, Sustainable Agriculture*, Stanford University Press (Stanford, 1993).

New London County Selectmen, *Agreeable to an Act of Assembly of the State of Connecticut, for Regulating the Prices of Labour, etc.* (New London, 1778).

New Zealand Ministry of Health, 'Eating and Activity Guidelines', 2 Aug. 2018, www.health.govt.nz/our-work/eating-and-activity-guidelines.

Nieto Olarte, Mauricio, *Orden Natural y orden social: Ciencia y política en el Semanario del Nuevo Reino de Granada*, Instituto Colombiano de Antropología e Historia (Madrid, 2007).

Nitti, Francesco S., 'The Food and Labour-Power of Nations', *Economic Journal* 6:21 (1896), 30–63.

Nuñez de Oria, Francisco, *Regimiento y aviso de sanidad, que trata de todos los generos de alimentos y del regimiento della* (Medina del Campo, 1586).

Nunn, Nathan and Nancy Qian, 'The Potato's Contribution to Population and Urbanization: Evidence from a Historical Experiment', *Quarterly Journal of Economics* 126 (2011), 593–650.

O'Connor, Kaori, 'The King's Christmas Pudding: Globalization, Recipes, and the Commodities of Empire', *Journal of Global History* 4 (2009), 127–155.

Ó Gráda, Cormac, 'Ireland's Great Famine: An Overview', in *When the Potato Failed: Causes and Effects of the 'Last' European Subsistence Crisis, 1845–1850*, ed. Cormac Ó Gráda, Richard Paping and Eric Vanhaute, Brepols (Turnhout, 2007), 43–57.

Ó Gráda, Cormac, Richard Paping and Eric Vanhaute, eds., *When the Potato Failed: Causes and Effects of the 'Last' European Subsistence Crisis, 1845–1850*, Brepols (Turnhout, 2007).

Occhiolini, Giovanni Battista, *Memorie sopra il meraviglioso frutto americano chiamato volgarmente patata* (Rome, 1784).

Ochoa, Carlos, 'Potato Collecting Expeditions in Chile, Bolivia and Peru, and the Genetic

Erosion of Indigenous Cultivars', in *Crop Genetic Resources for Today and Tomorrow*, ed. O. H. Frankel and J. G. Hawkes, Cambridge University Press (Cambridge, 1975), 167–172.

Olsson, Mats and Patrick Svensson, 'Agricultural Production in Southern Sweden1702–1864', in *Growth and Stagnation in European Historical Agriculture*, ed. Mats Olsson and Patrick Svensson, Brepols (Turnhout, 2011), 117–139.

Osborne, Michael, *Nature, the Exotic, and the Science of French Colonialism*, University of Indiana Press (Bloomington, 1994).

Overton, Mark, *Agricultural Revolution in England: The Transformation of the Agrarian Economy 1500–1850*, Cambridge University Press (Cambridge, 1996).

Pagden, Anthony, *The Languages of Political Theory in Early Modern Europe*, Cambridge University Press (Cambridge, 1987).

Pagden, Anthony, *Spanish Imperialism and the Political Imagination*, Yale University Press (New Haven, 1990).

Palanca Cañón, J. David, 'Introducción y Generalización del Cultivo y Consumo Alimentario y Médico de la Patata en el País Vasco', DPhil, Universidad del País Vasco, 2011.

Palmer, Catherine, 'From Theory to Practice: Experiencing the Nation in Everyday Life', *Journal of Material Culture* 3:2 (1998), 175–199.

Palomino Gonzales, María Mercedes, 'Gourmetización del alimento andino y la estetización del agricultor como parte del Boom Gastronómico Peruano', *Razón y Palabra* 20:94 (2016), 581–595.

Paping, Richard and Vincent Tassenaar, 'The Consequences of the Potato Disease in the Netherlands 1845–1860: A Regional Approach', in *When the Potato Failed: Causes and Effects of the 'Last' European Subsistence Crisis, 1845–1850*, ed. Cormac Ó Gráda, Richard Paping and Eric Vanhaute, Brepols (Turnhout, 2007), 149–184.

Paquette, Gabriel, ed., *Enlightened Reform in Southern Europe and Its Atlantic Colonies, c.1750–1830*, Palgrave Macmillan (London, 2016).

Paquette, Gabriel, *Enlightenment, Governance, and Reform in Spain and its Empire, 1759–1808*, Palgrave Macmillan (Basingstoke, 2008).

Parker, Geoffrey, *Global Crisis: War, Climate Change and Catastrophe in the Seventeenth Century*, Yale University Press (New Haven, 2013).

Parker, George, *The Gardeners Almanack* (London, 1702).

Parkinson, John, *Paradisi in Sole Paradisus Terrestris* (London, 1629).

Parliamentary *History of England, from the Earliest Period to the Year 1803* (London, 1818).

Parmentier, Antoine Augustin, *Manière de Faire le Pain de Pommes de Terre sans Mélange de Farine* (Paris, 1779).

Parmentier, Antoine Augustin, *Mémoire couronné le 25 aout 1784, par l'Académie Royale des Sciences, Belles Lettres et Arts de Bordeaux, sur cette question: Quel seroit le meilleur procédé pour conserver, le plus long-temps possible, ou en grain ou en farine, le maïs ou blé de Turquie, plus connu dans la Guienne sous le nom de blé d'Espagne? Et quels seroient les différens moyens d'en tirer parti, dans les années abondantes, indépendamment des usages connus & ordinaires dans cette province?* (Bordeaux, 1785).

Parmentier, Antoine Augustin, *Observations on Such Nutritive Vegetables as May be Substituted in the Place of Ordinary Food, in Times of Scarcity* (London, 1783).

Parmentier, Antoine Augustin, *Les pommes de terre, considérées relativement à la santé & àl'économie: ouvrage dans lequel on traite aussi du froment & du riz* (Paris, 1781).

Parmentier, Antoine Augustin, *Traité sur la culture et les usages des pommes de terre, de la patate, et du topinambour* (Paris, 1789).

Parque de la Papa, www.parquedelapapa.org/esp/03parke_01.html.

Parry, John, 'Plantations and Provision Grounds: An Historical Sketch of the Introduction of Food Crops to Jamaica', *Revista de historia de América* 19 (1955), 1–20.

Pauer, Erich, 'Neighbourhood Associations and Food Distribution in Japanese Cities in World War II', in *Agriculture and Food Supply in the Second World War*, ed. Bernd Martin and Alan Milward, Scripta Mercaturae Verlag (Ostfildern, 1985), 219–241.

Pedersen, Susan, *Family Dependence, and the Origins of the Welfare State: Britain and France, 1914–1945*, Cambridge University Press (Cambridge, 1993).

Penn, William, 'A Further Account of the Province of Pennsylvania and its Improvements,' *Pennsylvania Magazine of History and Biography* 9:1 (1885), 62–81.

Pennell, Sara, 'Recipes and Reception: Tracking New World Foodstuffs in Early Modern British Culinary Texts, c.1650–1750', *Food and History* 7:1 (2010), 11–33.

Persson, Karl Gunnar, *Grain Markets in Europe, 1500–1900: Integration and Deregulation*, Cambridge University Press (Cambridge, 2009).

Petersen, Christian, *Bread and the British Economy, c.1770–1870*, ed.Andrew Jenkins, Scholar Press (Aldershot, 1995).

Petránová, Lydia, 'The Rationing System in Bohemia during the First World War', in *The Origins and Development of Food Policies in Europe*, ed. John Burnett and Derek Oddy, Leicester University Press (London, 1994), 23–38.

Petrie, Hazel, *Chefs of Industry: Maori Tribal Enterprise in Early Colonial New Zealand*, Auckland University Press (Auckland, 2013).

Petty, William, *Tracts; Chiefly Relating to Ireland* (Dublin, 1769).

Petyt, William, *Britannia Languens: or, a Discourse of Trade* (London, 1689).

Phillips, Denise, *Acolytes of Nature: Defining Natural Science in Germany, 1770–1850*, University of Chicago Press (Chicago, 2012).

Phillips, Edward, *The New World of Words, or A General English Dictionary* (London, 1678).

Piktin, Donald, *The House that Giacomo Built: History of an Italian Family, 1898–1978*, Cambridge University Press (Cambridge, 1985).

Pilcher, Jeffrey, *¡Que Vivan los Tamales! Food and the Making of Mexican Identity*, University of New Mexico Press (Albuquerque, 1998).

Pingali, P. L., 'Green Revolution: Impacts, Limits, and the Path Ahead', *Proceedings of the National Academy of Sciences* 109:31 (2012), 12302–12308.

Piqueras Haba, Juan, 'La difusión de la patata en España (1750–1850): El papel de las Sociedades Económicas y del clero rural', *Ería: revista cuatrimestral de geografía* 27 (1992), 80–89.

Pissling, Wilhelm Franz, *Gesundheitslehre für das Volk* (Vienna, 1856).

Platell, Amanda, 'Sorry, Why Should the NHS Treat People for Being Fat?', Mail Online, 27 Feb. 2009, www.dailymail.co.uk/debate/article-1156678/AMANDAPLATELL-Sorry-NHS-treat-people-fat.html.

Ploeg, Jan Douwe van der, 'Potatoes and Knowledge', in *An Anthropological Critique of Development: The Growth of Ignorance*, ed. Mark Hobart, Routledge (London, 1993), 209–27.

Pohl-Valero, Stefan, 'Food Science, Race, and the Nation in Colombia', *Oxford Research Encyclopedia of Latin American History* (2016), http://oxfordre.com/latinamericanhistory/view/10.1093/acrefore/9780199366439.001.0001/acrefore9780199366439-e-321?print=pdf.

Pohl-Valero, Stefan, '"La raza entra por la boca": Energy, Diet, and Eugenics in Colombia, 1890–1940', *Hispanic American Historical Review* 94:3 (2014), 455–486.

Poissonnier Desperrières, [Antoine], *Mémoire sur les avantages qu'il y aurait a changer absolument la nourriture des gens de mer* (Versailles, 1772).

Poissonnier Desperrières, Antoine, *Traité des maladies des gens de mer* (Paris, 1767).

Pollock, Griselda, 'Van Gogh and the Poor Slaves: Images of Rural Labour as Modern Art', *Art History* 11:3 (1988), 408–432.

'Pomme de terre, Topinambour, Batate, Truffe blanche, Truffe rouge', 1765, in *Encyclopédie, ou dictionnaire raisonné des sciences, des artes et des métiers, etc.*, ed. Denis Diderot and Jean le Rond d'Alembert, University of Chicago: ARTFL

Encyclopédie Project, ed. Robert Morrissey, http://encyclopedie.uchicago.edu.

Popplow, Marcus, 'Economizing Agricultural Resources in the German Economic Enlightenment', in *Materials and Expertise in Early Modern Europe: Between Market and Laboratory*, ed. Ursula Klein and E. C. Spary, University of Chicago Press (Chicago, 2010), 261–287.

Porras Barrenechea, Raúl, ed., *Cartas del Perú (1524–1543)*, Colección de Documentos Inéditos para la Historia del Perú (Lima, 1959).

Porter, Andrew, *Religion versus Empire? British Protestant Missionaries and Overseas Expansion, 1700–1914*, Manchester University Press (Manchester, 2004).

Post, John, 'Nutritional Status and Mortality in Eighteenth-Century Europe', in *Hunger in History: Food Shortage, Poverty and Deprivation*, ed. Lucille Newman, Wiley-Blackwell (Oxford, 1995), 241–280.

Potato Pete's Recipe Book, Ministry of Food ([London], [1940?]).

Pottier, Johan, *The Anthropology of Food: The Social Dynamics of Food Security*, Polity Press (Cambridge, 1999).

Prain, Gordon, Sam Fujisaka and Michael Warren, eds., *Biological and Cultural Diversity: The Role of Indigenous Agricultural Experimentation in Development*, Intermediate Technology Publications (London, 1999).

Premo, Bianca, *The Enlightenment on Trial: Ordinary Litigants and Colonialism in the Spanish Empire*, Oxford University Press (Oxford, 2017).

Procter, Robert, *The Nazi War on Cancer*, Princeton University Press (Princeton, 1999).

Pullan, Brian, *Poverty and Charity: Europe, Italy, Venice, 1400–1700*, Variorum (Aldershot, 1994).

Purchas, Samuel, *Hakluytus Posthumus or Purchas his Pilgrimes* (Glasgow, 1905).

Qu, Dongyu and Kaiyun Xie, eds., *How the Chinese Eat Potatoes*, World Scientific (Hackensack, NJ, 2008).

Quesnay, François, 'Fermiers', 1756, in *Encyclopédie, ou dictionnaire raisonné des sciences, des arts et des métiers*, ed. Denis Diderot and Jean le Rond d'Alembert, VI:537, University of Chicago, ARTFL Encyclopédie Project (Autumn 2017 Edition), ed. Robert Morrissey and Glenn Roe, http://encyclopedie.uchicago.edu.

Rabinbach, Anson, *The Human Motor: Energy, Fatigue and the Origins of Modernity*, University of California Press (Berkeley, 1992).

Raj, Kapil, 'Colonial Encounters and the Forging of New Knowledges and National Identities: Great Britain and India, 1760–1850', *Osiris* 15 (2000), 119–134.

Raj, Kapil, *Relocating Modern Science: Circulation and the Construction of Knowledge in South Asia and Europe, 1650–1900*, Palgrave Macmillan (Basingstoke, 2007).

Rath, Eric, *Food and Fantasy in Early Modern Japan*, University of California Press (Berkeley, 2010).

Ray, Utsa, *Culinary Culture in Colonial India: A Cosmopolitan Platter and the Middle-Class*, Cambridge University Press (Delhi, 2015).

Reader, John, Potato: *The History of the Propitious Esculent*, Yale University Press (New Haven, 2008).

Reagin, Nancy, *Sweeping the German Nation: Domesticity and National Identity in Germany, 1870–1945*, Cambridge University Press (Cambridge, 2007).

Redlich, Fritz, 'Science and Charity: Count Rumford and his Followers', *International Review of Social History* 16:2 (1971), 184–216.

'Review of *The Nation's Larder*', *Public Health* 54 (Oct. 1940–Sept. 1941), 14–16.

Ricciardi, V., N. Ramankutty, Z. Mehrabi, L. Jarvis and B. Chookolingo, 'How Much of the World's Food do Smallholders Produce?', *Global Food Security* 17 (2018), 64–72.

Rich, Barnabe, *True Report of a Late Practice Enterprised by a Papist* (London, 1582).

Richardson, Sarah, *The Political Worlds of Women: Gender and Politics in Nineteenth Century Britain*, Routledge (New York, 2013).

Richthofen, Ferdinand von, *Ferdinand von Richthofen's Tagebücher aus China*, ed. E. von Tiessen, 2 vols., Dietrich Reimer (Berlin, 1907).

Rickman, Geoffrey, *Corn Supply of Ancient Rome*, Clarendon Press (Oxford, 1980).

Riera Climent, Luis and Juan Riera Palmero, 'Los alimentos americanos en los Extractos de la Bascongada (1768–1793): el maíz y la patata', *Llull* 30 (2007), 525–540.

Ries, Nancy, 'Potato Ontology: Surviving Postsocialism in Russia', *Cultural Anthropology* 24:2 (2009), 181–212.

Rigaud, Lucas, *Cozinheiro moderno ou nova arte de cozinha* (Lisbon, 1785).

Riley, H. T., ed., *Memorials of London and London Life in the 13th, 14th and 15th Centuries* (London, 1868), British History Online, www.british-history.ac.uk.

Riley, James, *Population Thought in the Age of the Demographic Revolution*, Carolina Academic Press (Durham, 1985).

Riquetti, Victor de, *Philosophie rurale: ou, Économie générale et politique de l'agriculture*, 3 vols. (Amsterdam, 1763).

Roberts, Lissa, 'Situating Science in Global History: Local Exchanges and Networks of Circulation', *Itinerario* 33:1 (2009), 9–30.

Rockefeller Foundation, 'Accelerating Agricultural Modernization in Developing Nations', Bellagio, 2–6 Feb. 1970.

Rodríguez de Campomanes, Pedro, *Discurso sobre el fomento de la industria popular* (Madrid, 1774).

Roncaglia, Alessandro, *The Wealth of Ideas: A History of Economic Thought*, Cambridge University Press (Cambridge, 2001).

Rosenhane, Schering, *Oeconomia*, ed. Torsten Lagerstedt, Almqvist &Wiksells Boktryckeri (Uppsala, 1944 [1662]).

Rowntree, B. Seebohm, *Poverty: A Study of Town Life*, Macmillan (London, 1908).

Roy, Parama, 'A Dietetics of Virile Emergency', *Women's Studies International Forum* 44 (2014), 255–265.

Roy, Parama, 'Meat-Eating, Masculinity, and Renunciation in India: A Gandhian Grammar of Diet', *Gender & History* 14 (2002), 62–91.

Royer, Johann, *Eine gute Anleitung wie man ... Garten-Gewächse ... nützen sole* (Braunschweig, 1651).

Rubner, Max, *Volksernährungsfragen* (Leipzig, 1908).

Rudenschöld, Ulrik, *Almanach* (Gothenburg, 1750).

Rule, John and John Wells, *Crime, Protest and Popular Politics in Southern England, 1740–1850*, Hambledon Press (London, 1997).

Rumpolts, Marx, *Ein new Kochbuch* (Frankfurt am Main, 1581).

Runge, C. Fore, 'Economic Consequences of the Obese', *Medscape* 56:11 (2007), 2668–2672.

Rusnock, Andrea, *Vital Accounts: Quantifying Health and Population in Eighteenth-Century England and France*, Cambridge University Press (Cambridge, 2002).

Safier, Neil, *Measuring the World: Enlightenment Science and South America*, University of Chicago Press (Chicago, 2008).

Safley, Thomas Max, ed., *The Reformation of Charity: The Secular and the Religious in Early Modern Poor Relief*, Brill (Boston, 2003).

Said, Edward, *Culture and Imperialism*, Vintage (London, 1994).

Saint Pierre, J. H. B. de, *A Voyage to the Isle of France, the Isle of Bourbon, and the Cape of Good Hope: with Observations and Reflections upon Nature and Mankind* (London, 1800).

Sala, Giovanni Dominici, *De Alimentis et Eorum Recta Administratione Liber* (Padua, 1628).

Sala, Ilaria María, 'Tudou for the Tuhao: Can "Sister Potato," a Singing Peasant, Convince the Chinese to Eat More of the Lowly Spud?', *Quartz*, 24 Feb. 2016, https://qz.com/622594/can-potato-sister-a-singing-peasant-convince-the-chinese-to-eat-more-of-the-lowly-spud/.

Salaman, Redcliffe, *History and Social Influence of the Potato*, ed. J. G. Hawkes, Cambridge University Press (Cambridge, 2000 [1949]).

Salinas y Cordova, Buenaventura de, *Memorial, Informe y Manifiesto* ([Madrid?], c.1646).

Salmon, William, *The Family-Dictionary, or, Household Companion* (London, 1710).

Salomon, Frank and George Urioste, *The Huarochirí Manuscript: A Testament of Ancient and Colonial Andean Religion*, University of Texas Press (Austin, 1991).

Santo Thomas, Domingo de, *Grammatica, o Arte, que ha compuesto de la lengua general de los indios, del Peru* (Valladolid, 1560).

Saraiva, Tiago, *Fascist Pigs: Technoscientific Organisms and the History of Fascism*, MIT Press (Cambridge, MA, 2016).

Sauer, Carl O., *Agricultural Origins and Dispersals: The Domestication of Animals and Foodstuffs*, MIT Press (Cambridge, MA, 1969 [1952]).

Schanbacer, William, *The Politics of Food: The Global Conflict between Food Security and Food Sovereignty*, Praeger (Santa Barbara, 2010).

Schiebinger, Londa, *Plants and Empire: Colonial Bio-prospecting in the Atlantic World*, Harvard University Press (Cambridge, MA, 2009).

Schiebinger, Londa and Claudia Swan, eds., *Colonial Botany: Science, Commerce and Politics in the Early Modern World*, University of Pennsylvania Press (Philadelphia, 2005).

Schlüter, Regina G., 'Promoting Regional Cuisine as Intangible Cultural Heritage in Latin America', in *Food and the Tourism Experience: The OECD–Korea Workshop*, Organisation for Economic Co-operation and Development Publishing (Paris, 2012), 89–99.

Schmalzer, Sigrid, *Red Revolution, Green Revolution: Scientific Farming in Socialist China*, University of Chicago Press (Chicago, 2016).

Scholliers, Peter, 'The Policy of Survival: Food, the State and Social Relations in Belgium', 1914–1921, in *The Origins and Development of Food Policies in Europe*, ed.John Burnett and Derek Oddy, Leicester University Press (London, 1994), 39–53.

Schuler, Johann Melchior, *Geschichte und Beschreibung des Landes Glarus* (Zurich, 1837).

Schulze-Gävernitz, Gerhart von, *The Cotton Trade in England and on the Continent:A Study in the Field of the Cotton Industry*, trans. Oscar Hall (London, 1895).

Scott, Gregory, 'Plants, People, and the Conservation of Biodiversity of Potatoes in Peru', *Natureza & Conservação: Brazilian Journal of Nature Conservation* 9:1 (2011), 21–38.

Scott, James C., *Against the Grain: A Deep History of the Earliest States*, Yale University Press (New Haven, 2017).

Scott, James C., *The Art of Not Being Governed: An Anarchist History of Upland Southeast Asia*, Yale University Press (New Haven, 2009).

Scott, James C., *Seeing Like a State: How Certain Schemes to Improve the Human*

Condition Have Failed, Yale University Press (New Haven, 1998).

Serrano, Elena, 'Making *Oeconomic* People: The Spanish *Magazine of Agriculture and Arts for Parish Rectors* (1797–1808)', *History and Technology* 30:3 (2014), 149–176.

Serres, Michel, 'Theory of the Quasi-Object', in *The Parasite*, trans. Lawrence Schehr, intro. Cary Wolfe, University of Minnesota Press (Minneapolis, 2007), 224–234.

Serres, Olivier de, *Le theatre d'agriculture et mesnage des champs* (Paris, 1603).

SERVINDI, 'Perú: Hoy se celebra el Día Nacional de la Papa', SERVINDI, 31 May 2006, www.servindi.org/actualidad/686.

Shammas, Carole, 'The Eighteenth-Century English Diet and Economic Change', *Explorations in Economic History* 21 (1984), 254–269.

Shammas, Carole, *The Pre-Industrial Consumer in England and America*, Clarendon Press (Oxford, 1990).

Shapiro, Laura, *Perfection Salad: Women and Cooking at the Turn of the Century*, Farrar, Straus & Giroux (New York, 1986).

Sharp, Buchanan, *Famine and Scarcity in Late Medieval and Early Modern England: The Regulation of Grain Marketing, 1256–1631*, Cambridge University Press (Cambridge, 2016).

Shaw, David John, *World Food Security: A History since 1945*, Palgrave Macmillan (Basingstoke, 2007).

Sheehan, Jonathan and Dror Wahrman, *Invisible Hands: Self-Organization and the Eighteenth Century*, University of Chicago Press (Chicago, 2015).

Shepherd, Chris, 'Imperial Science: The Rockefeller Foundation and Agricultural Science in Peru, 1940–1960', *Science as Culture* 14:2 (2005), 113–137.

Sheridan, Richard, 'Captain Bligh, the Breadfruit and the Botanic Gardens of Jamaica', *Journal of Caribbean History* 23:1 (1989), 28–50.

Sheridan, Richard, 'The Crisis of Slave Subsistence in the British West Indies during and after the American Revolution', *William and Mary Quarterly* 33:4 (1976), 615–641.

Sherman, Sandra, *Imagining Poverty: Quantification and the Decline of Paternalism*, Ohio State University Press (Columbus, 2001).

Sherwood, Joan, *Poverty in Eighteenth-Century Spain: The Women and Children of the Inclusa*, University of Toronto Press (Toronto, 1988).

Silva, Renan, *La ilustración en el Virreinato de Nueva Granada*, La Carreta (Bogotá, 2005).

Silverblatt, Irene, *Moon, Sun, and Witches: Gender Ideologies and Class in Inca and Colonial Peru*, Princeton University Press (Princeton, 1987).

Simmons, Amelia, *American Cookery* (Hartford, 1796).

Simmons, Dana, *Vital Minimum: Need, Science and Politics in Modern France*, University of Chicago Press (Chicago, 2015).

Simonen, Seppo, *Raivaajia ja rakentajia. Suomen maatalouden historiaa*, Kirjayhtymä (Helsinki, 1964).

Simonton, Deborah, *A History of European Women's Work: 1700 to the Present*, Routledge (London, 2003).

Skytte, Carl, 'Ron at utaf potatoes brånna brånnavin', in *Konglig Svenska Vetenskaps Academiens Handlingar* 8 (Stockholm, 1747), 252–253.

Slicher Van Bath, B. H., *The Agrarian History of Western Europe. A.D. 500–1850*, trans. Olive Ordish, Edward Arnold (London, 1966).

Smith, Adam, *An Inquiry into the Nature and Causes of the Wealth of Nations (1776), The Glasgow Edition of the Works and Correspondence of Adam Smith*, vol. II, ed. William B. Todd, Oxford University Press (Oxford, 1975).

Smith, Adam, *The Theory of Moral Sentiments* (1759), *The Glasgow Edition of the Works and Correspondence of Adam Smith*, vol. I, ed. D. D. Raphael and A. L. Macfie, Oxford University Press (Oxford, 1976).

Smith, E., *The Compleat Housewife* (London, 1739).

Smith, Pamela, *The Body of the Artisan: Art and Experience in the Scientific Revolution*, University of Chicago Press (Chicago, 2004).

Smith, R. E. F. and David Christian, *Bread and Salt: A Social and Economic History of Food and Drink in Russia*, Cambridge University Press (Cambridge, 1984).

Smith, Thomas and John Choules, *Origin and History of Missions; Containing Faithful Accounts of the Voyages, Travels, Labors and Successes of the Various Missionaries, Who Have Been Sent Forth to Evangelize the Heathen*, 2 vols. (Boston, 1832).

Smith, William, *Sure Guide in Sickness and Health, in the Choice of Food, and Use of Medicine* (London, 1779).

Sokolow, Jayme, 'Count Rumford and Late Enlightenment Science, Technology and Reform', *The Eighteenth Century: Theory and Interpretation* 21:1 (1980), 76–86.

Solander, Daniel, *Daniel Solander: Selected Correspondence, 1753–1782*, ed. Edward Duyker and Per Tingbrand, Miegunyah Press (Melbourne, 1995).

Sood, S., V. Bhardwaj, S. K. Pandey and S. K. Chakrabarti, 'History of Potato Breeding: Improvement, Diversification, and Diversity', in *The Potato Genome: Compendium of Plant Genomes*, ed. S. Kumar Chakrabarti, C. Xie and J. Kumar Tiwari, Springer (Cham, 2017), 31–72.

Soule, Emily Berquist, *The Bishop's Utopia: Envisioning Improvement in Colonial Peru*, University of Pennsylvania Press (Philadelphia, 2014).

Spalding, Karen, *Huarochirí: An Andean Society under Inca and Spanish Rule*, Stanford University Press (Stanford, 1984).

Spang, Rebecca, *The Invention of the Restaurant: Paris and Modern Gastronomic Culture*, Harvard University Press (Cambridge, MA, 2000).

Spary, Emma, *Eating the Enlightenment: Food and the Sciences in Paris, 1670–1760*, University of Chicago Press (Chicago, 2012).

Spary, Emma, *Feeding France: New Sciences of Food, 1760–1815*, Cambridge University Press (Cambridge, 2014).

Spary, Emma, *Utopia's Garden: French Natural History from Old Regime to Revolution*, University of Chicago Press (Chicago, 2000).

Spary, Emma and Paul White, 'Food of Paradise: Tahitian Breadfruit and the Autocritique of European Consumption', *Endeavour* 28:2 (2004), 75–80.

Stapelbroek, Koen and Jani Marjanen, eds., *The Rise of Economic Societies in the Eighteenth Century*, Palgrave Macmillan (Basingstoke, 2012).

Staples, Amy L. S., *The Birth of Development: How the World Bank, Food and Agriculture Organization, and the World Health Organization Changed the World, 1945–1965*, Kent State University Press (Kent, OH, 2006).

Stedman Jones, Gareth, *An End to Poverty? A Historical Debate*, Columbia University Press (New York, 2004).

Stedman Jones, Gareth, *Karl Marx: Greatness and Illusion*, Penguin (London, 2016).

Steel, Carolyn, *Hungry City: How Food Shapes Our Lives*, Vintage (London, 2013).

Stern, Philip and Carl Wennerlind, eds., *Mercantalism Reimagined: Political Economy in Early Modern Britain and its Empire*, Oxford University Press (Oxford, 2014).

Stern, Walter, 'The Bread Crisis in Britain, 1795–96', *Economica*, new series, 31:122 (1964), 168–187.

Stewart, John, *An Account of Jamaica and its Inhabitants* (London, 1808).

Struys, Jan Janszoon, *Drie aanmerkelyke en zeer rampspoedige Reizen, door Italien, Griekenland, Lystland, Moscovien, Tartaryen, Meden, Persien, Oostindien, Japan, en verscheiden andere Gewesten* (Amsterdam, 1676).

Styles, John, 'Custom or Consumption? Plebeian Fashion in Eighteenth-Century England', in *Luxury in the Eighteenth Century: Debates, Desires and Delectable Goods*, ed. Maxine Berg and Elizabeth Eger, Palgrave Macmillan (Basingstoke, 2003), 103–115.

Swislocki, Mark, 'Nutritional Governmentality: Food and the Politics of Health in Late Imperial and Republican China', *Radical History Review* 110 (2011), 9–35.

Switzer, Stephen, *The Practical Kitchen Gardiner* (London, 1727).

Symner, Miles, 'Notes on Natural History in Ireland', c.1656–1660, University of Sheffield,

Hartlib Papers 62/45/6A, www.hrionline.ac.uk/hartlib/context.

Tabáres de Ulloa, Francisco, *Observaciones prácticas sobre el cacahuete, o maní de América: su producción en España, bondad del fruto, y sus varios usos, particularmente para la extracción de aceyte; modo de cultivarle y beneficiarle para bien de la nación* (Valencia, 1800).

Talve, Ilmar, 'The Potato in Finnish Food Economy', in *Food in Perspective:Proceedings of the Third International Conference on Ethnological Food Research, Cardiff, Wales, 1977*, ed. Alexander Fenton and Trefor Owen (Edinburgh, 1981), 277–282.

Tapwell, Thomas, *A Friendly Address to the Poor of Great Britain on the Present Scarcity of Wheat and Dearness of Wheaten Bread* (London, 1796).

Targioni-Tozzetti, Antonio, *Cenni storici sulla introduzione di varie piante nell'agricoltura ed Orticoltura Toscana* (Florence, 1853).

Tavárez Simó, Fidel José, 'La invención de un imperio comercial hispano, 1740–1765', *Magallánica, Revista de Historia Moderna* 3 (2015), 56–76.

Taylor, Charles, 'Atomism', in *Philosophical Papers*, Cambridge University Press (Cambridge, 1985), II, 187–210.

Temple, Richard, 'The Agri-Horticultural Society of India', *Calcutta Review* 22 (1854), 341–359.

Tennant, William, *Indian Recreations; Consisting Chiefly of Strictures on the Domestic and Rural Economy of the Mahommedans & Hindoos*, 2 vols. (Edinburgh, 1803–1804).

Terrón, Eloy, *España, encrucijada de culturas alimentarias: su papel en la difusión de los cultivos americanos*, Ministerio de Agricultura, Pesca y Alimentación, Secretaría General Técnica (Madrid, 1992).

Terry, Edward, *A Voyage to East-India, &c.* (London, 1777).

Teuteberg, Hans Jürgen, 'Der Verzehr von Nahrungsmitteln in Deutschland pro Kopf und Jahr seit Beginn der Industrialisierung (1850–1975). Versuch einer quantitativen Langzeitanalyse', *Archiv für Sozialgeschichte* 19 (1979), 331–388.

Thaler, Richard and Cass Sunstein, *Nudge: Improving Decisions about Health, Wealth and Happiness*, Yale University Press (New Haven, 2008).

Thane, Pat, 'The Working Class and State "Welfare" in Britain, 1890–1914', *Historical Journal*, 27:2 (1984), 877–900.

Thick, Malcolm, 'Root Crops and the Feeding of London's Poor in the Late Sixteenth and Early Seventeenth Centuries', in *English Rural Society, 1500–1800: Essays in Honour of Joan Thirsk*, ed. John Chartres and David Hey, Cambridge University Press (Cambridge, 1990), 279–296.

Thirsk, Joan, *The Agrarian History of England and Wales, 1640–1750*, Cambridge

University Press (Cambridge, 2011).

Thirsk, Joan, *Food in Early Modern England: Phases, Fads, Fashions 1500–1760*, Hambledon (London, 2007).

Thomopoulos, Elaine, *The History of Greece*, Greenwood (Santa Barbara, 2012).

[Thompson, Benjamin], *Ensayos, políticos, económicos y fisosóficos del Conde de Rumford, traducidos de órden de la Real Sociedad Económica de esta corte*, trans. Domingo Agüero y Neira, 2 vols. (Madrid, 1800–1801).

Thompson, Benjamin, *Essays, Political, Economical and Philosophical*, 3 vols. (London, 1797–1803).

Thompson, E. P., *The Making of the English Working Class*, Penguin (Harmondsworth, 1984).

Thompson, E. P., 'The Moral Economy of an English Crowd in the Eighteenth Century', *Past & Present* 50:1 (1971), 76–136.

Tiedemann, Friedrich, *Geschichte des Tabaks und anderer ähnlicher Genußmittel* (Frankfurt, 1854).

Tilly, Charles, 'Food Supply and Public Order in Modern Europe', in *The Formation of National States in Western Europe*, ed. Charles Tilly, Princeton University Press (Princeton, 1975), 380–455.

Tobin, Beth, *Colonizing Nature: The Tropics in British Arts and Letters, 1760–1820*, University of Pennsylvania Press (Philadelphia, 2004).

Tomaselli, Sylvana, 'Moral Philosophy and Population Questions in Eighteenth-Century Europe', *Population and Development Review* 14 (1988), 7–29.

Touchet, Julien, *Botanique & Colonisation en Guyane française (1720–1848)*, Ibis Rouge (Cahors, 2004).

Toussaint-Samat, Maguelonne, *A History of Food: A New Expanded Edition*, trans. Anthea Bell, Blackwell (Oxford, 2009).

Transactions of the Agricultural and Horticultural Society of India, 2 vols. (Calcutta, 1838).

Treitel, Corinna, 'Food Science/Food Politics: Max Rubner and "Rational Nutrition" in Fin-de-Siècle Berlin', in *Food and the City in Europe since 1800*, ed.Peter Lummel, Derek Oddy and Peter Atkins, Ashgate (Aldershot, 2012), 51–62.

Treitel, Corinna, 'Max Rubner and the Biopolitics of Rational Nutrition', *Central European History* 41 (2008), 1–25.

Trentmann, Frank, *Empire of Things: How We Became a World of Consumers, from the Fifteenth Century to the Twenty-First*, Allen Lane (Milton Keynes, 2016).

Trentmann, Frank and Flemming Just, eds., *Food and Conflict in Europe in the Age of the*

Two World Wars, Palgrave Macmillan (Basingstoke, 2006).

Trevelyan, Charles, *The Irish Crisis* (London, 1848).

Tribe, Keith, 'Cameralism and the Sciences of the State', in *The Cambridge History of Eighteenth-Century Political Thought*, ed. Mark Goldie and Robert Wokler, Cambridge University Press (Cambridge, 2006), 525–546.

Tribe, Keith, 'Continental Political Economy from the Physiocrats to the Marginal Revolution', in *The Cambridge History of Science*, vol. VII: The Modern Social Sciences, ed. Roy Porter, Theodore M. Porter and Dorothy Ross, Cambridge University Press (Cambridge, 2003), 154–170.

Tribe, Keith, Governing Economy: *The Reformation of German Economic Discourse, 1750–1840*, Cambridge University Press (Cambridge, 1988).

Tribe, Keith, 'Henry Sidgwick, Moral Order and Utilitarianism', *Journal of the History of Economic Thought* 24:4 (2017), 907–930.

Tribe, Keith, *Land, Labour and Economic Discourse*, Routledge (London, 1978).

Tribe, Keith, *Strategies of Economic Order: German Economic Discourse, 1750–1950*, Cambridge University Press (Cambridge, 1995).

Truman, Harry S., 'Inaugural Address', 20 Jan. 1949, Harry S. Truman Presidential Library & Museum, https://trumanlibrary.org/whistlestop/50yr_archive/inagural20jan1949.htm.

Turnbull, William, *The Naval Surgeon Comprising the Entire Duties of Professional Men at Sea. To Which are Subjoined, a System of Naval Surgery and a Compendious Pharmacopoeia* (London, 1806).

Turner, William, *An Almanack for the Year of our Lord God 1701* (London, 1701).

Turton, Thomas, *An Address to the Good Sense and Candour of the People in behalf of the Dealers in Corn: with Some Observations on a Late Trial for Regrating* (London, 1800).

Ugent, Donald, Tom Dillehay and Carlos Ramírez, 'Potato Remains from a Late Pleistocene Settlement in Southcentral Chile', *Economic Botany* 41:1 (1987), 17–27.

UK Department of Health, 'Choosing Health: Making Healthy Choices Easier', 16 Nov. 2004, http://webarchive.nationalarchives.gov.uk and www.dh.gov.uk/en/Publicationsandstatistics/Publications/PublicationsPolicyAndGuidance/DH_4094550.

Unanue, Hipólito, *Observaciones sobre el clima del Lima y sus influencias en los seres organizados* (Lima, 1806).

Underrättelse om potatoës eller jord-pärons plantering och bruk, hämtad af flere årens försök, uti Rambergs bergslag (Stockholm, 1757).

United Nations, *Report of the Forty-Ninth Session of the Committee on Commodity Problems*, Rome, 14–25 Oct. 1974.

Uriz, Joaquín Xavier de, *Causas prácticas de la muerte de los niños expósitos en sus primeros años: remedio en su origen de un tan grave mal: y modo de formarlos útiles a la religión y al estado con notable aumento de la población, fuerzas, y riqueza de España*, 2 vols. (Pamplona, 1801).

US Department of Health and Human Services, 'A Healthier You', 2005, https://health.gov/dietaryguidelines/dga2005/healthieryou/contents.htm.

Usoz, Javier, 'Political Economy and the Creation of the Public Sphere during the Spanish Enlightenment', in *The Spanish Enlightenment Revisited*, ed. Jesús Astigarraga, Voltaire Foundation (Oxford, 2015), 117–122.

Valle y Caviedes, Juan del, *Obra completa*, ed. Daniel Reedy, Ayacucho (Caracas, 1984).

Valles Garrido, J. M., 'La distribución de sopas económicas del Conde Rumford en la Segovia de comienzos del siglo XIX', *Estudios Segovianos* 92 (1995), 143–176.

Van de Ven, Hans J., *War and Nationalism in China, 1925–1945*, Routledge (Abingdon, 2003).

Van Gogh, Vincent, Vincent van Gogh: The Letters, www.vangoghletters.org.

Vandenbroeke, Christian, 'Cultivation and Consumption of the Potato in the 17th and 18th Century,' *Acta Historiae Neerlandica* 5 (1971), 15–39.

Vanhaute, Eric, '"So Worthy an Example to Ireland": The Subsistence and Industrial Crisis of 1845–1850 in Flanders', in *When the Potato Failed: Causes and Effects of the 'Last' European Subsistence Crisis, 1845–1850*, ed. Cormac Ó Gráda, Richard Paping and Eric Vanhaute, Brepols (Turnhout, 2007), 123–148.

Vanhaute, Eric, Richard Paping and Cormac Ó Gráda, 'The European Subsistence Crisis of 1848–1850: A Comparative Perspective', in *When the Potato Failed: Causes and Effects of the 'Last' European Subsistence Crisis, 1845–1850*, ed. Cormac Ó Gráda, Richard Paping and Eric Vanhaute, Brepols (Turnhout, 2007), 1–31.

Vardi, Liana, *The Physiocrats and the World of the Enlightenment*, Cambridge University Press (Cambridge, 2012).

Varenne de Béost, Claude-Marc-Antoine, *La cuisine des pauvres ou Collection des meilleurs Mémoires qui ont parus depuis peu* (Dijon, 1772).

Verhandlungen und Schriften der hamburgischen Gesellschaft zur Beförderung der Künste und nützlichen Gewerbe 1 (Hamburg, 1790).

Vernon, James, 'The Ethics of Hunger and the Assembly of Society: The Techno-Politics of the School Meal in Modern Britain', *American Historical Review* 110:3 (2005), 693–725.

Vernon, James, *Hunger: A Modern History*, Harvard University Press (Cambridge, MA, 2007).

'Vice Minister of the Party at Ministry of Agriculture Emphasizes: Strengthen Potato

Industry and Make it the Aircraft Carrier to Protect Food Safety of the Country', Institute of Food Science and Technology (CASS), 2017, http://iappst.caas.cn/en/news/74612.htm.

Vico, Giambattista, *The New Science of Giambattista Vico. Unabridged Translation of the Third Edition, 1744*, trans. Thomas Goddard Bergin and Max Harold Fisch, Cornell University Press (Ithaca, 1984).

Viqueira Albán, Juan Pedro, *Propriety and Permissiveness in Bourbon Mexico*, trans. Sonya Lipsett-Rivera and Sergio Rivera Ayala, SR Books (Wilmington, 1999).

Virey, Julien Joseph, 'De la vie et des ouvrages d'Antoine-Augustin Parmentier', *Bulletin de Pharmacie et des Sciences Accessoires* 2 (1814), 60–61.

Virey, Julien Joseph, 'Pomme-de-terre, ou papas des Américas: Recherches sur son origine et l'epoque de son introduction en Europe', in *Nouveau dictionnaire d'histoire naturelle, appliquée aux arts, a l'agriculture, a l'économie rural et domistique, a la médecine, etc, par une société de naturalistes et d'agriculteurs* (Paris, 1818), XXVII, 526–549.

Voekel, Pamela, *Alone before God: The Religious Origins of Modernity in Mexico*, Duke University Press (Durham, 2002).

Vos, Paula de, 'Natural History and the Pursuit of Empire in Eighteenth-Century Spain', *Eighteenth-Century Studies* 40:2 (2007), 209–239.

Wahnbaeck, Till, *Luxury and Public Happiness: Political Economy in the Italian Enlightenment*, Oxford University Press (Oxford, 2004).

Walter, John, 'The Social Economy of Dearth in Early Modern England', in *Famine, Disease and the Social Order in Early Modern Society*, ed. John Walter and Roger Schofield, Cambridge University Press (Cambridge, 1989), 75–128.

Wang, Xiaoying, 'The Post-Communist Personality: The Spectre of China's Capitalist Market Reforms', *China Journal* 47 (2002), 1–17.

Ward, Bernardo, *Obra pía y eficaz modo para remediar la miseria de la gente pobre de España* (Madrid, 1787 [1750]).

Ward, Bernardo, *Proyecto económico, en que se proponen varias providencias, dirigidas á promover los intereses de España, con los medios y fondos necesarios para su plantificación* (Madrid, 1779).

Warner, Jessica, *Craze: Gin and Debauchery in an Age of Reason*, Random House (New York, 2003).

Washington, George, *The Writings of George Washington from the Original Manuscript Sources, 1745–1799*, ed. John Fitzpatrick, Government Printing Office (Washington, DC, 1931–1944).

Watt, George, *Dictionary of the Economic Products of India*, 6 vols. (London and Calcutta, 1891).

Watts, Michael, 'Development and Governmentality', *Singapore Journal of Tropical Geography* 24:1 (2003), 6–34.

Webb, Samantha, 'Not So Pleasant to the Taste: Coleridge in Bristol during the Mixed Bread Campaign of 1795', *Romanticism* 12:1 (2006), 5–14.

Webb, Sidney and Beatrice Webb, 'The Assize of Bread', *The Economic Journal* 14:54 (1904), 196–218.

Weismantel, Mary, *Food, Gender and Poverty in the Ecuadorian Andes*, University of Pennsylvania Press (Philadelphia, 1988).

Wells, Roger, *Wretched Faces: Famine in Wartime England, 1793–1801*, Sutton Press (Gloucester, 1988).

West-India Planter, *Remarks on the Evidence Delivered on the Petition presented by the West-India Planters and Merchants, to the Hon. the House of Commons, on the 16th of March 1775* (London, 1777).

Wheeler, Erica, 'To Feed or to Educate? Labelling in Targeted Nutrition Intervention', *Development and Change* 16 (1985), 475–483.

Whitaker, William, 'Food Entitlements', in *The Cambridge World History of Food*, ed. Kenneth Kiple and Kriemhild Coneè Ornelas, 2 vols. Cambridge University Press (Cambridge, 2001), II, 1585–1592.

White, Gilbert, *Natural History of Selborne* (Edinburgh, 1833).

Wiegelmann, Günter, *Alltags- und Festspeisen in Mitteleuropa. Innovation, Strukturen und Regionen vom späten Mittelalter bis zum 20. Jahrhundert*, Waxmann (Marburg 1967).

Wilk, Richard, 'Beauty and the Feast: Official and Visceral Nationalism, in Belize', *Ethnos* 58 (1993), 294–316.

Will, Pierre-Etienne and R. Bin Wong, *Nourish the People: The State Civilian Granary System in China, 1650–1850*, University of Michigan Press (Ann Arbor, 1991).

Wills, C. J., *The Land of the Lion and the Sun, or Modern Persia, Being Experiences of Life in Persia from 1866 to 1881* (London, 1891).

Wilse, J. N., *Physisk, oeconomisk og statistisk Beskrivelse over Spydeberg Præstegield og Egn i Aggershuus-Stift udi Norge* (Christiania, 1779).

Wilson, Mary Tolford, 'Americans Learn to Grow the Irish Potato', *New England Quarterly* 32:3 (1959), 333–350.

Winter, J. M., 'Military Fitness and Civilian Health in Britain during the First World War', *Journal of Contemporary History* 15 (1980), 211–244.

Wirilander, Kaarlo, *Savon Historia: Savo Kaskisavujen Kautena 1721–1870*,

Kustannuskiila Oy (Kuopio, 1960).

Withers, Charles, *Placing the Enlightenment: Thinking Geographically about the Age of Reason*, University of Chicago Press (Chicago, 2007).

Woolley, Hannah, *The Queen-Like Closet* (London, 1672).

Worboys, Michael, 'The Discovery of Colonial Malnutrition between the Wars', in *Imperial Medicine and Indigenous Societies*, ed. David Arnold, Manchester University Press (Manchester, 1988), 208–225.

Yan, Yunxiang, 'The Chinese Path to Individualization', *British Journal of Sociology* 61:3 (2010), 489–512.

Yang, Mei-ling, 'Creating the Kitchen Patriot: Media Promotion of Food Rationing and Nutrition Campaigns on the American Home Front during World War II', *American Journalism* 22:3 (2005), 55–75.

Ying, Li, 'The Great Potato Debate', *Global Times*, 22 Jan. 2015, www.globaltimes.cn/content/903438.shtml.

Young, Arthur, *Farmer's Letters to the People of England*, second edition (London, 1768).

Young, Arthur, *The Question of Scarcity Plainly Stated, and Remedies Considered* (London, 1800).

Young, Arthur, *A Tour in Ireland; with General Observations on the Present State of that Kingdom: Made in the Years 1776, 1777, and 1778. And Brought Down to the End of 1779*, 2 vols. (London, 1780).

Zanon, Antonio, *Della coltivazione, e dell'uso delle patate e d'altre piante commestibili* (Venice, 1767).

Zárate, Agustín de, *Historia del descubrimiento y conquista del Perú*, ed. Franklin Pease and Teodoro Hampe Martínez, Pontífica Universidad Católica del Perú (Lima, 1995 [1555]).

Zeta Quinde, Rosa, *El pensamiento ilustrado en el Mercurio peruano, 1791–1794*, Universidad de Piura (Piura, 2000).

Zilberstein, Anya, 'Inured to Empire: Wild Rice and Climate Change', *William and Mary Quarterly* 72:1 (2015), 127–158.

Zweiniger-Bargielowska, Ina, *Managing the Body: Beauty, Health and Fitness in Britain, 1880–1939*, Oxford University Press (Oxford, 2010).

Zwinger, Theodor, *Theatrum Botanicum* (Basle, 1696).

Zylberberg, David, 'Fuel Prices, Regional Diets and Cooking Habits in the English Industrial Revolution (1750–1830)', *Past & Present* 229:1 (2015), 91–122.

索引

academies of science 科学院
　promote potatoes 推广土豆, 70, 202
Acurio, Gastón 加斯顿·阿克瑞奥, 197
Africa 非洲, 48, 123–124, 125, 181
　potato promotion 土豆推广, 124
　West African agricultural expertise 西非农业专业知识, 111, 121
Agamben, Giorgio 吉奥乔·阿甘本, 210
Agricultural and Horticultural Society of India 印度农业和园艺学会, 118–119
Agriculture 农业
　African 非洲的, 111, 124
　Andean 安第斯山区的, 14, 194, 199–201, 206
　Chinese 中国的, 132, 133–134
　early modern agricultural revolution 近代早期农业革命, 48–49
　European 欧洲的, 48–49, 147, 149, 152
　Indian 印度的, 117, 119
　wartime policies 战时政策, 174, 180, 181, 183–184
Alcohol 酒类

beer 啤酒, 66, 148
consumption 消费, 55, 65, 66, 148, 170
distilled from potatoes 马铃薯蒸馏而成, 71, 73, 143
Aldini, Tobias 托拜厄斯·阿尔迪尼, 31
Allotments 配额, 181
Alsace 阿尔萨斯
　spread of potatoes 土豆的传播, 29, 76
Alströmer, Jonas 乔纳斯·阿尔斯托姆, 70
Ambrosoli, Mauro 莫罗·安布罗索利, 49
American Revolution 美国革命, 93, 123
Andagoya, Pascual de 帕斯夸尔·德·安达戈亚, 24
Andes 安第斯山脉 See also 另见 Peru 秘鲁
　dietary importance of potatoes 土豆在饮食上的重要性, 14, 23–24, 128
　potato cultivation 土豆栽培, 14, 199–201, 206
　religious significance of potatoes 土豆的宗教意义, 128–129

Appleby, Joyce 乔伊斯·阿普尔比, 105
Arch, Joseph 约瑟夫·阿奇, 150
Argentina 阿根廷, 180
artisans 工匠, 42, 44
assizes of bread 对面包的审判, 143
Atwater, Wilbur 威尔伯·阿特沃特, 159, 174
Australia 澳大利亚, 122, 227
Austria 奥地利, 38

bakers 面包师, 7, 68, 143, 146, 176
Baldini, Filippo 菲利普·巴尔迪尼, 70
Banks, Joseph 约瑟夫·班克斯, 114, 122
barley 大麦, 96, 108, 143, 144
 eaten by the poor 为穷人所食, 100, 144
 in soup 加入汤中, 93, 97, 98, 99
Bauhin, Gaspard 加斯帕德·鲍欣, 30, 31
Baur, Erwin 埃尔文·鲍尔, 178
Bavetta, Sebastiano 塞巴斯蒂安·巴韦塔, 17
Baxter, Richard 理查德·巴克斯特, 43
beans 豆类, 24, 36, 38, 97, 98, 174
Becher, Johann Joachim 约翰·约阿希姆·贝歇耳, 59
Beinart, William 威廉·贝纳特, 138
Belgium 比利时, 29, 30, 74, 124, 155, 164
Benevolence 善心
 dietary 饮食的, 17, 55, 113–115, 116–119
Bengal 孟加拉, 8, 117, 118, 121
Berry, Wendell 温德尔·拜瑞, 212
Bible《圣经》, 25, 26
Bignami, Pietro Maria 彼得罗·马利亚·贝格美, 64, 82
Blackpool 布莱克浦, 165

Blair, Tony 托尼·布莱尔, 209
Bligh, William 威廉·布莱, 112, 113
Bloomberg, Michael 迈克尔·布隆伯格, 1, 6
Board of Agriculture 英国农业委员会, 67–69, 75, 92, 96
 Committee on Potatoes 土豆委员会, 68, 143
Boer War 布尔战争, 13, 172
Bolivia 玻利维亚, 178
Bologna 博洛尼亚, 64, 82
botanical gardens 植物园, 71, 73, 113, 126
 acclimatisation gardens 适应性菜园, 118, 121
 Calcutta 加尔各答, 114
Botero, Giovanni 乔万尼·博泰罗, 9, 217
Boutelou, Claudio 克劳迪奥·博特罗, 73
Braunschweig 不伦瑞克, 37
bread 面包, 174
 black and brown 黑面包, 43, 161
 in working-class diet 在工人阶级的饮食中, 43, 146, 148, 163, 164
 mixed 混合面包, 68, 143, 146
 price 价格, 45, 90
 rye 黑麦面包, 43, 161, 164
 white 白面包, 65, 66, 166
breadfruit 面包果, 75, 111–114
Buchan, William 威廉·巴肯, 13, 66–67, 209
 recommends potatoes 推荐土豆, 11, 52, 67
buckwheat 荞麦, 74, 75
Burke, Edmund 埃德蒙·伯克, 14
Burkina Faso 布基纳法索, 123

Bush, George W. 乔治·W. 布什, 208

cabbages 卷心菜, 26, 43, 125, 142, 161
Cadet de Vaux, Antoine-Alexis 安托万-亚历克西斯·卡代·德沃克斯, 65, 73, 95–96
Calcutta 加尔各答, 114, 119
calories 卡路里, 141, 183 *See also* 另见 nutritional science 营养学
Campini, Antonio 安东尼奥·坎皮尼, 70
Canada 加拿大, 180
Canary Islands 加那利群岛, 38
Capitalism 资本主义
　　agrarian 农业, 48, 149
　　nineteenth-century 19 世纪, 19, 140–141, 147, 149, 150–151, 152, 156–157, 158–160, 165–166
　　state 国家, 19, 134–137, 139
Captain Swing 斯温上尉起义 *See* 参见 riots and public unrest 暴动和公众骚乱
Carleson, Edvard 爱德华·卡尔森, 59
carrots 胡萝卜, 31, 38, 43, 48, 88, 142, 222
cashew nuts 腰果, 115
Casteau, Lancelot de 兰斯洛特·德·卡斯特, 37
cauliflowers 花椰菜, 118
Cecil, William 威廉·塞西尔, 6
chakitaqlla 脚犁, 15, 196
Charles III 卡洛斯三世, 70, 71, 87
Chartism 宪章运动, 165
Chastellux, Marquis de 沙特吕侯爵, 10, 91, 92
chestnuts 栗子, 39
Chile 智利, 14, 23, 24, 178, 196
chilli peppers 辣椒, 4, 24, 26, 33, 115, 121
China 中国, 134–135, 191
　　Mao era 毛泽东时代, 133–134
　　nationalist health policies 民族主义健康政策, 132–133
　　potato promotion 土豆推广, 19, 134–137
　　potato research 土豆研究, 133, 135, 179
　　potatoes in 当地的土豆, 3, 4, 19, 110, 131–137
chips 薯片 *See* 参见 potatoes, processed 土豆加工食品
choice 选择, 17, 21, 80, 100, 137
　　'choosing subjects' "选择主体", 81, 208, 209
　　dietary 饮食的, 5, 17, 104, 105, 139, 185, 186, 207, 208, 210, 212
chuño 冻土豆, 24, 128
Cienfuegos, Bernardo de 贝尔纳多·德·西恩弗戈斯, 31, 39
Cieza de Leon, Pedro 佩德罗·谢萨·德莱昂, 24
Clusius, Carolus 卡罗勒斯·克卢修斯, 30, 34
Cobbett, William 威廉·科贝特, 146–150, 151
　　opinion of potatoes 对土豆的看法, 147–149
coffee 咖啡, 121
Collingham, Lizzie 莉齐·科林厄姆, 180, 182
Colombia 哥伦比亚, 178, 179, 191
colonialism 殖民主义, 123, 124, 139
　　in Africa 在非洲, 124, 181
　　in the antipodes 在澳大利亚和新西兰,

122

French 法国殖民主义, 122

in India 在印度, 116–120

in Ireland 在爱尔兰, 50, 151–157, 163

in North America 在北美, 123

in Spanish America 在西属美洲, 3, 14, 23, 126–131

in West Indies 在西印度群岛, 111–114

Colquhoun, Patrick 帕特里克·科尔克霍恩, 75

Columbian exchange 哥伦布大交换, 18, 35, 121

Committee on Food Habits 饮食习惯委员会, 2, 185, 207

complexion 气色, 31, 60

consumerism 消费主义, 105, 139, 154, 157

cookery books and recipes 烹饪书和菜谱, 13, 92 See also 另见 potato bread 土豆面包

 potato recipes 土豆菜谱, 36–38, 71, 220

  Britain 英国, 70, 168

  China 中国, 136

  England 英格兰, 34, 38

  France 法国, 70

  Germany 德国, 43, 184

  India 印度, 120

  Iran 伊朗, 108

  Italy 意大利, 30, 33

  Japan 日本, 182

  Russia 俄罗斯, 162

  Spain 西班牙, 31, 72

  USA 美国, 123, 175, 187, 211

 soup recipes 汤的食谱, 88–89, 91, 97

cookery classes 烹饪课, 13, 171

for working-class girls 针对工人阶级女孩, 13, 171, 173

cottage gardens 农舍菜园 See also 另见 potato gardens potatoes grown in 种植土豆的土豆菜园, 27, 29, 44, 45, 142

Count Rumford 拉姆福德伯爵 See 参见 Thompson, Benjamin 本杰明·汤普森

Coveney, John 约翰·柯文尼, 208

Crosby, Alfred 阿尔弗雷德·克罗斯比, 121

croutons 面包丁, 93–95, 99

Cuba 古巴, 122

Cullather, Nick 尼克·库拉瑟, 188, 190

Cyprus 塞浦路斯, 181

Czech Republic 捷克共和国, 79

dairy products 乳制品

 cheese 奶酪, 36, 43, 142

 eaten with potatoes 和土豆一起吃, 10, 16, 36, 43, 52, 67, 155

Davalos, José Manuel 何塞·曼纽尔·达瓦洛斯, 129

Dean, Mitchell 米切尔·迪恩, 199

Dechambre, Amédée 阿梅代·德尚布尔, 47, 157

Democratic Republic of the Congo 刚果民主共和国, 124

Denmark 丹麦, 70, 161, 197

De Soucey, Michaela 米卡拉·德索西, 196

development programmes 发展计划, 188–192, 194, 199

 criticise traditional agriculture 批判传统农业, 20–21, 193–194, 205

 ignore tubers 忽略块茎植物, 133–134, 186, 191, 192

索引 323

dietary advice 饮食建议
- aimed at the poor 针对穷人, 9, 65–67
- government guidelines 政府指导方针, 1, 2, 4, 6, 168, 185, 207, 210, 211
- in humoral medicine 在体液医学, 66–67

digestion 消化, 83, 146, 159, 174

Doyle, Henry 亨利·道尔, 71, 76–77

Drake, Francis 弗朗西斯·德雷克, 128

East India College 东印度学院, 152, 156

East India Company 东印度公司, 115, 117

eating 吃
- as a political act 作为政治行为, 84, 100, 146

eating habits 饮食习惯
- economic, military and political impact 经济、军事、政治影响, 2, 5, 14, 20, 21
- of elites 精英的, 9, 49–50, 129, 132
- of industrial workers 产业工人的, 11–13, 141, 158–160, 161, 163–165
- of ordinary people 普通人的
  - criticised 被批判, 11, 20, 66–67, 146, 164, 173
  - economic, military and political impact 经济、军事、政治影响, 56, 199
  - eighteenth-century debates 18 世纪的讨论, 59, 62, 63, 66, 69
  - nineteenth-century debates 19 世纪的讨论, 141, 164, 165, 167
  - twentieth-century debates 20 世纪的讨论, 133, 172, 173, 180
  - politically irrelevant 无关政治, 9, 14, 132, 207
  - politically relevant 有关政治, 10–11, 17, 20, 26, 132, 139, 207

ecological imperialism 生态帝国主义, 121

economic botany 经济植物学, 75, 121

economic societies 经济学会, 55, 65, 71, 73, 74, 75, 82
- Lima 利马, 127–128, 129
- Madrid 马德里, 97–98
- promote potatoes 推广土豆, 97–98
- Valencia 巴伦西亚, 75, 97

économistes 经济学家 *See* 参见 physiocracy 重农主义

Ecuador 厄瓜多尔, 178, 197

eddoes 芋头, 113

Eden, Frederick Morton 弗雷德里克·莫顿·伊登, 42–43, 65–66, 82, 90

Edwards, Bryan 布莱恩·爱德华兹, 111

efficiency 效率, 4–5, 20, 141, 158, 173, 188

Egypt 埃及, 181

El Niño effect 厄尔尼诺效应, 89

emulation 模仿
- of elite eating habits 精英的饮食习惯, 132, 143–144, 146

Engel, Samuel 塞缪尔·恩格尔, 65

England 英格兰 *See also* 另见 Great Britain 英国
- agriculture 农业, 27–29, 38, 48–49, 147, 149
- hostility towards potatoes 对土豆的敌意, 46, 50, 140, 147–149
- potato consumption 土豆消费, 27–29, 34, 36, 38, 39, 41, 43, 142, 149–150, 161, 164–165
- potato nomenclature 土豆命名, 33
- tithe disputes 什一税争议, 27–29, 222

Enlightenment 启蒙运动, 5, 9–10, 56,

91–92, 104–105, 113

Hispanic 西班牙, 126–127

Evelyn, John 约翰·伊夫林, 38

famine and dearth 饥荒和匮乏, 6, 9, 202
  See also 另见 Great Famine (Ireland) 爱尔兰大饥荒, hunger 饥饿
  China 中国, 132, 134
  Europe 欧洲, 55, 59, 64, 66, 90, 100, 181
  India 印度, 114–115, 117, 119
  potato famines 土豆饥荒, 155

fasting 禁食, 121

fatigue 疲劳, 160, 166

Feng, Xiaoyan 冯小燕, 136

Fenton, Ioulia Evgenyevna 尤莉亚·芬顿, 205

Ferrières, Madeleine 马德琳·费里埃尔, 44

Fielding, Henry 亨利·菲尔丁, 61

Finland 芬兰, 79
  potato promotion 土豆推广, 70, 74, 224
  potatoes absent 未出现土豆, 41

First World War 第一次世界大战, 20
  government food agencies 政府食品管理部门, 173
  importance of potatoes 土豆的重要性, 173–176
  in Britain 在英国, 174
  in Germany 在德国, 175–177
  in USA 在美国, 174–175

Flanders 佛兰德斯
  spread of potatoes 土豆的传播, 29, 41, 161

Fleischhacker, Samuel 塞缪尔·弗莱施哈克尔, 105

fodder crops 饲料作物, 48–49

Food and Agriculture Organization of the United Nations 联合国粮农组织, 5, 188, 191–193, 194, 199, 206

food security 粮食安全, 2–3, 5, 133, 179, 189–90, 191, 192, 195, 198, 199, 205, 206 See also 另见 peasants and small farmers 农民与小农场主
  contribution of potatoes 土豆的贡献, 134–135, 183, 192–193, 194–195, 201, 204
  in China 在中国, 19, 110, 134–135, 136–137

food sovereignty 粮食主权, 6

Foucault, Michel 米歇尔·福柯, 10, 57, 92, 210

foundlings 弃婴, 62, 71, 99

France 法国, 158
  potatoes promoted 土豆推广, 10–11, 53, 65, 70, 73, 83
  spread of potatoes 土豆传播, 29, 38, 39, 41

Frederick II 腓特烈二世, 25, 26, 28, 54, 55, 91, 212

free trade 自由贸易, 17, 81–82, 85–87, 101

freedom 自由, 1, 2, 17, 79–80, 104–105, 207–208, 212
  from hunger 免于挨饿, 188
  'nutritional' "营养"自由, 183

French Revolutionary Wars 法国革命战争, 59, 90, 140, 144

fries 薯条 See 参见 potatoes, processed 土豆加工食品

fruit 水果, 115, 118, 122, 135, 138

索引 325

Fuel and Food Administration 燃料与食品管理局, 173, 174
fuel supply 燃料供应, 41, 43, 45, 69

Galiani, Ferdinando 费迪南多·加利亚尼, 87
Galicia (Eastern Europe) 加里西亚（东欧）, 44
Galicia (Spain) 加里西亚（西班牙）, 29, 45
Gandhi, Mahatma 圣雄甘地, 120
gastronationalism 饮食民族主义, 196–198, 201
genetic erosion 遗传侵蚀, 192
geographical indicators 地理标志, 197
George III 乔治三世, 113, 140
Germany 德国, 158, 175–177, 178, 183–184 *See also 另见* Prussia 普鲁士
  Nazi food policy 纳粹粮食政策, 183–184
  Nazi potato breeding programmes 纳粹土豆培育计划, 43
  potato consumption 土豆消费, 38, 43, 161, 164–165, 175–176
  potato nomenclature 土豆命名, 34
  potato promotion 土豆推广, 74, 175–176, 183–184
  spread of potatoes 土豆传播, 30, 36, 41, 42, 44
grain trade 粮食贸易, 7, 81–82, 86–87
granaries 粮仓, 7, 132
Great Britain 英国, 65–69, 82
  1790s, 100, 142–146
  industrialisation 工业化, 158, 164–167
  military forces 军队, 13, 61, 62–63
  potato consumption 土豆消费, 62, 69, 102, 174, 180–181
  potato promotion 土豆推广, 66–69, 70, 142–146, 168, 174, 180–181, 227
  public health 公共卫生, 57, 171, 172–173, 180, 209
Great Famine (Ireland) 爱尔兰大饥荒, 155–157
Greece 希腊, 54
Green Revolution 绿色革命, 134, 190
Guaman Poma de Ayala, Felipe 费利佩·瓜曼·波马, 15, 129
Guan, Jia-Ji 管家骥, 179
Guatemala 危地马拉, 205
guilds 行会, 7
guinea pigs 豚鼠, 24, 129

hachis Parmentier 帕蒙蒂埃烤土豆泥, 203, 204
Hall, Douglas 道格拉斯·霍尔, 42
Hancock, Simon 西蒙·汉考克, 177
hand-loom weavers 手工织布机, 100, 165
Hanway, Jonas 乔纳斯·汉威, 60
happiness 幸福, 17
  as political 政治上, 79, 91, 100
  eighteenth-century notions 18世纪观念, 80–81, 82, 83, 87, 91–93, 101, 102, 210
  in colonial ideology 在殖民意识形态中, 116–117, 119
  potatoes promote 推广土豆, 64, 81, 82–83, 91, 97, 98, 135–136
Harcourt, Bernard 伯纳德·哈考特, 5
Haudenosaunee 豪德诺索尼, 122
health manuals 健康手册, 31–33, 60–61
Henry, David 大卫·亨利, 65
Henry, Diana 戴安娜·亨利, 1

Henry VIII 亨利八世, 8

herbals 本草书, 29–31

herrings 红鲱鱼, 140

Hertzberg, Peter Harboe 彼得·哈伯·赫茨伯格, 12

Hildesheim, Wilhelm 威廉·希尔德斯海姆, 160

Hobbes, Thomas 托马斯·霍布斯, 9

Hobsbawm, Eric 艾瑞克·霍布斯鲍姆, 149

Hoffman, Philip 菲利普·霍夫曼, 48

horsemeat scandal 马肉丑闻, 1

horticultural manuals 农艺手册, 25–26, 34, 38, 42, 69

hospitals 医院, 56, 62, 97

House of Commons 下议院, 91, 103, 144

Howard, John 约翰·霍华德, 70

humoral medicine 体液医学, 31, 61

hunger 饥饿 See also 另见 famine and dearth 饥荒和匮乏
  and public order 和公共秩序, 9, 64, 89–90, 132
  during the First World War 第一次世界大战期间, 176
  during the Second World War 第二次世界大战期间, 89–90, 181
  potatoes vanquish 土豆征服, 52, 71, 82

Hutcheson, Francis 弗兰西斯·哈奇森, 92

Imperial List of Approved Varieties 帝国批准品种名录, 184

improvement 改良, 116–119, 123–124, 139, 156–157

Inca empire 印加帝国, 3, 7, 23–24, 129

India 印度, 8, 115–121, 179, 190

colonial policies 殖民政策, 114–119, 172
potatoes promoted 推广土豆, 118–119
spread of potatoes 土豆的传播, 116, 119–122

individualism 个人主义, 19, 186, 208

Indonesia 印度尼西亚, 137

industrialisation 工业化, 19, 141, 152, 160, 165

infant feeding 喂养婴儿, 62, 71, 77

infant mortality 婴儿死亡率, 55, 62

International Potato Center 国际马铃薯中心, 195

International Year of the Potato 国际土豆年, 3, 20, 136, 192–193, 194, 195

Iran 伊朗, 107–108, 124
  potatoes promoted 土豆推广, 107

Iraq 伊拉克, 181

Ireland 爱尔兰, 42, 45, 50 See also 另见 Great Famine (Ireland) 爱尔兰大饥荒
  1641 Rebellion 1641年叛乱, 39
  Cobbett's views of 科贝特对其的看法, 147–148
  in nineteenth-century thought 在19世纪思想中, 147–148, 151–157, 163
  Malthus' views of 马尔萨斯对其的看法, 153–154
  potato consumption condemned 土豆消费受到谴责, 19, 50, 76, 141, 147–148, 152, 153–154, 163
  potato consumption praised 土豆消费收到赞赏, 19, 64–5, 77, 151–2
  widespread potato consumption 广泛的土豆消费, 25, 39, 43, 50, 64–5, 76, 155

Irish Lumper 爱尔兰大土豆, 155

Iroquois 易洛魁 See 参见 Haudenosaunee 豪德诺索尼
Italy 意大利, 48, 49, 87, 182
    potato nomenclature 土豆命名, 34
    potato promotion 土豆推广, 53, 64, 70, 82
    spread of potatoes 土豆传播, 30, 42

Japan 日本, 7, 172, 182
Jefferson, Thomas 托马斯·杰斐逊, 110–111
Jerusalem artichokes 耶路撒冷洋蓟, 26, 33, 34, 35
Jews 犹太人, 182, 183
Jones-Brydges, Harford 哈福德·琼斯-布里奇斯, 107

Kant, Immanuel 伊曼努尔·康德, 17, 104–105
Kapodistrias, Ioannis 依昂尼斯·卡波迪思特里亚斯, 54
K-Brot K 面包 *See* 参见 potato bread 土豆面包
Kenya 肯尼亚, 124, 181, 191
Khan, Asaf 阿萨夫汗, 115
Kyd, Robert 罗伯特·基德, 114

labour supply 劳动力供应, 56, 58–59, 60, 61–62, 64–65, 149, 158–159
Lagrange, Jean-Louis 约瑟夫-路易斯·拉格朗日, 59
Lancashire 兰开夏郡, 164, 165
    potatoes in 当地的土豆, 28, 39, 76
land 土地
    access to 获得土地, 45, 131, 147, 155, 176

landlords 地主
    encouraged to provide potato gardens 被鼓励提供土豆菜园, 67
    unenthusiastic about potatoes 不热衷于土豆, 46, 50
late blight 晚疫病, 155, 177, 184
Laurell, Axel 阿克塞尔·劳雷尔, 70, 224
Lavoisier, Antoine-Laurent 安托万-洛朗·拉瓦锡, 71
L'Écluse, Charles de 查尔斯·德·埃克卢斯 *See* 参见 Clusius, Carolus 卡罗勒斯·克卢修斯
Legrand d'Aussy, Pierre-Jean-Baptiste 皮埃尔-让-巴蒂斯特·勒格兰德·奥西, 52
Leon Pinelo, Antonio 安东尼奥·莱昂·皮内洛, 39
leprosy 麻风
    caused by potatoes 土豆引起的, 25, 26, 47, 53
Li, Keqiang 李克强, 135
liberalism 自由主义, 5, 17, 151, 157
Linnaeus, Carl 卡尔·林奈, 46
llamas 骆驼, 129, 130
Lloyd, David 大卫·劳埃德, 152
Lloyd George, David 大卫·劳合·乔治, 172
Lobb, Theophilus 西奥菲勒斯·洛布, 91
Locke, John 约翰·洛克, 116
London 伦敦, 97, 100, 146, 157
Lorraine 洛林
    potatoes in 当地的土豆, 41, 76
Louis XVI 路易十六, 53

Machiavelli, Niccolò 尼可罗·马基雅维

利, 9

Madrid 马德里, 64
- potatoes in 当地的土豆, 32, 39, 77
- soup kitchens 施粥所, 97–98

Maimone, Dario 达里奥·迈蒙, 17

maize 玉米, 24, 49, 138
- breeding 育种, 177, 192
- in Africa 在非洲, 47
- in Andes 在安第斯山区, 15, 24, 128
- in China 在中国, 4, 131, 135
- in Europe 在欧洲, 29, 47–48, 143, 144
- in mixed breads 在混合面包中, 143, 144

Malabar coast 马拉巴尔海岸, 115

Málaga 马拉加, 34

Malawi 马拉维, 24

Malcolm, John 约翰·马尔科姆, 107

Malthus, Thomas Robert 托马斯·罗伯特·马尔萨斯, 57, 141, 152–154

Manchester 曼彻斯特, 142, 164, 165, 172

Mandeville, Bernard 伯纳德·曼德维尔, 84

manioc 木薯, 113, 129

Māori 毛利, 124

Marsham, Robert Bullock 罗伯特·布洛克·马舍姆, 151

Marx, Karl 卡尔·马克思, 157, 245

McCulloch, John Ramsay 约翰·雷姆赛·麦克库洛赫, 154, 157

McMahon, Darrin 达林·麦马翁, 91

McNeill, William 威廉·麦克尼尔, 42

McVitie's 麦维他, 81

meat 肉类, 30, 63, 66, 88, 129, 135, 142, 158, 165 See also 另见 guinea pigs 豚鼠, pigs 猪

bacon 培根, 10, 36, 148
- in Indian diet 在印度饮食中, 120
- superior to potatoes 优于土豆, 163, 164, 166

Mehtab, Sheikh 谢赫·梅赫塔布, 120

Melon, Jean-François 让-弗朗索瓦·梅隆, 60

Mencius 孟子, 8

merchants 商人
- spread potatoes 传播土豆, 115, 116, 121, 131

Mexico 墨西哥, 49, 171

Middleton, Karen 凯伦·米德尔顿, 138

migration 移民, 155, 164

military 军队
- diet 饮食, 63, 77, 96, 123, 172
  - nutritional evaluations 营养评估, 160
- health of soldiers linked to diet 将士兵健康和饮食相关联, 2, 13, 61
- 'poor physique' 健康状况不佳, 2, 13, 173

Mill, James 詹姆斯·穆勒, 57

millet 小米, 14, 43, 48

Ministry of Food 食品部, 168
- Potato Division 马铃薯司, 181

Mirabeau, Marquis de 米拉波侯爵 See 参见 Riqueti, Victor 维克多·里克蒂

missionaries 传教士, 125, 137 See also 另见 potato priests 土豆牧师
- spread potatoes 传播土豆, 17, 108, 123

Moheau, Jean-Baptiste 让-巴蒂斯特·莫霍, 58

Moleschott, Jacob 雅各布·摩莱萧特, 162–163

Molokhovets, Elena 埃琳娜·莫洛科韦兹, 162

monoculture 单一栽培

　risks of 的风险, 155

Monte Verde 蒙特佛得角, 23

Montesquieu 孟德斯鸠

　(Charles-Louis de Secondat 查尔斯-路易·德·塞斯塔), 85

More, Hannah 汉娜·莫尔, 70, 144

Moscow 莫斯科

　potatoes in 当地的土豆, 181

Munich 慕尼黑, 95

Murphy, Margot 玛格特·墨菲, 187

nanny state 保姆式国家, 79, 232

Naples 那不勒斯, 64

Napoleon 拿破仑, 96

national potato days 全国土豆日, 196, 197

nationalist movements 民族主义运动, 120, 132–133, 188

Navarra, Pietro 彼得罗·纳瓦拉, 17

neoliberalism 新自由主义, 3, 17

Netherlands 荷兰, 48, 155

　potatoes in 当地的土豆, 29, 161

Netting, Robert 罗伯特·内廷, 45, 205

New Andean Cuisine 新安第斯菜系, 197

New Guinea 新几内亚, 182

New York City 纽约

　soda ban 汽水禁令, 1, 6

New Zealand 新西兰, 125, 209

newspapers and journals 报纸杂志

　discuss cheap soup 讨论廉价汤, 88, 97, 99

　discuss potatoes 讨论土豆, 52, 57, 70, 71–73

　dispense horticultural and dietary advice 提供农艺和饮食建议, 181, 182, 183, 185

Niemcewicz, Julian Ursyn 朱利安·尤尔辛·聂姆策维奇, 73

Nine Years War 九年战争, 41

Nitti, Francesco 弗朗西斯科·尼蒂, 159, 164, 166, 167

Norway 挪威

　potato promotion, 12, 70

nutrition transition 营养转变, 135

nutritional governmentality 营养治国, 5, 110

nutritional science 营养学, 120, 132–133, 159–160, 171

　evaluation of potatoes 对土豆的评估, 20, 141, 160, 161–165, 174

oats 燕麦, 26, 43, 44

　healthful 有益健康, 144, 174

　in mixed bread 在混合面包中, 68, 143

obesity 肥胖, 2, 81, 209

Occhiolini, Giovanni Battista 乔瓦尼·巴蒂斯塔·奥奇奥利尼, 53

Ochoa, Carlos 卡洛斯·奥乔亚, 178, 179, 192, 200

oeconomy 经济, 74–75

Oldham 奥尔德姆, 100, 147, 164

olla podrida 西班牙什锦菜, 36

onions 洋葱, 43, 62, 63

Ottoman empire 奥斯曼帝国, 7

Palestine 巴勒斯坦, 181

Paris 巴黎

　soup kitchens 施粥所, 95–96, 99

Parkinson, John 约翰·帕金森, 38

Parmentier, Antoine Augustin 安托万·奥古斯丁·帕蒙蒂埃, 53, 59, 76, 83,

202–205
parsnips 防风萝卜, 38, 43, 222
peanuts 花生, 4, 75, 131
peasants and small farmers 农民与小农场主, 7
 condemned as backward 被谴责为落后, 27, 47, 119, 124, 194
 contributions to food security 对粮食安全的贡献, 6, 20–21, 192, 194, 195
 contributions to potato cultivation 对土豆栽培的贡献, 49, 130, 200, 201, 205, 206
 custodians of biodiversity 生物多样性保护者, 20–21, 192, 195, 201, 206
 experimentalists 实验主义者, 48, 205, 206
 innovative practices 创新实践, 16, 18, 22, 44, 49, 131
Peru 秘鲁, 126–31, 178, 194, 196 See also 另见 Inca empire 印加帝国
 gastronationalism 饮食民族主义, 126–131
 plant breeding 职务育种, 179
 potatoes disdained 鄙视土豆, 128, 130
Petty, William 威廉·配第, 50, 76, 151
Petyt, William 威廉·培提特, 60
Philippines 菲律宾, 131
physiocracy 重农主义, 57, 85–87
pigs 猪
 in Māori culture 在毛利文化中, 125
 in Nazi ideology 在纳粹意识形态中, 184
Pitt, William, the Younger 小威廉·皮特, 140, 144, 145
Pizarro, Francisco 弗朗西斯科·皮萨罗, 23

plant breeding 植物育种, 177–179, 183, 190
 in China 在中国, 133
 in Nazi ideology 在纳粹意识形态中, 184
 in India 在印度, 179, 190
 in Latin America 在拉丁美洲, 178–179
 potatoes 土豆, 20, 133, 135, 177–179, 192
 in the USA 在美国, 177, 178
Poissonnier Desperrières, Antoine 安托万·普瓦索尼耶·德斯佩里埃, 63
Poland 波兰, 44, 73, 161, 184
polenta 玉米粥, 43, 48
political economy 政治经济学, 17, 55, 83, 207 See also 另见 physiocracy 重农主义; Smith, Adam 亚当·斯密
 dietary dimensions 饮食维度, 5, 80, 81, 82, 101–104, 212
poor houses 济贫院, 93, 146
poor laws 济贫法, 119, 148, 153
 New Poor Law of 1834 1834 年新济贫法, 148, 152
population 人口
 eighteenth-century debates 18 世纪的讨论, 10, 55, 56–65
 in Foucault 在福柯的论述中, 9–10, 210
 nineteenth-century debates 19 世纪的讨论, 141, 154
 twentieth-century debates 20 世纪的讨论, 188
population growth 人口增长, 131
 eighteenth-century 18 世纪, 89, 143
 linked to diet 与饮食相关联, 62
 nineteenth-century 19 世纪, 158

potatoes encourage 土豆的刺激, 19, 45, 65, 70, 77, 151, 154
'population quality' "人口质量", 134
Portugal 葡萄牙, 39, 42
potato bread 土豆面包
    eighteenth-century experiments 18 世纪实验, 68, 71, 74, 143, 204
    in Britain 在英国, 68, 69, 144, 145
    K-Brot K 面包, 176
    praised 受到赞扬, 69, 71, 73, 143, 144
    recipes 菜谱, 69, 73, 123, 142–146, 175
    rejected by the poor 被穷人拒绝, 142–146
potato-collecting expeditions 土豆采集考察, 178, 179, 192
potato gardens 土豆菜园, 67, 117, 148, 154
potato germplasm collections 土豆种质资源库, 179
potato heroes 土豆英雄, 25, 54, 202–205, 224
Potato Park 土豆公园, 198
Potato Pete 土豆皮特, 168, 181
potato priests 土豆牧师, 71
potato research institutes 马铃薯研究所, 177, 179 *See also* 另见 International Potato Center 国际马铃薯中心
potato soups 土豆汤, 100, 144, 182
    Rumford soup 拉姆福德汤, 93–99
    supposedly relished by the poor 据说穷人喜欢, 91, 95, 98, 100
potato starch 土豆淀粉, 67, 143, 202, 204
potatoes 土豆

'America's best gift' "美洲的最佳礼物", 70, 73, 204
animal feed 喂养动物, 30, 47, 49, 71, 75, 184, 202
botanical classification 生物学分类, 25, 26, 29, 33, 129
commercial varieties 商业品种, 20, 177, 178, 184, 198
    Renacimiento 复兴, 178, 200
cradle area 发源地, 14, 23, 178
cultivation techniques 培育技术, 15, 25, 68, 70, 71, 124, 194
demonstrate European superiority 证明欧洲优越性, 108, 117, 123, 124, 138
diseases 疾病, 67, 155, 184, 195 *See also* 另见 late blight 晚疫病
domestication 驯化, 14, 23
early descriptions 早期描述, 23, 24, 30, 34
    likened to truffles 和松露相联系, 23, 24, 34, 36, 37
eaten by ordinary people 为普通人所食
    Andes 安第斯山区, 14, 23–24, 128
    China 中国, 110, 131, 134
    Europe 欧洲, 16, 20, 29, 30, 44, 46, 47, 69, 142, 149–150, 161, 164
    India 印度, 119–120
    North America 北美, 123
eaters beautiful 使食用者貌美, 102, 123
encourage autarky 鼓励自给自足, 19, 152, 157, 193
facilitate laziness 促进懒惰, 14, 50, 76
as field crop 作为庄稼, 28, 41, 43, 50, 200
freeze-dried 冻干 *See* 参见 *chuño* 冻土

豆

global spread 全球传播, 3, 14, 18, 24, 108–110, 138–139

grown on mountainous land 在山地生长, 15, 23, 45, 131

healthful 有益健康, 11, 45, 52, 76–77, 136

heritage food 遗产食物, 198, 201

historiographic treatment 史学处理, 25, 51, 55, 142, 202–205

impede formation of political consciousness 阻碍政治意识的形成, 154, 157

impede industrial development 阻碍工业发展, 160–165

in debates about capitalism 关于资本主义的讨论, 19, 140, 141, 142, 146, 147–149, 151, 152, 156–157, 166

in herbals 在本草书中, 29–31

in military diet 在军队饮食中, 41, 62, 68, 77, 123

in Swing Riots 在斯温上尉起义中, 149–150

incite lust 刺激性欲, 31, 41

increase grain exports 提高粮食出口, 67, 77, 96, 102

'lazy potato blood' "懒惰的土豆之血", 163

nomenclature 命名, 33–35

per capita consumption 人均消费量, 3, 24, 174, 209

popular with children 受儿童欢迎, 75, 83

popularised by military 被军队推广, 119, 182

preservation techniques 保存技术, 24, 68, 71

processed 加工, 1, 135, 178, 209, 210

promote backwardness 加剧落后, 148, 154, 157, 163

promoted in eighteenth century 在18世纪被推广, 11, 16, 17, 51, 52–56, 64–65, 69–77, 204

research 研究, 68, 74, 118, 229 See also 另见 plant breeding 植物育种

short- and long-day varieties 短日照和长日照, 30, 44

state evading 逃避国家, 16, 19, 44, 45–46, 50, 134, 151, 152, 157

substitute for bread 替代面包, 146, 149, 151, 204

supposedly shunned 认为应被回避, 27, 142

trade 贸易, 32, 38, 39, 40, 50, 122, 123, 125, 130, 174, 175, 176

traditional landraces 传统地方品种, 178, 192, 201

unhealthy 不健康, 129–130, 141, 148, 160, 161–165

unmentioned in Bible 《圣经》未提及, 25

wild 野生, 14, 23, 178

yield 收获, 16, 44, 45, 46, 75, 155, 160, 177

poverty 贫困, 151

in Britain 在英国, 100, 147, 150, 165, 166, 173

in Ireland 在爱尔兰, 148, 154

leads to poor diet 导致饮食状况不佳, 11, 103, 150

political threat 政治威胁, 170, 188

Prentice, Thomas 托马斯·普伦蒂斯, 43

prisons 监狱, 97
prizes 奖项
　　promoting potatoes 推广土豆, 11, 68, 73–74, 118
　　promoting soup 推广汤, 88
productivity 生产力
　　of agriculture 农业的, 13, 118, 205
　　of workers 工人的, 13, 61
prostitutes 妓女, 236
protein 蛋白质, 120, 163–164, 171, 174
　　in potatoes 土豆中, 44, 162, 166
Prussia 普鲁士, 55, 155, 160 See also 另见 Germany 德国
public health 公共卫生, 2, 57, 59, 80, 172
　　and public order 和公共秩序, 141, 159
　　responsibility for ensuring 确保公共卫生的责任, 1, 5, 14, 119, 132, 171, 173, 177

quantification 量化, 92–93
Quechua 盖丘亚语, 34
Quesnay, François 弗朗索瓦·魁奈, 85, 87
quinoa 藜麦, 15, 24, 75, 128, 197

Rabinbach, Anson 安森·拉宾巴赫, 159
Raffles, Stamford 斯坦福德·莱佛士, 137
railroads 铁路, 122, 173
Raleigh, Walter 沃尔特·罗利, 128
Rationing 定量配给
　　First World War 第一次世界大战, 175, 176, 177
　　Second World War 第二次世界大战, 168, 181, 183, 185

Rawls, John 约翰·罗尔斯, 79
reason of state 国家理性, 9
refugees 难民
　　spread new agricultural techniques 传播新的农业技术, 48
religion 宗教
　　and diet 和饮食, 8, 121, 125, 148
reproduction 生育, 62, 65
responsibilisation 责任化, 209, 210
restaurants 餐馆, 96, 174
rice 水稻, 7, 24, 63, 75
　　hybrid 杂交, 133, 190, 192
　　in China 在中国, 132, 133, 135
　　in India 在印度, 8, 117, 120
　　in Iran 在伊朗, 108, 109
　　in North America 在北美, 111
　　in soup 在汤中, 88, 99
　　West-African expertise 西非专业知识, 111
Richthofen, Ferdinand von 费迪南德·冯·里希特霍芬, 131
riots and public unrest 暴动和公众骚乱, 13
　　1766 Madrid riot 1766 年马德里暴动, 64, 87
　　in Britain 在英国, 90, 140, 142, 144, 145
　　Swing Riots 斯温上尉起义, 149–50
　　Swedish 'potato revolution' 瑞典"土豆革命", 176
Riqueti, Victor 维克多·里克蒂, 85, 99, 106
Rockefeller Foundation 洛克菲勒基金会, 189, 191, 192, 194, 195
Roman empire 罗马帝国, 7
Roncaglia, Alessandro 阿列桑德洛·荣卡

格利亚, 153
Rotherham 罗瑟勒姆, 1
Rowbottom, William 威廉·罗伯顿, 100
Rowntree, Seebohm 西博姆·朗特里, 13, 173
Royal Navy 英国海军, 62–63
Royal Society 英国皇家学会, 50, 114
Rubner, Max 马克斯·鲁布纳, 11, 160, 176
Rudé, George 乔治·鲁德, 149
Rumpolts, Marx 马克斯·朗波茨, 35–36
Russia 俄罗斯
 potato consumption 土豆消费, 161, 162, 181–182
 potato promotion 推广土豆, 73
 potatoes absent 未出现土豆, 41
rye 黑麦, 45, 74, 143, 155, 161 See also 另见 bread 面包

sago 西米, 75, 114
Said, Edward 爱德华·萨义德, 116
sailors and fishermen 水手和渔民, 62–63
 spread new commodities 传播新商品, 42, 116, 125
Salaman, Redcliffe 瑞德克里夫·沙勒曼, 25
Scandinavia 斯堪的纳维亚, 34
Schmalzer, Sigrid 西格丽德·施马尔泽（舒喜乐）, 133
school meals 学校伙食, 4, 11, 171, 176
Schulze-Gävernitz, Gerhart von 格哈特·冯·舒尔茨-格弗尼茨, 164–166
Scotland 苏格兰
 potato promotion 土豆推广, 74
 potatoes in 当地的土豆, 39, 42–43, 65, 69

Scott, James C. 詹姆斯·C. 斯科特, 45, 170
scurvy 坏血病, 63, 66
Second World War 第二次世界大战, 20, 179–180, 182
 importance of potatoes 土豆的重要性, 168, 179–186
 in Britain 在英国, 168, 180–181
 in Germany 在德国, 183–184
 in Japan 在日本, 182
 in Soviet Union 在苏联, 181–182
 in USA 在美国, 2, 184–186
self-interest 个人利益, 17, 87
 dietary 饮食的, 83, 84, 87–88, 91, 100, 105, 137, 139, 207
 in Adam Smith 在亚当·斯密的论述中, 101–102
 leads to potato consumption 导致土豆消费, 81, 83
self-organisation 自我组织, 17, 77–78, 103, 106
 in Adam Smith 亚当·斯密的论述中, 101–102
Serres, Olivier de 奥利维尔·德·塞尔, 38
Seven Years War 七年战争, 123, 202
Shaanxi Province 陕西省, 131
Shah Fath-Ali 法特赫-阿里沙, 107
Shanxi Province 山西省, 135
Sherman, Sandra 桑德拉·谢尔曼, 100, 103
Simmons, Amelia 阿米莉亚·西蒙斯, 123
Sinclair, John 约翰·辛克莱, 67, 68
Singha, Baneswar 班恩思瓦·辛哈, 120
Skytte, Carl 卡尔·斯基特, 72

slavery 奴隶制, 111, 112–113, 156
　　diet of the enslaved 奴隶的饮食, 112–114
Slicher van Bath, Bernard Hendrik 伯纳德·亨德里克·斯利歇尔·范巴斯, 47
Smith, Adam 亚当·斯密, 16, 104, 189, 236
　　on happiness 论幸福, 101–104
　　praises potatoes 赞扬土豆, 103
　　views on diet 对饮食的看法, 61
Solanum tuberosum 马铃薯 See 参见 potatoes 土豆
sorghum 高粱, 14, 133
soup kitchens 施粥所, 76, 88–89, 93–99
　　users 'ungrateful' 使用者"不感激", 90
South Africa 南非
　　potatoes in 当地的土豆, 125
Soviet Union 苏联 See 参见 Russia
Spain 西班牙, 29, 41, 63
　　population debates 人口争论, 57, 58–59
　　potato nomenclature 土豆命名, 34
　　potato promotion 土豆推广, 70, 73, 74, 76–77
　　soup kitchens 施粥所, 97–98
　　spread of potatoes 土豆传播, 35, 45
Spary, Emma 艾玛·斯帕里, 22, 99
Sri Lanka 斯里兰卡, 138
St Buryan 圣布里扬, 28
stooping crops 压弯的庄稼
　　grown by women 女性种植, 45
stoutness 体格强壮
　　a sign of health 健康的表现, 65, 67
sugar 糖
　　cultivation 培育, 113, 121, 122

　　in working-class diet 在工人阶级饮食中, 65
sumptuary laws 禁止奢侈的法律, 8
Sweden 瑞典, 176
　　hostility towards potatoes 对土豆的敌意, 46
　　population debates 人口争论, 57, 59
　　potato promotion 土豆推广, 70
　　spread of potatoes 土豆传播, 39, 161
sweet potatoes 红薯
　　in China 在中国, 131, 191
　　in Europe 在欧洲, 29, 32, 34, 35, 221
　　in India 在印度, 115
　　in West Indies 在西印度群岛, 113
Swislocki, Mark 马克·维斯罗基, 110
Switzerland 瑞士
　　potato promotion 土豆推广, 65
　　soup kitchens 施粥所, 90, 97
　　spread of potatoes 土豆传播, 26, 31, 45
Syria 叙利亚, 181

Tanzania 坦桑尼亚, 194
Tasmania 塔斯马尼亚, 122, 124
Tassoni, Giovanni 乔瓦尼·塔索尼, 182
Taylor, Charles 查尔斯·泰勒, 79
tea 茶, 121, 131, 165
　　in working-class diet 在工人阶级饮食中, 65, 148
technologies of statecraft 治国之道, 5, 170–171, 172, 173, 177, 199
　　mathematics 数学, 56
Tehran 德黑兰, 107, 124
Tennant, William 威廉·坦南特, 117
Terrón, Eloy 埃洛伊·特伦, 29
Terry, Edward 爱德华·特里, 115

thermodynamics 热力学, 159
Thompson, Benjamin 本杰明·汤普森, 73, 93–95, 96, 99
Tiedemann, Friedrich 弗里德里希·蒂德曼, 161
Tilly, Charles 查尔斯·蒂利, 8
tithes 什一税
　potato 土豆, 27–29, 39, 44
tobacco 烟草, 42, 118
tofu 豆腐, 7
tomatoes 西红柿, 29, 121
Törbel 托贝尔, 45
tourism 旅游业, 197
trade unionism 工会主义, 167
translations 翻译, 73, 74
　of Thompson's writings 汤普森的作品, 96, 97
Transylvania 特兰西瓦尼亚
　potato promotion 推广土豆, 74
Trevelyan, Charles 查尔斯·特里维廉, 155–156
Tribe, Keith 基思·特赖布, 104
Trieste 的里雅斯特, 99
tubers 块茎植物
　development programmes ignore 发展计划忽略, 133–134, 186, 191, 192
　state evading 规避国家, 45–46
Turkey 土耳其, 18
Turkey Twizzlers 油炸鸡肉卷, 1
turkeys 火鸡, 36
turnips 芜菁, 39, 43, 88, 142, 222
　potatoes prepared similarly 和土豆做法类似, 31
　rape-roots 油菜根, 30

Ukraine 乌克兰, 181

Unanue, Hipólito 希波利托·乌纳努埃, 130
United Nations 联合国, 3, 20, 75, 192
United States 美国, 122–123, 177, 178, 184–186
　potato promotion 推广土豆, 122–123, 174–175, 211
　public health policies 公共卫生政策, 122–123, 184–185, 210
　spread of potatoes 土豆的传播, 14, 23, 122–123
universities and institutes 大学和研究机构, 63, 122, 123, 129, 160, 163, 177, 178
　Cornell 康奈尔大学, 177, 190
upawasa 斋戒 *See* 参见 fasting 禁食
Utah 犹他州, 14, 23

Valverde, Vicente de 韦森特·德·瓦尔韦德, 23
Van Gogh, Vincent 文森特·梵高, 208
Vandenbroeke, Christian 克里斯蒂安·范登布鲁克, 29
Vavilov, Nikolai Ivanovich 尼古拉·伊万诺维奇·瓦维洛夫, 178
vegetables 蔬菜
　poor advised to eat 被推荐给穷人食用, 65, 66, 67
vegetarianism 素食主义, 120
Vernon, James 詹姆斯·弗农, 4, 170
Vico, Giambattista 詹巴蒂斯塔·维柯, 85
Victualling Board 食品储备局, 62–63
Virey, Julien-Joseph 朱利安-约瑟夫·维雷, 53
Virginia 弗吉尼亚州, 33

Vivarais 维瓦莱
  potatoes in 当地的土豆, 39, 44
Voit, Carl von 卡尔·冯·沃特, 159, 163

wages 薪水
  agricultural 农业的, 147, 149, 150, 156
  for urban workers 城市工人的, 174
Wakefield 韦克菲尔德, 140
Wales 威尔士, 61, 74, 174
Ward, Bernardo 伯纳多·沃德, 47, 58–59
Wargentin, Pehr 佩尔·沃占廷, 57
Webb, Beatrice and Sidney 阿特丽斯·韦伯和西德尼·韦伯, 144
welfarism 福利主义, 14, 171, 199
West Indies 西印度群岛, 69, 111–114, 156
  tuber nomenclature 块茎植物命名, 115
wheat 小麦, 24, 135, 180
  'chapati quality' "做面饼的品质", 191

compared with potatoes 和土豆比较, 44, 69
  multivalent meanings 多重含义, 49z
  shortages 短缺, 89, 143–146, 155
Wiegelmann, Günter 冈特·维格尔曼, 41
wild rice 野生水稻, 75
women 女性, 13
  agricultural contributions 农业贡献, 45, 195, 206
workers' canteens 工人食堂, 171, 176, 177, 182
World Bank 世界银行, 189

yams 山药, 14, 115, 191
York 约克郡, 11, 173
Young, Arthur 亚瑟·杨, 57, 91, 151–152

Zanon, Antonio 安东尼奥·詹农, 82
Zylberberg, David 戴维·齐尔贝尔博格, 41

图书在版编目(CIP)数据

土豆帝国/(英)丽贝卡·厄尔(Rebecca Earle)
著;刘媺译.—上海:上海人民出版社,2023
(历史·文化经典译丛)
书名原文:Feeding the People: The Politics of
the Potato
ISBN 978 - 7 - 208 - 17780 - 2

Ⅰ.①土… Ⅱ.①丽… ②刘… Ⅲ.①马铃薯-历史
-世界 Ⅳ.①S532-91

中国版本图书馆CIP数据核字(2022)第125217号

责任编辑 张晓玲 张晓婷
封面设计 王小阳

历史·文化经典译丛
土豆帝国
[英]丽贝卡·厄尔 著
刘 媺 译

出 版	上海人氏出版社
	(201101 上海市闵行区号景路159弄C座)
发 行	上海人民出版社发行中心
印 刷	上海商务联西印刷有限公司
开 本	720×1000 1/16
印 张	22
插 页	5
字 数	263,000
版 次	2023年6月第1版
印 次	2023年6月第1次印刷
ISBN 978 - 7 - 208 - 17780 - 2/K · 3213	
定 价	118.00元

This is a simplified Chinese edition of the following title published by Cambridge University Press:

Feeding the people: The Politics of the Potato
ISBN 978-1-108-48406-0 (Hardback)
By Rebecca Earle, first published by Cambridge University Press 2020.

This simplified Chinese edition for the People's Republic of China (excluding Hong Kong, Macau and Taiwan) is published by arrangement with the Press Syndicate of the University of Cambridge, Cambridge, United Kingdom.

© Cambridge University Press and Shanghai People's Publishing House 2023

This simplified Chinese edition is authorized for sale in the People's Republic of China (excluding Hong Kong, Macau and Taiwan) only. Unauthorized export of this simplified Chinese edition is a violation of the Copyright Act. No part of this publication may be reproduced or distributed by any means, or stored in a database or retrieval system, without the prior written permission of Cambridge University Press and Shanghai People's Publishing House.

Copies of this book sold without a Cambridge University Press sticker on the cover are unauthorized and illegal.

本书封面贴有 Cambridge University Press 防伪标签，无标签者不得销售。
此版本仅限在中华人民共和国境内（不包括香港、澳门特别行政区及台湾省）销售。

上海人民出版社·独角兽

"独角兽·历史文化"书目

[英]佩里·安德森著作
《从古代到封建主义的过渡》
《绝对主义国家的系谱》
《新的旧世界》

[英]李德·哈特著作
《战略论:间接路线》
《第一次世界大战战史》
《第二次世界大战战史》
《山的那一边:被俘德国将领谈二战》
《大西庇阿:胜过拿破仑》
《英国的防卫》

[美]洛伊斯·N.玛格纳著作
《生命科学史》(第三版)
《医学史》(第二版)
《传染病的文化史》

《欧洲文艺复兴》
《欧洲现代史:从文艺复兴到现在》
《非洲现代史》(第三版)
《巴拉聚克:历史时光中的法国小镇》
《语言帝国:世界语言史》
《鎏金舞台:歌剧的社会史》
《铁路改变世界》
《棉的全球史》
《土豆帝国》
《伦敦城记》
《威尼斯城记》

《工业革命(1760—1830)》
《世界和日本》
《激荡的百年史》
《论历史》
《论帝国:美国、战争和世界霸权》
《社会达尔文主义:美国思想潜流》
《法国大革命:马赛曲的回响》

阅读,不止于法律。更多精彩书讯,敬请关注:

微信公众号

微博号

视频号